# Textbooks in Telecommunication Engineering

## Era of the Internet and Network Science/Engineering

**Series Editor**

Tarek S. El-Bawab, Ph.D., Boston, MA, USA

Telecommunication and networks have evolved to embrace all aspects of our everyday life, including smart cities and infrastructures, healthcare, banking and businesses, manufacturing, space and aviation, meteorology and climate change, oceans and marine life, Internet of Things, defense, homeland security, education, research, social media, entertainment, and many others. Network applications and services continue to expand, virtually without limits. Therefore, specialized telecommunication and network engineering programs are recognized as a necessity to accelerate the pace of advancement in this field, and to prepare a new generation of engineers for imminent needs in our modern life. These programs need curricula, courses, labs, and textbooks of their own.

The IEEE Communications Society's Telecommunication Engineering Education (TEE) movement, led by Tarek S. El-Bawab- the editor of this Series, resulted in recognition of this field of engineering by the Accreditation Board for Engineering and Technology (ABET) on November 1, 2014. This Springer Series was launched to capitalizes on this milestone. The Series goal is to produce high-quality textbooks to fulfill the education needs of telecommunication and network engineering, and to support the development of specialized undergraduate and graduate curricula in this regard. The Series also supports research in this field and helps prepare its scholars for global challenges that lay ahead. The Series have published innovative textbooks in areas of network science and engineering where textbooks have been rare. It is producing high-quality volumes featuring innovative presentation media, interactive content, and online resources for students and professors.

Book proposals are solicited in all topics of telecommunication and network engineering including, but not limited to: network architecture and protocols; traffic engineering; network design, dimensioning, modeling, measurements, and analytics; network management and softwarization; cybersecurity; synchronization and control; applications of artificial intelligence in telecommunications and networks; applications of data science in telecommunications and networks; network availability, reliability, protection, recovery and restoration; wireless communication systems; cellular technologies and networks (through 5G, 6G, and beyond); satellite and space communications and networks; optical communications and networks; heterogeneous networks; broadband access and free-space optical communications; MSO/cable networks; storage networks; optical interconnects; and data centers; social networks; transmission media and systems; switching and routing (from legacy to today's paradigms); network applications and services; telecom economics and business; telecom regulation and policies; standards and standardization; and laboratories.

Proposals of interest shall be for textbooks that can be used to develop university courses, either in full or in part. They should include recent advances in the field while capturing whatever fundamentals that are necessary for students to understand the bases of the topic and appreciate its evolution trends. Books in this series will provide high-quality illustrations, examples, end-of-chapters' problems/exercises and case studies.

This series is indexed in Scopus. For further information and to submit proposals, please contact the Series Editor, Dr. Tarek S. El-Bawab, telbawab@ieee.org; or Mary James, Executive Editor at Springer, mary.james@springer.com.

Wei Chen · Zhijin Qin

# Semantic Communications

Springer

Wei Chen
School of Electronic and Information Engineering
Beijing Jiaotong University
Beijing, China

Zhijin Qin
Department of Electronic Engineering
Tsinghua University
Beijing, China

ISSN 2524-4345 ISSN 2524-4353 (electronic)
Textbooks in Telecommunication Engineering
ISBN 978-3-032-11004-6 ISBN 978-3-032-11005-3 (eBook)
https://doi.org/10.1007/978-3-032-11005-3



This Springer imprint is published by the registered company Springer Nature Switzerland AG
The registered company address is: Gewerbestrasse 11, 6330 Cham, Switzerland

# Note from the Series Editor

This Series has its roots in the Telecommunication Engineering Education (TEE) movement in the USA, which I had the honor of leading from 2008 to 2015. The movement was a result of remarkable impacts telecom networks had already started to have on society, and many more that were looming on the horizon at that time. Major changes in the theoretical foundations of this field of engineering, its design strategies, and its practical applications were unfolding at the same time. The industries we had known as the Telecom, Cable, and Data-Communication industries through the twentieth century were going through technical innovations and were reshuffling, repositioning themselves in the market, and converging, in preparation for the global economy of the twenty-first century.

The TEE movement was prompted by a personal experience I went through as I moved from industry to academia. *Telecom engineering*, or *Network Engineering*, since the network is centric to the evolution telecom has been through for three to four decades so far, became a distinct multidisciplinary field of study. This field blends ingredients from electrical engineering,computer engineering, computer science, and other disciplines, but may not be taught merely as a branch of any one of them, per se. It became evident that society would be better served by adding specialized undergraduate/graduate courses and programs in telecom/network engineering to our higher-education system, especially in the USA where this was not the norm.

The story of the TEE movement has been told and also discussed within broader contexts in the literature (e.g., *IEEE Communications Magazine, Vol. 53, No. 11, pp. 35–9, November 2015*). The movement had accomplished its main goal of adding telecom/network engineering to ABET's portfolio of recognized engineering disciplines. ABET is the Accreditation Board of Engineering and Technology, Inc., which accredits engineering programs in the USA and more than 40 other countries. A lot of work had to be done to capitalize on that milestone, however. The technical literature of this discipline was rich in scholarly research papers, but not as rich in textbooks designed to support the development of new university courses and curricula. These textbooks are needed to facilitate rigorous academic training for a new generation of network engineers. The idea of this Series came up subsequently to help the telecom/ networking community overcome this shortcoming. Since its inception, Springer's TTE (*Textbooks in Telecommunication Engineering*) has been doing this job. The proximity of the acronyms herein—between TEE and TTE—is coincidental, but worth noting!

I am proud of the inroads TTE has made in publishing quality textbooks and in fulfilling the goals and objectives the Series has had since its beginning. By the end of 2024, we approach the milestone of having 35 books, including 20 published and about 15 in the making (whether in the writing, reviewing, or production phases). More than 50 authors have contributed to this Series so far and they come from about 20 different countries and cultures all over the world.

The topics covered by this Series so far scan the entire landscape of telecom and network engineering, including wireless, optical, satellite; access and core; devices, systems, and networks; space-aerial-terrestrial integration; queuing theory and traffic engineering; network management; Internet and Internet of Things (IoT); signal processing; applied electromagnetics; and economy/business aspects of telecom. The Series includes a cellulargenerations primer and has a Wi-Fi primer project under progress. It covers the most cutting-edge topics with textbooks that are carefully designed to strike the right balance between the need to create new courses thereto today and the fact that progress is imminent in each of these topics, and in the short term. These include books that discuss the roles of edge and cloud computing; of blockchains, and of artificial intelligence and machine learning in network engineering. Special topics such as under-water communications, railways telecom, networked navigation systems, semantic communications, video analytics, and broadband to combat the digital divide are also covered by TTE. I have tried my best to make this Series a top de facto global home for quality telecom/network engineering textbooks, and am hoping that many in our community can start to consider it as such now. The fruits of the hard work put by many people into fulfilling this goal are now handy to pick by our community.

I want to thank my wife Shahira and my daughter Nadine for their support of my work in this Series over nearly ten years. I have put a lot of time into this task and they have understood how this work is important to me. I owe a big deal to all the authors of this Series. Words cannot describe my appreciation of their contributions and talent. I would like to also express my thanks to Springer in general, and to Mary James in particular. Mary has been my main partner throughout this journey, and I couldn't have thought of a better partner than her. Finally, I thank all my colleagues in the field of telecom/network engineering and our readership for their continued support of the mission and the endeavor of this great project.

December 2024 Tarek S. El-Bawab

# Foreword

As wireless communication networks continue to evolve, the demand for more intelligent, efficient, and adaptive communication paradigms has never been more pressing. Traditional communication systems have long focused on syntactic accurate communication to guarantee the transmitted bits are received correctly regardless the content of the underlying data. Semantic communication systems focus on efficient communication of the meaning behind the raw data. With the rapid advancement of artificial intelligence and the ever-growing complexity of network applications, a shift toward semantic communication has become essential.

It is my great pleasure to write the foreword for this insightful and timely book on semantic communications. This book provides a structured and accessible introduction to the principles of semantic communication, making complex concepts easier to understand and apply to several emerging systems and applications. What sets this book apart is its balanced blend of theory and practice. The authors not only elucidate mathematical frameworks but also provide actionable insights into implementing semantic communication systems. This book traces the evolution from conventional to semantic communication, introduces key components and techniques in semantic communication, and highlights real-world applications, which are particularly noteworthy, reflecting the cutting-edge advancements in the field.

I am confident that this book will serve as an invaluable resource for university students, academics, researchers, and engineers seeking to understand and advance the field of semantic communications, and undoubtedly facilitate learning and inspire innovation in the next generation of communication professionals.

I wholeheartedly recommend this book and congratulate the authors on their outstanding contribution to the wireless communication and artificial intelligence education community.

Singapore
April 2025

Dusit Niyato

# Preface

Since its emergence, communication technology has evolved continuously. Recent breakthroughs in artificial intelligence have accelerated the development of semantic communication as a transformative technological paradigm, driving significant advances in the field and facilitating its deployment in practical applications. This book systematically documents the evolution and key research achievements of semantic communication in recent years, comprehensively covering the most cutting-edge semantic communication technologies.

There are more than 300 pages of new material, covering

- First systematic integration of deep learning and semantic communication theoretical frameworks
- In-depth analysis of AI technologies in semantic communication
- Standardization-ready solutions for 6G-oriented semantic communication
- Practical implementation cases and performance optimization techniques

These contributions are particularly valuable for graduate students and researchers seeking to grasp the latest developments in the field. Much of the material is ideally suited for core modules in advanced graduate-level courses on wireless communications and intelligent networks.

By striking a balance between theoretical depth and practical guidance, this book serves not only as a graduate-level textbook but also as a technical reference for communication engineers. It is our hope that the work presented here will provide readers with a clear and effective pathway to understanding and deploying the core technologies of semantic communication, thereby contributing to the development of next-generation intelligent communication systems.

Beijing, China
April 2025

Wei Chen
Zhijin Qin

**Acknowledgements** The development of this new material represented a significant scholarly endeavor, made possible through the generous support and contributions of several institutions and individuals. We are particularly indebted to Beijing Jiaotong University and Tsinghua University for their institutional support. This work was also supported by the State Key Laboratory of Advanced Rail Autonomous Operation and the State Key Laboratory of Space Network and Communications.

We would like to express our sincere appreciation to the students whose time, dedication, and expertise greatly enriched this work. From Beijing Jiaotong University: Lei Guo, Yukun Cheng, Jiangyuan Guo, Weiran Jiang, Hao Chen, Mengmeng Zhu, Jiaming Cheng, and Songling Gao; and from Tsinghua University: Zijing Wang, Jingkai Ying, Chengbin Liang, Hao Cao, Yinhuan Huang, and Tianwei Zuo.

This work was also supported by several national research programs, including the Fundamental Research Funds for the Central Universities under Grant 2022JBQY004 and the National Natural Science Foundation of China under Grants 62293484, U2468201, W2421083, and 62221001 as well as Beijing Natural Science Foundation (F251001).

Their support has been instrumental in enabling the research and development documented in this book.

**Competing Interests** The authors have no competing interests to declare that are relevant to the content of this manuscript.

# Contents

## Part II Joint Coding for Semantic Communication

# Acronyms

| | |
|---|---|
| 1G | The First-Generation Cellular Networks |
| 2G | The Second-Generation Cellular Networks |
| 3G | The Third-Generation Cellular Networks |
| 4G | The Fourth-Generation Cellular Networks |
| 5G | The Fifth-Generation Cellular Networks |
| 6G | The Sixth-Generation Cellular Networks |
| ADN | Autonomous Driving Network |
| AEP | Asymptotic Equipartition Property |
| AI | Artificial Intelligence |
| AoI | Age of Information |
| AoII | Age of Incorrect Information |
| AR | Augmented Reality |
| ASK | Amplitude Shift Keying |
| AVC | Advanced Video Coding |
| AWGN | Additive White Gaussian Noise |
| BEV | Bird's Eye View |
| BLEU | Bilingual Evaluation Understudy |
| BP | Brevity Penalty |
| BPP | Bits per Pixel |
| BPSK | Binary Phase-Shift Keying |
| BSC | Binary Symmetric Channel |
| CAV | Connected Autonomous Vehicle |
| CDL | Clustered Delay Line |
| CDMA | Code Division Multiple Access |
| CNN | Convolutional Neural Network |
| CSI | Channel State Information |
| DCT | Discrete Cosine Transform |
| DDPM | Denoising Diffusion Probabilistic Model |
| DGM | Deep Generative Model |
| DJSCC | Deep Joint Semantic-Channel Coding |
| DL | Deep Learning |
| DLM | Detail Loss Metric |
| DMOS | Difference Mean Opinion Score |
| DoF | Degree of Freedom |
| DPCM | Differential Pulse Coded Modulation |
| DWT | Discrete Wavelet Transform |
| eMBB | Enhanced Mobile Broadband |
| FDMA | Frequency Division Multiple Access |
| FEC | Forward Error Correction |
| FID | Frechet Inception Distance |
| FSK | Frequency Shift Keying |
| GAN | Generative Adversarial Network |

| | |
|---|---|
| G-PCC | Geometry-based Point Cloud Compression |
| GPS | Global Positioning System |
| GSC | Generative Semantic Communication |
| HARQ | Hybrid Automatic Repeat reQuest |
| HEVC | High Efficiency Video Coding |
| HMDs | Head-Mounted Devices |
| HR-HSI | High-Resolution HSI |
| HSI | Hyperspectral Imaging |
| HTC | Human-Type Communication |
| i QAM | i Quadrature Amplitude Modulation |
| ICT | Information and Communication Technology |
| IMT | International Mobile Telecommunications |
| IoT | Internet of Things |
| ISAC | Integrated Sensing and Communication |
| ITU | International Telecommunication Union |
| JSCC | Joint Semantic-Channel Coding |
| JSCD | Joint Source-Channel Decoding |
| KL | Kullback-Leibler |
| LDPC | Low-Density Parity-Check |
| LEO | Low Earth Orbit |
| LLM | Large Language Model |
| LLR | Log-Likelihood Ratio |
| LPIPS | Learned Perceptual Image Patch Similarity |
| LR-HSI | Low-Resolution HSI |
| LTI | Linear Time-Invariant |
| MAC | Medium Access Control |
| MCPD | Mean Co-Located Pixel Difference |
| MCS | Modulation and Coding Scheme |
| MEPG | Moving Pictures Experts Group |
| MIMO | Multiple Input Multiple Output |
| ML | Machine Learning |
| MMSE | Minimum Mean Square Error |
| mMTC | Massive Machine-Type Communications |
| mmWave | Millimeter Wave |
| MOS | Mean Opinion Score |
| MSE | Mean Squared Error |
| MS-SSIM | Multi-Scale Structural Similarity Index |
| MTC | Machine-Type Communication |
| NIC | Neural Image Compression |
| NLOS | Non-Line of Sight |
| NLP | Natural Language Processing |
| NOMA | Non-Orthogonal Multiple Access |
| NTC | Nonlinear Transform Coding |
| NTSCC | Nonlinear Transform Semantic-Channel Coding |
| OFDM | Orthogonal Frequency Division Multiplexing |
| OSI | Open System Interconnection |
| PAPR | Peak-to-Average Power Ratio |
| PDSCH | Physical Downlink Shared CHannel |
| PESQ | Perceptual Evaluation of Speech Quality |
| PET | Priority Encoding Transmission |
| PSK | Phase Shift Keying |
| PSNR | Peak Signal to Noise Ratio |
| PTSC | PoinT cloud Semantic Communication |

| | |
|---|---|
| QAA | Question Answering Accuracy |
| QAM | Quadrature Amplitude Modulation |
| QoE | Quality of Experience |
| RIS | Reconfigurable Intelligent Surface |
| RMS | Root Mean Square |
| SDR | Signal-to-Distortion Ratio |
| SISO | Single-Input Single-Output |
| SMS | Short Message Service |
| SNR | Signal-to-Noise Ratio |
| SSCC | Separate Source and Channel Coding |
| SSIM | Structural Similarity Index |
| STBC | Space-Time Block Coding |
| STC | Space-Time Coding |
| SVD | Singular Value Decomposition |
| SVM | Support Vector Machine |
| TDL | Tapped Delay Line |
| TDMA | Time Division Multiple Access |
| UAV | Unmanned Aerial Vehicle |
| UEP | Unequal Error Protection |
| UMa | Urban Macro-Cell |
| UMi | Urban Micro-Cell |
| URLLC | Ultra-Reliable Low-Latency Communication |
| V2I | Vehicle-to-Infrastructure |
| V2V | Vehicle-to-Vehicle |
| V2X | Vehicle-to-Everything |
| VAE | Variational Autoencoder |
| VideoQA | Video Question Answering |
| VIF | Visual Information Fidelity |
| ViT | Vision Transformer |
| VLC | Visible Light Communication |
| VMAF | Visual Multimethod Assessment Fusion |
| VoI | Value of Information |
| V-PCC | Video-based Point Cloud Compression |
| VR | Virtual Reality |
| WCDMA | Wideband Code Division Multiple Access |
| XR | Extended Reality |
| ZF | Zero-Forcing |

# Part I

# From Conventional Communication to Semantic Communication

# 1 Development of Wireless Communication

**Abstract**

Wireless communication has advanced through successive generations, each transforming how people connect and exchange information. The first generation (1G) enabled analog voice services, followed by the digital leap of 2G and the mobile data era of 3G. With 4G came the smartphone revolution, powering mobile applications, video streaming, and social networking. The arrival of 5G further expanded capabilities with ultra-high-speed connectivity, low latency, and massive device integration, supporting applications from autonomous driving to industrial automation. Looking ahead, 6G promises to deliver not only faster speeds but also intelligent, adaptive, and multimodal services through the integration of artificial intelligence. A key paradigm shift is semantic communication, which focuses on transmitting and interpreting meaning rather than raw data, reducing redundancy and aligning communication with task requirements. This chapter reviews the historical evolution of wireless systems and introduces the emerging techniques that will shape efficient, intelligent communication in the 6G era.

**Keywords**

Wireless communication • Three-level communication theory • Semantic communication • Artificial intelligence • Information theory

## 1.1 History of Wireless Communication

The development of wireless communication is one of the most remarkable achievements in the history of technology, revolutionizing how people interact, communicate, and access information across long distances without the need for physical connections. Wireless communication refers to the transfer of data via electromagnetic waves, such as radio waves, microwaves, and infrared signals, instead of relying on traditional cables or wires. It encompasses a broad range of technologies, from early systems designed to transmit voice to modern systems that support high-speed data and multimedia services.

### 1.1.1 Early Beginnings and Theoretical Foundations

Wireless communication has its origins in the groundbreaking contributions of scientists and inventors in early nineteenth century. The theoretical foundations were established by Michael Faraday and James Clerk Maxwell, who explored electromagnetic radiation. Their research demonstrated the existence of electromagnetic waves and highlighted their potential to transmit information across distances. Heinrich Hertz advanced these theories through his pioneering experiments in the 1880s, where he successfully generated and detected radio waves. Hertz's work validated Maxwell's predictions and set the stage for the development of practical wireless technologies.

The journey of wireless communications began in the late nineteenth century, with the invention of the radiotelegraphy system by Guglielmo Marconi, an Italian inventor, often considered the father of modern wireless communication. Marconi began experimenting with wireless telegraphy, developing a system that could transmit and receive Morse code signals over long distances. In 1895, Marconi achieved a major milestone by transmitting the first wireless signal over a short distance [1]. His work laid the foundation for the invention of long-distance wireless communication. This groundbreaking achievement marked the birth of modern wireless communication. The real breakthrough, however, came in 1901 when Marconi succeeded in transmitting a signal across the Atlantic Ocean, demonstrating for the first time the feasibility of long-range wireless communication [2]. This achievement demonstrated the feasibility of long-range wireless commu-


W. Chen and Z. Qin, *Semantic Communications*, Textbooks in Telecommunication Engineering,
https://doi.org/10.1007/978-3-032-11005-3_1

nication and opened the door for future wireless technologies, signaling the start of a new era of global communication.

Although early pioneers like Marconi showcased the practical potential of wireless communication, a comprehensive theoretical framework was still missing. It was not until the mid-twentieth century that the theoretical framework for modern communication systems began to take shape. Claude E. Shannon played a crucial role in this transformation. His landmark paper [3], "A Mathematical Theory of Communication," published in 1948, fundamentally changed the field of communication engineering. Prior to Shannon's contributions, the design of communication systems relied heavily on empirical methods and trial-and-error approaches. Shannon recognized the need for a more systematic, mathematical foundation to better understand communication processes.

Shannon's information theory [4] introduced several core concepts that are still vital to modern communication systems. The first key concept is channel capacity, which defines the maximum possible rate at which information can be transmitted reliably over a communication channel. Shannon's formula for channel capacity provides critical insights into the limitations and design principles for communication systems. The second key concept is information entropy, which quantifies the uncertainty or randomness in a set of data. Shannon's formulation of entropy allows for measuring the information content, which is fundamental for efficient data encoding and transmission. The third key concept is noise, which is a prevalent factor in communication systems to reduce signal quality and performance. Shannon developed a mathematical approach to analyze how noise affects communication and suggested methods to minimize its disruptive influence.

By introducing these essential concepts, Shannon's theory equipped engineers with the tools to design more efficient communication systems. It emphasized optimizing bandwidth use, understanding channel capacity, and managing noise, all of which are crucial for the development of advanced techniques like error correction coding, modulation, and equalization. His work laid the groundwork for modern wireless communication systems, enabling reliable and high-performance transmissions.

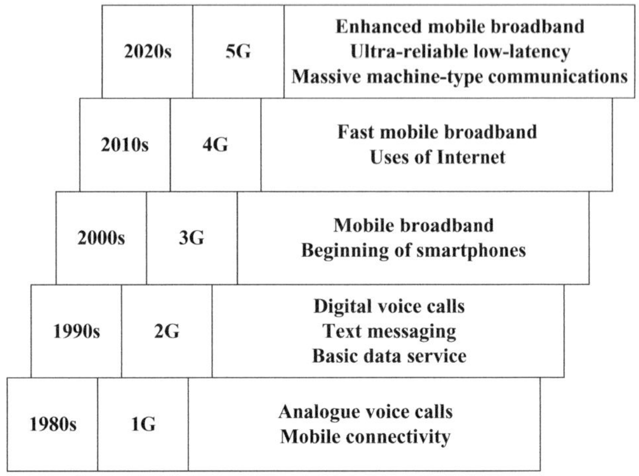


**Fig. 1.1** Development of cellular communications from 1G to 5G

### 1.1.2 From 1G to 5G: A Cellular Evolution

As shown in Fig. 1.1, the progression of cellular communication has been an extraordinary journey, marked by continual technological breakthroughs that have reshaped how we connect and communicate. From the early days of analog systems to today's high-speed, low-latency networks, each generation has built on the advancements of the previous one, sparking innovation and transforming industries.

As wireless communication systems advanced, the demand for commercial and industrial applications grew, prompting the development and deployment of first-generation (1G) cellular networks in the 1980s which marked the beginning of cellular communications. These networks were primarily designed to support analog voice communication. 1G systems relied on frequency division multiple access (FDMA), a technology that allowed multiple users to share the same frequency band by dividing it into smaller, non-overlapping sub-bands [5]. While 1G cellular networks represented a significant leap forward in communication, they suffered from various limitations, including poor spectrum efficiency, low security, and limited service quality.

These above challenges led to the development of second-generation (2G) cellular networks in the 1990s, which marked the transition from analog to digital communication. This transition brought about several improvements. The key advantage of 2G was its ability to support encrypted digital voice calls, improving security and providing better voice quality. 2G networks utilized technologies such as time division multiple access (TDMA) and code division multiple access (CDMA) to improve spectrum efficiency and support multiple users simultaneously. Moreover, 2G networks introduced data services, such as text messaging (SMS), as well as basic data transmission capabilities for applications like email [6].

With the increasing demand for mobile data and Internet services in the early 2000s, third-generation (3G) cellular networks emerged. 3G was a significant step forward in terms of data speeds and capacity. Wideband CDMA (WCDMA) is one of the key technologies to enable 3G, using spread spectrum techniques to improve capacity and reduce interference. Compared with 2G, 3G networks offer significantly higher data rates and enable faster mobile services, such as internet browsing, email, and the global positioning system (GPS) services [7]. The network capacity has also been increased

**Table 1.1** Capability comparison between IMT-2030 and IMT-2020

| Capabilities | IMT-2020 | IMT-2030 |
|---|---|---|
| Connection density (devices/km$^2$) | $10^6$ | $10^6 - 10^8$ |
| Mobility (km/h) | 500 | 500 – 1000 |
| Latency (ms) | $1 - 10^{-5}$ | $1 - 10^{-5}$ to $1 - 10^{-7}$ |
| Reliability | 1 | 0.1 – 1 |
| Positioning (cm) | NaN | 1 – 10 |

to accommodate more users in the age of 3G, facilitating seamless global roaming and allowing users to stay connected while traveling internationally.

The introduction of smartphones further fueled the need for faster, more efficient communication networks, leading to the rollout of fourth-generation (4G) networks in the 2010s. 4G networks, based on technologies like orthogonal frequency-division multiplexing (OFDM) and multiple-input multiple-output (MIMO), provided faster data rates and lower latency, which made it possible to stream high-definition videos, play online games, and participate in real-time video conferences from mobile devices [8].

Fifth-generation (5G) networks were commercialized in the 2020s which represent the latest evolution in wireless technology. 5G offers unprecedented speeds, significantly lower latency, and greater capacity compared to its predecessors, supporting ultra-reliable and low-latency communications (URLLC), machine-type communications (mMTC), and enhanced mobile broadband (eMBB) communications. By employing advanced technologies such as massive MIMO, millimeter-wave (mmWave) communication, and network slicing, 5G is capable of supporting a wide array of new applications and technologies, including the Internet of Things (IoT), autonomous driving, virtual reality (VR), and augmented reality (AR) [9]. The enhanced capabilities of 5G are essential to meet the demands of a highly connected world, where billions of devices are expected to be interconnected. In fact, it is predicted that global internet protocol (IP) data traffic will grow by 55% annually between 2020 and 2030, reaching a staggering 5016 exabytes per year, and there will be 25.9 billion connected mobile devices worldwide by 2025 [10].

### 1.1.3 The Promise of 6G: A New Era of Connectivity

The evolution of wireless communications is driven by the need to provide faster and more efficient communication services to meet the growing demands of society. Each new generation of wireless technology typically emerges about every ten years, with each generation providing significant advancements in terms of data rates, connectivity, and the variety of services offered. As we look toward the future, the next major step in wireless communication development is the emergence of sixth-generation (6G) cellular networks.

The 6G cellular network is poised to revolutionize wireless communication, building upon the foundation laid by 5G. It promises to deliver unprecedented levels of speed, latency, and capacity, enabling a wide range of innovative applications and services. Table 1.1 compares the capabilities between 5G and 6G. Compared to 5G, the key features of 6G include ultra-high speeds, with data rates reaching the terabits per second range, enabling near-instantaneous data transfer and high-quality streaming of immersive content. Additionally, 6G will significantly reduce latency, facilitating real-time applications such as remote surgery, autonomous vehicles, and virtual reality experiences. It will also support massive connectivity, accommodating a vast number of connected devices to drive IoT and smart city initiatives. Enhanced reliability is another critical feature, ensuring seamless connectivity even in challenging environments. Furthermore, 6G will incorporate advanced security features to protect sensitive data and safeguard network integrity, making it a robust and transformative technology for the future.

Figure 1.2 illustrates the evolution of communication usage scenarios from 5G to 6G, as proposed by IMT-2030. It builds upon the three foundational scenarios of 5G defined by IMT-2020 and expands them into six advanced scenarios for 6G. These include integrated sensing and communication, massive communications, ubiquitous connectivity, hyper-reliable and low-latency communication, artificial

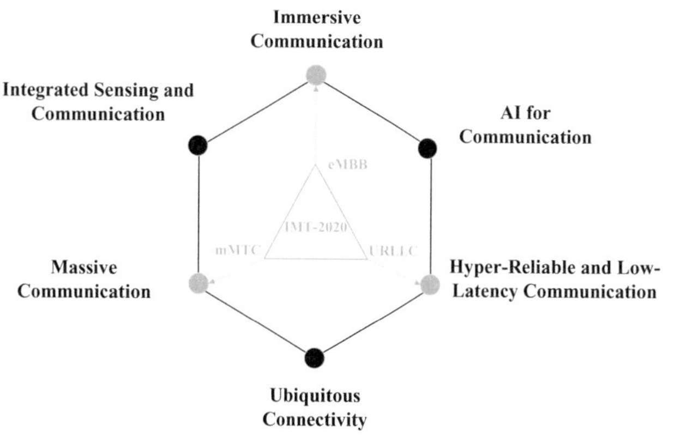


**Fig. 1.2** Usage scenarios of IMT-2030

intelligence (AI) for communications, and immersive communications. The figure highlights the transformative potential of 6G, showcasing its ability to support a broader range of applications, from ultra-high-speed data transfer and real-time control to seamless integration of sensing and communication functionalities. This evolution underscores the shift toward a more connected, intelligent, and adaptive wireless ecosystem, enabling groundbreaking use cases in areas such as autonomous systems, smart cities, and immersive digital experiences.

Early research on 6G is already underway, with a focus on exploring novel technologies such as Terahertz communication, visible light communications (VLC) [11], integrated sensing and communication (ISAC) [12], reconfigurable intelligent surfaces (RIS) [13], low Earth orbit (LEO) satellite networks [14], semantic communications [15], and so on. Terahertz communication, which utilizes higher-frequency bands, offers significantly higher bandwidth and data rates. VLC, on the other hand, uses visible light to transmit data by modulating the intensity of light-emitting diodes (LEDs), enabling high-speed information transmission, particularly beneficial in indoor environments where RF signals may be weak or unreliable. ISAC combines sensing and communication capabilities into a single system, enabling new applications such as autonomous vehicles and smart environments. LEO satellites, orbiting relatively close to Earth, provide global communication services with relatively low latency, revolutionizing global connectivity and playing a crucial role in the future of wireless communication. RIS dynamically adjusts the wireless environment to improve signal propagation and coverage. These technologies hold the potential to further enhance the performance and capabilities of wireless networks, enabling faster speeds, more reliable connections, and support for a wider range of applications in fields such as healthcare, transportation, and entertainment. For instance, 6G could support remote surgery, telemedicine, and personalized healthcare, as well as enable smart traffic management for autonomous vehicles. Industrial and agricultural IoTs are also key components in the 6G era. As research and development continue to advance, 6G promises to shape the future of wireless communication, driving innovation and improving our quality of life.

## 1.2 Three-Level Communication

The concept of semantic communications was introduced by Shannon and Weaver in their work [16], where they outlined a three-tier framework for communication. By moving beyond the traditional focus on bit-level accuracy, semantic communication holds the promise of enabling more intelligent, efficient, and personalized wireless communication systems.

- Level I is known as the technical level, which is concerned with ensuring the accurate transmission of each bit of information regardless of their underlying content. This level focuses on the accuracy and efficiency of transmitting symbols (e.g., bits, words, or signals) from a sender to a receiver. It deals with the technical challenges of communication, such as encoding, decoding, noise reduction, and channel capacity. The primary concern here is ensuring that the transmitted symbols are received correctly, without errors or distortions. Shannon's mathematical theory of communication primarily addresses this level, emphasizing the optimization of information transfer over a noisy channel.
- Level II is referred to as the semantic level, which emphasizes the accurate conveyance of the meaning behind the symbols being transmitted. The semantic level shifts the focus from the transmission of symbols to the meaning conveyed by those symbols. It addresses whether the receiver interprets the message in the way the sender intended. This level considers the clarity, context, and interpretation of the information, ensuring that the meaning is preserved and understood. Challenges at this level include ambiguity, cultural differences, and the complexity of language.
- Level III is termed the effectiveness level, which focuses on how well the transmitted meaning influences actions or behavior in the desired manner. The highest level of communication concerns the impact or effectiveness of the message on the receiver. It evaluates whether the communication achieves its intended purpose, such as influencing behavior, eliciting a response, or driving decision-making. This level goes beyond mere understanding and focuses on the practical outcomes of communication, such as persuasion, motivation, or action.

Existing wireless communication systems, spanning from 1G to 5G, align with the first level of Shannon's framework, where the primary goal is to reliably transmit raw data, often in the form of bits or symbols, across a communication channel. These systems, grounded in classical information theory, have historically focused on optimizing parameters such as channel capacity, spectrum efficiency, and delay to ensure the accurate and reliable transmission of information. From the earliest generations of cellular networks to the more advanced 5G systems, the objective has been to maximize the bit rate and to reduce error rates within the constraints of available bandwidth. The underlying assumption has been that as long as the bit sequence is transmitted correctly, the communication process is successful.

The advent of modern communication technologies has led to rapid advancements in performance, pushing the capabilities of wireless networks closer to the Shannon limit, where maximum channel capacity is achieved under ideal conditions. However, while these traditional systems have made remarkable strides in improving raw data transmission, they have inherently been designed without regard for the actual meaning of the transmitted symbols. That is, they focus on ensuring the delivery of bits, rather than the relevance, context, or task-specific utility of the transmitted data. As a result, the existing systems, despite their technical excellence in terms of data throughput and error correction, struggle to handle the increasing demand for semantic or context-aware data exchange in modern applications, particularly in the era of big data and the IoT.

The exponential growth of data generated by diverse sources—ranging from sensors, mobile devices, social media, to multimedia content—has highlighted the limitations of conventional communication systems. As wireless traffic expands, traditional systems face significant challenges in managing the spectrum and bandwidth constraints while maintaining efficiency. With the sheer volume of data being transmitted, much of it may be redundant, irrelevant, or unnecessary for the receiver to fully understand the intended message. This inefficiency is especially problematic in real-time applications, such as emergency response systems, autonomous vehicles, or remote healthcare, where the value of the information is highly context-dependent. Traditional communications, designed to simply transmit bits, fail to differentiate between what is important or relevant to the task at hand and what is extraneous. Thus, the need for a new communication paradigm that focuses on the meaning of the data (rather than merely its transmission) has become evident.

This is where the concept of semantic communication comes into play. As the next logical step beyond traditional information-theoretic communication systems, semantic communication represents a fundamental shift in how communication is conceptualized and implemented. Unlike traditional wireless communication systems, which focus solely on the accurate transmission of bits, semantic communication seeks to optimize the delivery of meaningful content. This paradigm considers the context of the communication and aims to convey only the most relevant information for a specific task or application, thereby reducing redundancy and improving efficiency.

At its core, semantic communication is a form of task-oriented or goal-oriented communication. Rather than transmitting entire bit sequences or raw data that may require substantial post-processing to extract meaningful insights, semantic communication systems aim to transmit data that is directly tied to achieving a specific objective. This objective could vary depending on the application, such as real-time object recognition in autonomous driving, emergency detection in disaster scenarios, or image classification in medical diagnostics. By focusing only on the information needed to perform a specific task, semantic communication can significantly reduce the amount of data that needs to be transmitted, thereby alleviating the burden on limited spectrum and bandwidth resources.

For instance, in the context of emergency detection, a traditional communication system would likely transmit a large volume of raw sensor data, which might include numerous readings that do not immediately indicate an emergency. However, in a semantic communication framework, only the data that signifies an anomaly or emergency situation would be transmitted, ensuring that the communication is not only more efficient but also more effective in achieving the goal. This shift toward relevance-based communication allows for more intelligent and context-aware networks, which can dynamically adjust to the needs of the application and the network conditions.

Semantic communication also represents a potential post-Shannon communication paradigm, extending beyond the first level of Shannon's framework, which focuses on error-free transmission of bits, to the second and third levels, which account for the extraction, transmission, and utilization of meaningful information. At these higher levels, the aim is no longer merely the reliable transmission of bits but the efficient and purposeful communication of information that has inherent value in achieving a given task. This approach is increasingly important as the demands on communication networks evolve, particularly with the rise of AI-powered applications, IoT systems, and big data analytics, which require more intelligent, context-aware, and efficient data exchange mechanisms.

The emergence of semantic communication is, therefore, seen as a critical solution to breaking through the spectrum and bandwidth bottlenecks that have long plagued traditional communication systems. By focusing on transmitting only the most relevant and task-specific information, semantic communication systems can provide significant improvements in both efficiency and performance, especially as we transition toward more complex, data-intensive communication ecosystems. The potential applications of semantic communication are vast, ranging from improving wireless resource management in satellite networks to enabling more efficient data exchange in next-generation mobile networks, thereby paving the way for smarter, more responsive communication systems that are better aligned with the needs of modern society.

In conclusion, the shift toward semantic communication represents a fundamental evolution in how we approach communication systems, especially as the demand for more intelligent, efficient, and context-aware networks grows. By prioritizing the meaning and relevance of information over the raw transmission of bits, semantic communication promises to reshape the landscape of wireless communication, opening up new possibilities for applications that require real-time, high-quality, and task-oriented data exchange.

## 1.3 AI for Next-Generation Wireless Communication

### 1.3.1 AI-Empowered Wireless Communication at Bit Level

Although 6G remains in the conceptual phase and has not yet been commercialized, there is a pressing need to develop more intelligent approaches to realize the next generation of wireless communication systems. AI and machine learning (ML) have shown tremendous potential in redefining wireless communications and are expected to be an indispensable part of future wireless networks. As wireless communication technologies continue to evolve, AI and ML are becoming increasingly essential to address the growing complexity of next-generation wireless systems and meet the demands of high-performance communication.

One of the most significant contributions of AI/ML in traditional wireless communications is in signal processing at the physical layer. Traditional wireless communication systems rely on predefined, fixed mathematical models to estimate and predict channel characteristics. These models often struggle in highly dynamic environments where factors such as mobility, interference, and varying atmospheric conditions can dramatically impact the communication channel. AI-based techniques, especially machine learning, can overcome these limitations by learning from vast datasets and adapting to real-time network conditions. For example, AI can significantly enhance channel estimation by leveraging historical data to better predict the behavior of the channel. Unlike traditional methods that might rely on static models or predefined assumptions, machine learning algorithms can learn continuously from the environment, allowing for more accurate and adaptive estimation in complex and unpredictable scenarios. This improvement leads to better signal quality, reduced interference, and more efficient data transmission in real time, which is especially important in environments such as urban areas with dense infrastructure or rural areas with variable propagation conditions.

Another major advantage of AI in traditional wireless communication systems lies in data processing and decision-making. The rapid expansion of wireless devices and the increasing volume of transmitted data present a significant challenge for conventional communication systems. These systems often struggle with efficiently processing and managing the growing amounts of data. With machine learning, AI can aid in data interpretation by identifying patterns and correlations within large datasets, automating decisions about how data should be routed, processed, and stored. For example, AI can help identify redundant information, optimizing both data storage and transmission strategies. Traditional communication systems often treat all data as equal, transmitting raw data without considering its relevance or redundancy. AI, however, can optimize this process by focusing on important or high-priority data, filtering out less relevant information, and reducing the burden on network resources. This not only enhances network efficiency but also helps manage the growing load on existing communication infrastructure.

AI and ML also play a pivotal role in resource allocation and network optimization. As wireless networks scale and grow in complexity, traditional methods of resource allocation, such as bandwidth, power, and frequency spectrum management become increasingly inefficient and difficult to manage. AI offers a more flexible and adaptive approach to optimize resource distribution. In modern wireless systems, especially in large-scale networks, traditional optimization techniques often rely on static assumptions and predefined algorithms. As network conditions become more complex, these approaches can become computationally expensive and less effective. AI-based solutions can dynamically adjust resource allocation strategies by learning from real-time data, optimizing for performance factors such as signal-to-noise ratio (SNR), throughput, latency, and power efficiency. For instance, beamforming techniques, which involve adjusting antenna configurations to improve signal quality, can benefit from AI's ability to model environmental factors and optimize antenna patterns in real time. Machine learning algorithms can also facilitate dynamic power control, ensuring that each user or device receives the appropriate amount of power, thus improving energy efficiency and extending the lifespan of devices.

AI can also contribute to predictive analytics for managing network traffic. Wireless communication systems, especially those supporting massive numbers of devices (such as IoT networks), need to adapt to highly fluctuating traffic patterns. In this context, AI can predict future traffic demands based on historical data and current network conditions. By anticipating network congestion or bandwidth demand spikes, AI can adjust the allocation of resources before problems arise, ensuring a smooth and uninterrupted user experience. For example, in cellular networks, AI can predict when and where congestion is likely to occur, allowing the network to proactively reassign users to different cells or channels, optimizing overall network capacity. This proactive approach to network management can significantly reduce latency and packet loss, improving the overall efficiency and reliability of the network.

Localization and positioning are other critical aspects of modern wireless communication systems, especially in applications such as asset tracking, indoor navigation, and location-based services. Traditional GPS-based systems often struggle in indoor environments or urban canyons where satellite signals are weak or obstructed. AI can enhance localization accuracy by using machine learning algorithms to process data from various sources, such as Wi-Fi signals, Bluetooth beacons, or cellular networks, to infer the location of devices with greater precision.

Moreover, AI-driven solutions can improve geospatial data analysis, which can be used for applications ranging from urban planning and smart cities to autonomous driving. By continuously learning from environmental factors, AI can improve the accuracy of positioning systems in real-time, making it possible to achieve highly reliable and accurate localization in environments where traditional methods fall short. AI can also be applied to predictive maintenance and fault detection in wireless communication networks. By leveraging machine learning models to analyze data from network components (e.g., base stations, routers, and antennas), AI can predict when certain elements are likely to fail or require maintenance. This proactive approach reduces downtime, minimizes service disruptions, and ensures that the network operates smoothly and efficiently. For instance, AI can detect signal degradation, hardware failures, or performance bottlenecks before they impact the end-user experience, allowing for quick corrective action to be taken.

The integration of AI and ML into wireless communication systems marks a transformative shift toward more intelligent, adaptive, and efficient networks. From signal processing and resource allocation to network optimization and predictive analytics, AI-driven solutions are essential for overcoming the limitations of traditional methods and enhancing the performance of wireless communication systems. As network demands continue to grow and become more complex, the ability of AI to provide real-time, data-driven decision-making will be crucial to the evolution of next-generation wireless systems. AI's role in optimizing various aspects of wireless communication ensures more reliable, efficient, and scalable networks that are better suited to meet the challenges of the future.

### 1.3.2 AI-Empowered Wireless Communication at Semantic Level

AI plays a pivotal role in transitioning traditional wireless communications to semantic communications, a paradigm shift where systems understand and prioritize the meaning of the transmitted data, rather than just focusing on bit-level accuracy. Traditional communication systems are designed to transmit raw data, relying on bit-level optimization techniques for efficient and reliable data delivery. While this approach has served well for many years, the ever-growing complexity of applications, data traffic, and the demands of next-generation technologies like 6G require a more sophisticated communication model—one that can handle not just the transmission of bits but the interpretation and delivery of meaningful information.

The core idea of semantic communication is to transmit not only the bits that form the message but also the meaning or context behind those bits, ensuring that the communication is more aligned with the end user's needs. This approach is fundamentally different from traditional communication methods, which focus purely on ensuring that bits are transmitted accurately between sender and receiver. AI technologies, particularly those rooted in natural language processing (NLP), deep learning, and computer vision, are at the forefront of enabling semantic communication systems. These technologies enable the system to interpret and extract essential semantic content from multimodal data—whether it's text, images, audio, or sensor data—and then adapt the transmission to reflect only the most relevant information. Rather than sending an entire dataset or raw bits, AI systems can identify and prioritize what matters most for the communication task at hand. For example, in an emergency scenario, AI might prioritize sending information about critical conditions—such as the location of a fire or the presence of hazardous materials—rather than transmitting irrelevant background data. By learning from context and past interactions, AI-based systems can tailor their communication strategy, adapting to changing environments and user needs in real time. This adaptive communication is particularly crucial for applications such as autonomous driving, where fast and reliable decision-making is essential, or in smart cities, where real-time coordination of sensors, infrastructure, and devices must be done intelligently to optimize the city's functionality.

This transition to semantic communication, enabled by AI, brings a host of advantages, particularly in the context of 6G networks and beyond. One of the primary advantages of semantic communication is its ability to compress information by transmitting only the essential semantic content. In traditional systems, all bits are transmitted, even if much of the information is redundant or irrelevant. Semantic communication, on the other hand, extracts the most relevant semantic content and encodes it efficiently, reducing the data load. This reduction in data transmission leads to significant improvements in network efficiency, reduces congestion, and enhances the overall capacity of the communication system. This is particularly important in high-density environments or remote areas, where bandwidth and spectrum are limited.

AI-empowered techniques can be employed to detect and correct errors for more reliable channel coding. Traditional error correction codes, while effective, often rely on the assumption that errors occur randomly across the transmitted bits. In contrast, AI can use contextual information, such as the meaning of the message or patterns in previous transmissions, to improve the accuracy of error detection and correction. For instance, if certain parts of the transmitted data are less important, AI can focus on error correction for the most critical parts, leading to more reliable and robust communication. Moreover, AI models can adapt dynamically to changing network conditions, improving the communication's reliability under different levels of interference or signal degradation.

With semantic communication, systems are not just transmitting raw data; they are delivering content that is highly relevant and task-specific. For example, in VR or AR applications, AI can ensure that only the most contextually significant elements of the experience are communicated, improving frame rates and reducing latency while preserving user experience. Furthermore, natural language interfaces—such as chatbots or voice assistants—benefit from semantic communication by enabling more intuitive, context-aware interactions. The system can focus on interpreting the intent behind user inputs, rather than merely processing the literal words, resulting in more fluid and accurate responses.

At the heart of semantic communication is the joint semantic-channel coding (JSCC). Traditional encoding methods focus on converting raw data into a format suitable for transmission, such as modulating bits into signals. In contrast, semantic encoding emphasizes understanding the meaning behind the data. AI algorithms can encode contextual information, such as the importance or relevance of specific data, into a more efficient format that requires less bandwidth. During decoding, AI ensures that the essential semantic content is extracted and interpreted correctly. This reduces redundancy and focuses communication on what truly matters to the task at hand. By focusing on the semantic content and not just the raw data, JSCC can reduce the amount of redundant data transmitted over the network. This reduces bandwidth usage and improves the overall efficiency of communication systems. Since only the essential parts of the message are transmitted, resources can be saved for other important tasks, making JSCC highly suitable for high-demand applications such as real-time video streaming, smart cities, or autonomous vehicles. Traditional communication systems rely on adding redundancy to protect the entire bitstream. However, in JSCC, redundancy is applied in a more targeted manner. By combining semantic prioritization with error protection, JSCC can offer enhanced reliability, especially in scenarios where errors in critical data must be minimized. This leads to more reliable communication in challenging environments, such as low-SNR conditions or wireless interference.

As AI continues to evolve, it will become increasingly integrated into the semantic communication, driving the next generation of 5G and 6G networks toward highly efficient, adaptive, and intelligent systems. In 6G networks, which are expected to support a diverse range of applications including autonomous vehicles, smart cities, virtual and augmented reality, and industrial automation, semantic communication will be essential for managing the vast and dynamic flow of data while ensuring that communication is not only efficient but also contextually relevant. This will allow for task-specific communication, where networks can prioritize the delivery of actionable, high-value information rather than focusing solely on transmitting large volumes of raw data. Ultimately, the integration of AI empowered semantic communication has the potential to revolutionize the way we communicate, moving beyond traditional bit-level transmission to a more meaningful, context-aware, and intelligent form of wireless communication. This shift toward semantic communication is poised to enable the next-generation wireless systems that are more resilient, efficient, and capable of addressing the increasing complexity and demands of modern applications.

## 1.4 Summary

This chapter provides an overview of the transformative journey of wireless communications, from analog voice service of 1G to the intelligent multimodal data service of future 6G. It highlights key milestones, including the digital revolution of 2G, the mobile data era of 3G, the smartphone revolution of 4G, and the ultra-high-speed 5G networks. This chapter also introduces the future 6G which aims to deliver unprecedented connectivity and intelligence through artificial intelligence, as well as the emerging semantic communications. Overall, this chapter outlines both the existing development and the cutting-edge technologies driving the future of wireless communication systems.

## Problems

1.1 Which of the following is considered the fundamental concept behind wireless communication?

1. Electromagnetic wave propagation
2. Packet switching
3. Optical fiber transmission
4. Digital modulation

1.2 Who is credited with validating Maxwell's theories by experimentally demonstrating radio waves?

A. Guglielmo Marconi
B. Heinrich Hertz
C. Claude Shannon
D. Michael Faraday

1.3 Who is credited with developing the mathematical theory of communication, laying the foundation for modern wireless communication?

A. Guglielmo Marconi
B. Alexander Graham Bell
C. Claude Shannon
D. Nikola Tesla

1.4 Which generation of cellular networks introduced encrypted digital voice calls and SMS?

A. 1G
B. 2G
C. 3G
D. 4G

1.5 Which generation of wireless networks introduced packet-switched data services for the first time?

A. 1G
B. 2G
C. 3G
D. 4G

1.6 What is the primary advantage of using OFDM (Orthogonal Frequency Division Multiplexing) in 4G and beyond?

A. Increases signal bandwidth
B. Improves battery life of devices
C. Eliminates the need for modulation
D. Enhances spectral efficiency and reduces interference

1.7 Which of the following is NOT a major feature of 5G networks?

A. Ultra-low latency
B. High-speed broadband connectivity
C. High energy consumption
D. Massive machine-type communication

1.8 When is the 6G network expected to be commercially deployed?

A. 2025
B. 2030
C. 2035
D. 2040

1.9 Semantic communication differs from traditional communication in that it primarily focuses on:

A. Extracting semantic meaning
B. Minimizing transmission error
C. Reducing bandwidth consumption only
D. Ensuring only encrypted transmission

1.10 AI improves channel estimation in wireless systems by:

A. Using static models
B. Reducing bandwidth
C. Ignoring noise
D. Learning from real-time data

1.11 Which of the following best describes the Shannon-Weaver three-level theory of communication?

A. It defines communication as a process of encoding and decoding semantic meaning
B. It models communication in terms of source, channel, and receiver with noise
C. It emphasizes the importance of linguistic structures in message transmission
D. It is specifically designed for cognitive radio networks

1.12 In Shannon and Weaver's framework, Level III communication refers to:

A. Bit-level accuracy
B. Semantic interpretation
C. Influence on receiver behavior
D. Channel capacity optimization

1.13 Which of the following best describes the role of AI in semantic communications?

A. It assists in reducing data redundancy without affecting meaning
B. It increases transmission power for better connectivity
C. It replaces traditional communication structure
D. It focuses solely on error correction

1.14 Which of the following best describes Weaver's achievement?

A. Information theory
B. Automatic control theory
C. Theory of relativity
D. Electromagnetic radiation theory

1.15 Which technology is expected to most likely play a major role in enabling semantic communications?

A. Blockchain
B. DNA computing
C. Quantum cryptography
D. Large language model

## References


1. G. Marconi, Wireless telegraphic communication. Nobel Lect. **11**, 198–222 (1909)
2. R. Simons, Guglielmo marconi and early systems of wireless communication. GEC Rev. **11**(1), 37–55 (1996)
3. C.E. Shannon, A mathematical theory of communication. Bell Syst. Tech. J. **27**(3), 379–423 (1948)
4. T.M. Cover, *Elements of information theory*. Wiley (1999)
5. P. Nicopolitidis, M.S. Obaidat, G.I. Papadimitriou,A.S. Pomportsis, *Wireless networks*. Wiley (2003)
6. M. Zeng, A. Annamalai, V.K. Bhargava, Recent advances in cellular wireless communications. IEEE Commun. Mag. **37**(9), 128–138 (1999)
7. C. Smith, D. Collins, *3G Wireless Networks*. McGraw-Hill Education (2002)
8. U. Varshney, R. Jain, Issues in emerging 4G wireless networks. Computer **34**(6), 94–96 (2001)
9. M. Agiwal, A. Roy, N. Saxena, Next generation 5G wireless networks: a comprehensive survey. IEEE Commun. Surv. Tuts. **18**(3), 1617–1655 (2016)
10. F. Tariq, M.R. Khandaker, K.-K. Wong, M.A. Imran, M. Bennis, M. Debbah, A speculative study on 6G. IEEE Wirel. Commun. **27**(4), 118–125 (2020)
11. A. Jovicic, J. Li, T. Richardson, Visible light communication: opportunities, challenges and the path to market. IEEE Commun. Mag. **51**(12), 26–32 (2013)
12. Y. Cui, F. Liu, X. Jing, J. Mu, Integrating sensing and communications for ubiquitous IoT: applications, trends, and challenges. IEEE Netw. **35**(5), 158–167 (2021)
13. Q. Wu, R. Zhang, Towards smart and reconfigurable environment: intelligent reflecting surface aided wireless network. IEEE Commun. Mag. **58**, 106–112 (2020). (Jan.)
14. G. Maral, J.-J. de Ridder, B.G. Evans, M. Richharia, Low earth orbit satellite systems for communications. Int. J. Satell. Commun. Netw. **9**(4), 209–225 (1991)
15. Z. Qin, L. Liang, Z. Wang, S. Jin, X. Tao, W. Tong, G.Y. Li, AI empowered wireless communications: from bits to semantics. Proc. IEEE **112**(7), 621–652 (2024)
16. C.E. Shannon, W. Weaver, *The Mathematical Theory of Communication* (University of Illinois Press, Urbana, IL, 1949)

# Introduction of Semantic Communication

# 2

**Abstract**

Classic Shannon information theory provides the mathematical framework for understanding and quantifying information in communication systems. It forms the backbone of modern telecommunications, data compression, and error correction. Shannon's key insights revolve around defining the limits of information transmission and developing methods for efficiently encoding, transmitting, and decoding data, even in the presence of noise. Semantic communication is still in its early stage of research, in which the unified mathematical theory and performance evaluation system with generalization value are crucial to the development of semantic communication. This chapter discusses the basic fundamental mathematical framework and theory of semantic communications and the role of the performance evaluation metrics in the design and implementation of semantic communication systems.

**Keywords**

Semantic information theory • Semantic entropy • Semantic rate distortion • Semantic channel capacity • Performance evaluation metric • Benchmark dataset

## 2.1 Fundamental Theory of Semantic Communication

### 2.1.1 Shannon's Information Theory

Shannon Information Theory, founded by Claude Shannon in the mid-twentieth century, provides the mathematical framework for understanding and quantifying information in communication systems. It forms the backbone of modern telecommunications, data compression, and error correction. Shannon's key insights revolve around defining the limits of information transmission and developing methods for efficiently encoding, transmitting, and decoding data, even in the presence of noise. At the core of Shannon's theory are fundamental concepts such as entropy, mutual information, and channel capacity. These concepts are crucial for understanding how information is represented and transferred through communication channels. Shannon's work has had a profound impact, enabling the design of reliable communication systems, efficient data storage, and error-correcting codes. In this section, we explore the key metrics in Shannon Information Theory, including entropy, mutual information, divergence, and channel capacity. We also discuss Shannon's three fundamental theorems, which form the foundation of modern communication systems.

Entropy is a measure of the uncertainty or unpredictability of a random variable. It quantifies the amount of information required to describe the state of a system. For a discrete random variable $X$, the entropy $H(X)$ is defined as

$$H(X) = -\sum_{x \in X} p(x) \log p(x), \tag{2.1}$$

where $p(x)$ is the probability of the event $x$. Entropy measures the average number of bits needed to represent an outcome of $X$. Higher entropy means greater uncertainty and more information. For a deterministic (non-random) variable, entropy is zero, since there is no uncertainty.

Mutual information measures the amount of information that one random variable shares with another. It quantifies the reduction in uncertainty of one variable given the knowledge of the other. For two random variables $X$ and $Y$, the mutual information $I(X;Y)$ is defined as

$$I(X;Y) = H(X) + H(Y) - H(X,Y), \tag{2.2}$$

where $H(X,Y)$ is the joint entropy of $X$ and $Y$. Mutual information is symmetric, meaning $I(X;Y) = I(Y;X)$, and is always non-negative. It is zero if $X$ and $Y$ are independent, indicating that knowing $X$ provides no information about $Y$.


W. Chen and Z. Qin, *Semantic Communications*, Textbooks in Telecommunication Engineering,
https://doi.org/10.1007/978-3-032-11005-3_2

Kullback-Leibler (KL) divergence measures the difference between two probability distributions. It quantifies how much information is lost when using one distribution to approximate another. For two probability distributions $P$ and $Q$, the KL divergence is defined as

$$D_{\mathrm{KL}}(P \parallel Q) = \sum_{x \in X} p(x) \log \left( \frac{p(x)}{q(x)} \right). \tag{2.3}$$

The KL divergence is always non-negative and is zero if $P = Q$. It is not symmetric, meaning $D_{\mathrm{KL}}(P \parallel Q) \neq D_{\mathrm{KL}}(Q \parallel P)$. It can be interpreted as the "extra" entropy required when using $Q$ to approximate $P$.

Channel capacity is the maximum rate at which information can be reliably transmitted over a communication channel with a given noise level. It is defined in terms of mutual information between the input and output of the channel. For a discrete memoryless channel, the channel capacity $C$ is given by:

$$C = \max_{p(x)} I(X; Y), \tag{2.4}$$

where $p(x)$ is the probability distribution over the input symbols $X$, and $I(X; Y)$ is the mutual information between the input and output. Channel capacity represents the upper limit on the data rate that can be achieved with arbitrarily low error probability. It depends on the channel characteristics (e.g., noise level, bandwidth) and the encoding strategy used.

Shannon's work has had a profound impact, enabling the design of reliable communication systems, efficient data storage, and error-correcting codes. This theory is divided into three major parts:

- **Source Coding**: The process of efficiently representing information produced by a source.
- **Channel Coding**: The process of reliably transmitting information over a noisy communication channel.
- **Rate-Distortion Theory**: The trade-off between data compression and the distortion (loss of quality) in the transmitted information.

The first theorem establishes the minimum average number of bits required to represent a source of information. It states that if a source produces symbols with probabilities $p(x)$, the average number of bits required to encode the source is at least the entropy of the source. This is also known as the *optimal encoding*.

$$L_{\min} = H(X), \tag{2.5}$$

where $L_{\min}$ is the minimum average number of bits needed per symbol. This theorem underpins data compression algorithms, ensuring that no loss of information occurs in the process.

The second theorem describes the ability to transmit information over a noisy channel. It states that for any communication channel, there exists a coding scheme that allows information to be transmitted at rates below the channel capacity $C$ with an arbitrarily small probability of error. In other words, error-free communication is possible as long as the transmission rate does not exceed the channel capacity.

$$R \leq C, \tag{2.6}$$

where $R$ is the transmission rate (bits per second), and $C$ is the channel capacity. This theorem is foundational to error-correcting codes, which allow reliable communication even in noisy environments.

The third theorem defines the minimum number of bits required to represent a source at a specified level of distortion. It is concerned with the trade-off between compression and fidelity (i.e., the distortion between the original and compressed signals). Given a source $X$ and a distortion measure $d(x, \hat{x})$, the minimum rate $R(D)$ required to achieve an average distortion $D$ is

$$\begin{aligned} R(D) &= \min_{p(\hat{x})} I(X; \hat{X}) \\ \text{s.t.} \quad & \mathbb{E}[d(X, \hat{X})] \leq D, \end{aligned} \tag{2.7}$$

where $\hat{X}$ is the reconstructed version of $X$, and $D$ is the allowed distortion. This theorem is critical in lossy compression algorithms, such as those used in image or audio compression.

### 2.1.2 Mathematical Theory of Semantic Communication

From the technical level to the semantic level, the goal of communication has changed from accurate transmission of symbols to effective exchange of semantics. The classic Shannon information theory is based on probability and statistics, ignoring the semantic representation of specific tasks. It solves two basic problems from the technical level: the critical data compression value and the critical communication transmission rate value. The mathematical theory of semantic communication and the mathematical representation of semantics can be reduced to the problems of semantic information theory. Although there is no recognized and unified theoretical framework for semantic information theory, in recent years, academia and industry have increasingly conducted research on semantic information theory. In 2021, Huawei proposed the top ten challenges in the field of information and communication technology (ICT) in the post-Shannon era, listing semantic information theory as the top problem in basic theory.

The measurement of semantic information is the foundation of semantic information theory. Entropy is a measure of the uncertainty of random variables, and semantic entropy is a measure of semantic uncertainty or the amount of information. So far, the intuitive and universal mathematical description of semantic entropy is still a challenging problem: on the one hand, the basic connotation of semantics is not easy to define and thus difficult to measure; on the other hand, people do not understand the mechanism and process of semantic generation.

Semantic entropy originated in the pre-Shannon era and is related to the analysis of natural language. In 1952, the concept of semantic entropy was proposed for language task [1], using logical probability to measure the amount of semantic information contained in a sentence:

$$H_s(e) = -\log m(e), \tag{2.8}$$

where $m(e)$ refers to the probability that event $e$ is true in all possible cases, that is, the logical probability of event $e$. It can be seen that the greater the logical probability of a sentence being true, the less information it contains. However, this creates a paradox, that is, any fact that contradicts itself will have infinite information. For example, we have

$$H_s(e \wedge \neg e) \rightarrow \infty, \tag{2.9}$$

which indicates the probability of "event $e$ and event not $e$" occurring is infinite. In 2004, a stronger semantic information theory framework was proposed [2], which uses the distance from the real event to represent the amount of semantic information, thus solving this paradox. In 2011, semantic information was quantitatively described based on truth similarity [3]. These two methods measure the information of an event relative to a reference event, and the information value is always between 0 and 1. However, these measures all rely on the existence of a reference event. In essence, their work provides a measure of the semantic similarity between two sentences, rather than a measure of semantic uncertainty or the amount of information. In 2011, propositional logic was used to expand the representation of $m(e)$ [4], further expanding the meaning of semantic entropy in language tasks. In 2024, a general semantic communication theory framework based on synonymous mapping was proposed [5], which considered the impact of information meaning on semantic information measurement. This model probabilistically maps different source sequences with the same semantics from the grammatical space to the same element in the semantic space, and defines semantic entropy

$$H_s(\tilde{U}) = E_p\left[\log \frac{1}{p(U_{\tilde{u}})}\right], \tag{2.10}$$

and semantic mutual information

$$I_s(\tilde{U}; \tilde{V}) = H_s(\tilde{U}) + H_s(\tilde{V}) - H_s(\tilde{U}, \tilde{V}). \tag{2.11}$$

These concepts are redefined in the semantic space in comparison with the classic Shannon information theory. Here, $\tilde{U}$ represents the semantic alphabet.

For intelligent tasks, a method was proposed to measure the semantic entropy of words in text [6]. Specifically, for a word $w$, the semantic entropy can be expressed as

$$\begin{aligned} H_s(w) &= H(T|w) + N(w) \\ &= -\sum_{t\in T} p(t|w)\log p(t|w) + p(NULL|w)\log F(w). \end{aligned} \tag{2.12}$$

Here, $H(T|w)$ represents translation inconsistency, which indicates the uncertainty caused by a word being translated into different ways, $T$ represents the set of target words, $N(w)$ represents the contribution of the empty link of $w$, which indicates the possibility of encountering difficulties when translating from one language to another, and $F(w)$ is the frequency of $w$. For classification tasks, the concept of matching and membership was introduced to define semantic entropy [7]. Membership is a difficult concept in fuzzy set theory and is generally given based on experience. Define $\varsigma$ as a semantic concept and $\mu_\varsigma(x)$ as the membership of each $x \in X$. For a certain category $C_j$, the matching degree $D_j(\varsigma)$ is defined as

$$D_j(\varsigma) = \frac{\sum\limits_{x\in X_{C_j}} \mu_\varsigma(x)}{\sum\limits_{x\in X} \mu_\varsigma(x)}. \tag{2.13}$$

For category $C_j$, its semantic entropy is defined as

$$H_s(\varsigma) = -D_j(\varsigma)\log D_j(\varsigma). \tag{2.14}$$

Adding the semantic entropy of all classes in $X$, we can get the overall semantic entropy on set $X$. A method to quantify task-related semantic entropy was proposed [8], defining semantic entropy as the minimum number of semantic queries about data X whose answers are sufficient to predict task $V$:

$$\begin{aligned} &H_S(X; V) \triangleq \min_E E_S[|Code_Q^E(X)|], \\ &s.t. \;\; P(v|Code_Q^E(X)) = P(v|x), \quad \forall x, v, \end{aligned} \tag{2.15}$$

where $Code_Q^E(X)$ represents the query vector extracted from $X$ using the semantic encoder $E$.

Semantics is the understanding of signals. Knowledge plays an important role in the process of semantic generation, playing the role of semantic encoding and representation. The semantic entropy of a sentence was studied [9] from the perspective of knowledge base based on logical probability. Specifically, let the knowledge base be $K$, and use $p[K|-q]$ to represent the probability that sentence $q$ is correct relative to the knowledge base $K$, which can be simplified to $p_q = p[K|-q]$, then the semantic entropy of q relative to K is given as

$$H_s(q) = -P_q \log P_q + (1 - P_q) \log (1 - P_q). \tag{2.16}$$

It is well known that extended patterns can combine simple things into complex systems and even generate intelligence. In the human language system, the components of subject, predicate, object, attributive, adverbial and complement form sentences, which can express meanings that a single word cannot have. Inspired by this, the intelligent semantics was predicted to be extensible, and the representation of semantic entropy should reflect the concept of knowledge expansion [10]. In the expansion of semantics, a collision occurs at the knowledge level. Semantics is the product of the interaction between knowledge and signals under the collision. For example, Apple Inc. belongs to the category of business companies, and thirteen belongs to the category of numbers, but their collision may produce a word - iPhone, which belongs to the category of mobile communication products. Let $X_1$ and $X_2$ represent signals, and $K_A^1$ and $K_A^2$ represent two knowledge instances, then the semantic entropy can be expressed as

$$H_S(X_1 \oplus X_2, K_A^1 \odot K_A^2), \tag{2.17}$$

where $\oplus$ represents expansion and $\odot$ represents collision.

Different from logical probability or statistical probability models, semantic entropy is derived in different forms for different tasks and backgrounds. Semantic information was defined as grammatical information that describes the relationship between a system and its environment [11]. Semantic entropy based on a language understanding model was derived [12] with the aid of background knowledge. The generalized information theory was introduced [13] in which mathematical measurement was used to describe semantic information based on the Bayesian formula, logical probability and fuzzy set theories.

In semantic communication, information extraction and encoding at the semantic level can help further compress the redundancy of semantic information and extract the most important semantic features, thereby improving the effectiveness of semantic transmission. The semantic rate distortion was defined as [14]:

$$\begin{aligned} R(D_s, D_a) &= \min I(X; \hat{S}, \hat{X}) \\ \text{s.t.} \quad & E[\hat{d}_s(X, \hat{S})] \le D_s, \\ & E[\hat{d}_a(X, \hat{S})] \le D_a. \end{aligned} \tag{2.18}$$

Among them, $S$ and $\hat{S}$ are the understanding of semantics by the sender and receiver, and $X$ and $\hat{X}$ are semantic representations. $D_s$ represents semantic distortion, and $D_a$ represents signal distortion, which represent the distortion caused by semantic coding and channel noise, respectively. Through the above simplification, this problem becomes a Shannon rate-distortion problem with two distortion metrics. Moreover, the semantic rate-distortion of two users was further generalized from the perspective of side information, which can be expressed as [15]:

$$R(D_1, D_2, D_s) = \min I(X_1, X_2; \hat{X}_1, \hat{X}_2, \hat{S}|Y), \tag{2.19}$$

where $X_1$ and $X_2$ are the semantic representations of the two users, respectively, and $Y$ is the side information.

The capacity of a semantic channel was expressed as the maximum rate of semantic information transmitted through the semantic channel or the ratio of the maximum semantic communication volume to the communication data size [16]. On the one hand, at this rate, semantic information can be sent through the semantic channel and recovered at the output of the semantic channel; on the other hand, the transmission rate of semantic data on the channel cannot exceed the semantic capacity of the channel. In 2011, the semantic channel capacity was defined as [4]:

$$C_s = \sup_{P(Z|X)} \left\{ I(X; V) - H(Z|X) + \overline{H_S(V)} \right\}. \tag{2.20}$$

For any $\epsilon > 0$ and $R < C_s$, there exists a coding strategy that makes the maximum probability of semantic errors less than $\epsilon$. Where $X$ is the source, $V$ is the transmission task, $Z$ is the semantic representation, $I(X; V)$ is the mutual information between the source X and the transmission task $V$, $H(Z|X)$ represents the semantic uncertainty of the encoding, and $\overline{H_S(V)}$ is the average logical information of the received message of task $V$. The semantic channel capacity can be higher or lower than the Shannon channel capacity $I(X; V)$, depending on the semantic encoding strategy and the receiver's ability to interpret the received message. If $\overline{H_S(V)} - H(Z|X) > 0$, it means that the receiver can handle semantic ambiguity, the receiver can answer questions with the received sentences, and semantic compression can achieve a higher transmission rate. In this case, $C_s$ is higher than Shannon capacity. If $\overline{H_S(V)} - H(Z|X) < 0$, the receiver cannot resolve the original semantic ambiguity, the receiver cannot answer questions based on the received sentences, and $C_s$ is lower than Shannon capacity.

## 2.2 Performance Evaluation Metric

Performance evaluation is an important part of the design of semantic communication systems. The performance of the communication system can be measured from the two aspects of accuracy and timeliness. The evaluation systems used by existing semantic communication-related research have mainly evolved from the traditional communication system evaluation system, or migrated from the downstream semantic task evaluation system. For the situation evolved from the traditional communication system evaluation system, the corresponding accuracy evaluation index often uses the bit error rate, and the timeliness evaluation index often uses the bit transmission rate. The problem with this type of evaluation index is that it cannot effectively reflect the ability of the communication transceiver to transmit semantic information. For the situation migrated from the downstream semantic task evaluation system, since the system model mostly uses the communication module and the downstream semantic task module to determine the structure and the end-to-end training fixed parameters, this type of evaluation index has the characteristics of being highly related to the communication task scenario and information modality. For example, semantic communication systems for semantic reconstruction generally use the error function of the original information and the reconstructed information as the accuracy metric, and the compression ratio function as the timeliness metric; while semantic communication systems for semantic tasks such as text translation and image segmentation generally use the difference between the model output and the supervision label as the accuracy metric.

The definitions of the above evaluation indicators are different, and the range of values is significantly different, which leads to the inability to horizontally connect the system models of many studies. Even for the same evaluation indicators, due to differences in channels, interference, source semantic characteristics, etc., semantic transmission schemes in different scenarios cannot be effectively compared. A unified and generalized semantic communication evaluation system can intuitively evaluate the performance of the communication system from a holistic perspective, which can not only verify the rationality of the system, but also provide a reference for the optimization and improvement of the system, and then provide a direction for the evolution of the system. Relevant research on the semantic communication evaluation system needs to be carried out urgently to lay a solid foundation for the steady development of semantic communication.

### 2.2.1 Accuracy-Based Metric

The performance evaluation indicators of semantic communication are highly related to specific communication task scenarios and information modalities. Among them, information reconstruction tasks are widely used in application scenarios where humans are the final receiving users, such as audio and video calls, multimedia file cloud backup, etc. This type of task requires the information to be restored to the information modality of the sending end at the receiving end, and the accuracy of the information is guaranteed as much as possible. The performance evaluation indicators are defined based on the difference between the sent information and the reconstructed information; non-information reconstruction tasks often appear in human-machine and machine-machine interaction scenarios, such as intelligent driving and industrial IoT. This type of task requires that the key content of the sent information can be understood and applied, and the performance evaluation indicators are defined based on the difference between the output information and the corresponding supervision label.

For information modalities, humans use different key capture methods for modal information such as text, image/video, and voice, and the distortion of each modal information cannot be simply measured uniformly by methods such as Euclidean distance. There are great differences in the subsequent tasks derived from each modal information. Tasks such as text sentiment analysis, intelligent question and answer, image segmentation, and speech recognition all need to define their own performance evaluation indicators. Based on the division of reconstruction tasks and other representative artificial intelligence tasks (such as text classification, image object detection, and speech recognition), this section analyzes and summarizes the performance evaluation indicators of three common modal information transmission models: text, image/video, and speech, laying the foundation for the subsequent construction of a unified and generalized semantic communication performance evaluation system.

For text signals, the bilingual evaluation understudy (BLEU) is the most commonly used evaluation indicator for measuring the quality of text information reconstruction, representing the proportion of the number of phrases in the reconstructed text that appear in the original text. The BLEU is defined based on n-gram which is given as

$$\mathrm{BLEU} = \mathrm{BP} \cdot \exp\left(\sum_{n=1}^{N} \omega_n \log p_n\right). \tag{2.21}$$

Here, BP is the penalty term for the reconstructed text sentence being too short which is given as

$$\mathrm{BP} = \begin{Bmatrix} 1, & \text{if } l_{\hat{X}} > l_X \\ e^{1-\frac{l_X}{l_{\hat{X}}}}, & \text{if } l_{\hat{X}} \le l_X \end{Bmatrix}. \tag{2.22}$$

$l_{\hat{X}}$ and $l_X$ are the lengths of the reconstructed text and the original text sentence, respectively, $\omega_n$ is the normalized weight, and $N = 4$ is an empirical value. $p_n$ represents n-gram accuracy which is defined as

$$p_n = \frac{\sum\limits_{n-\text{gram} \in \hat{X}} \text{Count}_{clip}(n - \text{gram})}{\sum\limits_{n-\text{gram}' \in \hat{X}} \text{Count}_{clip}(n - \text{gram}')}. \tag{2.23}$$

$\text{Count}_{clip}(n - \text{gram})$ and $\text{Count}_{clip}(n - \text{gram}')$ represent the number of occurrences of n-length phrases in the original text $X$ and the reconstructed text $\hat{X}$, respectively. The physical meaning of BLEU is to measure the proportion of all n-length phrases in the reconstructed text that appear in the original text at the same time. Although it is the most commonly used indicator for text information reconstruction, BLEU still cannot achieve an understanding of lexical semantics and has shortcomings such as not considering the recall rate of the lexical corpus. In 2021, a text-oriented semantic channel joint coding framework was proposed based on natural language processing, and a semantic similarity indicator was designed based on the parameters of the neural network as a supplement to BLEU [17].

For speech signals, the mean opinion score (MOS) is a subjective metric used to evaluate the quality of speech signals based on human listeners' perceptions. It is one of the oldest and most widely used speech quality metrics. MOS is typically obtained by asking a panel of human listeners to rate the quality of a speech sample on a scale from 1 to 5, where

- 1 represents "bad" quality,
- 2 represents "poor" quality,
- 3 represents "fair" quality,
- 4 represents "good" quality, and
- 5 represents "excellent" quality.

The final MOS score is the average rating given by all participants in the subjective test. MOS is often used as a reference for validating objective metrics as it provides a direct correlation with human perception. However, since MOS requires human participation, it is time-consuming and may not be suitable for real-time or large-scale evaluations.

Perceptual evaluation of speech quality (PESQ) is a standardized indicator that can evaluate the reconstruction or transmission quality of speech signals under various conditions. PESQ is a widely adopted objective measure for evaluating the quality of speech signals in telecommunication systems. Developed by the International Telecommunication Union - Telecommunication Standardization Sector (ITU-T) in the recommendation P.862 as a replacement for the earlier models like the speech intelligibility index (SII) and the MOS, PESQ is designed to correlate well with human listeners subjective assessments. PESQ is an indicator to measure the intelligibility of the reconstructed speech signal, focusing on people's subjective feelings, and was widely used in early wireless communication systems. PESQ works by comparing the processed speech signal (after transmission or processing) to the original, unprocessed reference signal. The comparison is based on both distortion and the perceptual quality of the speech signal, considering the nonlinearities of human hearing. It incorporates a perceptual model that simulates how humans perceive changes in speech quality due to factors such as noise, compression, and packet loss.

Specifically, PESQ is defined as

$$\text{PESQ} = 4.5 - 0.1 d_{\text{SYM}} - 0.0309 d_{\text{ASYM}}. \tag{2.24}$$

$d_{\text{SYM}}$ and $d_{\text{ASYM}}$ are the symmetrical interference and asymmetrical interference of the speech information obtained by the algorithm, respectively, which reflects the prediction accuracy and generalization ability of the reconstruction model. PESQ provides a score between 1.0 (bad quality) and 4.5 (excellent quality). A PESQ score of 3.5 or higher indicates good speech quality, while scores below 3.0 generally reflect poor quality. PESQ was used as a system transmission performance evaluation index [18] in which a semantic channel joint coding framework was designed for speech based on this index.

The signal-to-distortion ratio (SDR) is a simple but effective metric for assessing the quality of a speech signal in the presence of distortion or noise. SDR compares the original speech signal to the distorted version, providing a measure of the overall degradation in signal quality due to transmission or processing. SDR is defined as the ratio of the power of the original signal to the power of the distortion (error or noise). Mathematically, it is given by

$$\text{SDR} = 10\log_{10}\left(\frac{\|x\|^2}{\|e\|^2}\right), \tag{2.25}$$

where $x$ is the original clean speech signal, and $e$ is the distortion or error signal (the difference between the processed signal and the original signal). SDR is expressed in dB, where a higher SDR value indicates better quality. Typical SDR values range from 0 dB (extremely distorted) to 30 dB (high quality). A value greater than 10 dB generally indicates an acceptable level of quality for speech communications. SDR is easy to compute and provides a straightforward way to assess signal degradation. Unlike PESQ, SDR does not consider the perceptual aspects of speech quality in detail, but rather provides an overall estimate of the distortion. It is commonly used in research to quantify the effects of speech enhancement or noise suppression techniques.

For image or video signals, peak signal-to-noise ratio (PSNR) and structural similarity index (SSIM) are the most commonly used indicators for evaluating image/video transmission quality. PSNR has a simple definition and low computational complexity, and can accurately reflect the pixel-level reconstruction error of image signals. PSNR is defined as the logarithmic function of the mean square error (MSE). The PSNR is given by the following formula:

$$\mathrm{PSNR} = 10 \cdot \log_{10}\left(\frac{\mathrm{MAX}_I{}^2}{\mathrm{MSE}}\right), \tag{2.26}$$

where $\mathrm{MAX}_I$ is the maximum possible pixel value of the image (e.g., 255 for 8-bit images), MSE is the Mean Squared Error, defined as

$$\mathrm{MSE} = \frac{1}{m \cdot n}\sum_{i=1}^{m}\sum_{j=1}^{n}(I(i,j) - K(i,j))^2, \tag{2.27}$$

where $I(i,j)$ is the pixel value of the original image, and $K(i,j)$is the pixel value of the distorted image. Higher PSNR values generally indicate better image quality, with values typically ranging from 20 to 50 dB. A higher PSNR suggests less distortion and a closer match to the original image. PSNR is an accurate reflection of the pixel-level reconstruction error of the image/video frame. It is widely used in early works because of its simple definition and easy calculation. A semantic communication framework was proposed for images [19], which solved the transmission problem when the knowledge base is mismatched and proved the performance gain of the scheme on PSNR. However, due to the lack of consideration of factors such as the physiological characteristics of the human eye, the evaluation accuracy of PSNR is not as good as that of SSIM.

SSIM is a perceptual metric that evaluates the similarity between two images based on their luminance, contrast, and structure. Unlike PSNR, which is based purely on pixel-wise differences, SSIM considers structural information, making it a more reliable metric for human perception. The SSIM between two images $I$ and $K$ is calculated as

$$\mathrm{SSIM}(I,K) = \frac{(2\mu_I\mu_K + C_1)(2\sigma_{IK} + C_2)}{(\mu_I^2 + \mu_K^2 + C_1)(\sigma_I^2 + \sigma_K^2 + C_2)}, \tag{2.28}$$

where

- $\mu_I$ and $\mu_K$ are the mean values of images $I$ and $K$,
- $\sigma_I^2$ and $\sigma_K^2$ are the variances of the two images,
- $\sigma_{IK}$ is the covariance between the images, and
- $C_1$ and $C_2$ are small constants to avoid division by zero.

SSIM ranges from $-1$ to 1, where 1 indicates perfect similarity between the two images, and -1 indicates completely different images. A higher SSIM score indicates that the two images are more similar in terms of structure and quality, which aligns better with human visual perception than PSNR. Compared with PSNR, SSIM further considers the impact of brightness, contrast, and structure on the quality of reconstructed images, and the evaluation accuracy is closer to human subjective feelings.

To further improve upon SSIM, multi-scale structural similarity index (MS-SSIM) [20] was introduced, which incorporates multi-scale information to better capture the perceptual quality of compressed images. MS-SSIM computes SSIM at multiple scales and combines the results to provide a more comprehensive assessment of image quality. It is defined as

$$\mathrm{MS\text{-}SSIM}(x,y) = \frac{1}{L}\sum_{i=1}^{L}\mathrm{SSIM}(x_i, y_i), \tag{2.29}$$

where $L$ is the number of scales, and $x_i$ and $y_i$ are the images at scale $i$. By evaluating structural similarity at different resolutions, MS-SSIM better reflects human perception of image quality, particularly for images with complex textures and varying levels of detail.

In addition to SSIM and MS-SSIM, the learned perceptual image patch similarity (LPIPS) [21] metric has been proposed as a deep learning-based approach to assess perceptual image quality. LPIPS measures the similarity between two images based on features extracted from pre-trained deep neural networks. Its core assumption is that neural networks pre-trained on large-scale image datasets have implicitly learned important perceptual features such as texture, color, and structure. In the feature space of these networks, the Euclidean distance between image features can better reflect the perceptual differences between images. LPIPS is calculated as

$$\mathrm{LPIPS}(x,y) = \sum_{l=1}^{L} w_l \cdot \|\phi_l(x) - \phi_l(y)\|_2^2, \tag{2.30}$$

where $\phi_l(x)$ represents the feature extracted from image $x$ at layer $l$, and $w_l$ is a channel weight. Compared to traditional metrics, LPIPS captures high-level perceptual differences that align more closely with human judgments.

The Frechet inception distance (FID) is a generative model evaluation metric that quantifies the statistical similarity between real and synthetic image distributions. It is widely used to assess the quality of outputs from generative adversarial networks (GANs) and diffusion models in tasks like image synthesis and style transfer. A lower FID value indicates smaller distributional divergence and higher perceptual fidelity. FID is computed as

$$\mathrm{FID} = |\mu_r - \mu_g|^2 + \mathrm{Tr}\left(\Sigma_r + \Sigma_g - 2(\Sigma_r\Sigma_g)^{1/2}\right), \tag{2.31}$$

where $\mu_r$, $\Sigma_r$ and $\mu_g$, $\Sigma_g$ denote the mean vectors and covariance matrices of Inception-v3 features extracted from real and generated images, respectively, and $\mathrm{Tr}(\cdot)$ calculates the trace of a matrix.

Although objective metrics such as PSNR and LPIPS provide efficient and reproducible quality evaluations, subjective metrics remain the gold standard for assessing video quality based on human perception. The most commonly used subjective metric is the MOS for images or videos, which is calculated as

$$\mathrm{MOS} = \frac{1}{N}\sum_{i=1}^{N} S_{i,j,k}, \tag{2.32}$$

where $S_{i,j,k}$ represents the score given by the $i$-th subject for the $k$-th video under the $j$-th testing condition. MOS reflects the average score given by all subjects for a specific video. Another subjective measure, the difference mean opinion score (DMOS), quantifies the perceptual difference between a reference video and a distorted version:

$$\mathrm{DMOS} = \mathrm{MOS}_{\mathrm{reference}} - \mathrm{MOS}_{\mathrm{distorted}}. \tag{2.33}$$

Traditional subjective evaluations are conducted in controlled environments with strict conditions, ensuring high reliability but making large-scale assessments costly and time-consuming. Recently, online crowdsourcing has emerged as an alternative approach, though it requires robust filtering to handle participant inconsistencies.

To bridge the gap between traditional objective metrics and subjective perception, the visual multimethod assessment fusion (VMAF) [22] model was developed. VMAF is a learning-based video quality metric that integrates multiple low-level visual features with subjective scores to provide a more accurate representation of perceived video quality. Unlike PSNR and SSIM, which primarily focus on spatial distortion within individual frames, VMAF accounts for perceptual quality across a broader range of video characteristics. The key components of VMAF include:

- Diverse content features: Covers various types of content, such as animations, indoor/outdoor scenes, facial close-ups, and high-motion sequences.
- Diverse low-Level features: Incorporates brightness, contrast, sharpness, color variations, and other visual properties.
- Various distortion types: Accounts for blocking artifacts, ringing artifacts, mosquito noise, and other compression-related distortions.

VMAF employs machine learning techniques, such as support vector machines (SVMs), to train a video quality prediction model based on a combination of fundamental metrics and subjective scores. Its core evaluation metrics include:

- Visual information fidelity (VIF): Based on information theory, it quantifies the amount of information preserved between the original and distorted images.
- Detail loss metric (DLM): Measures the loss of visual details due to compression.
- Mean co-located pixel difference (MCPD): Evaluates temporal consistency by analyzing brightness variations between adjacent frames.

VMAF is widely applied in streaming quality monitoring, offering an enhanced alignment with subjective perception. By integrating multiple evaluation dimensions, it provides a robust and adaptable framework for video quality assessment, improving the optimization of video compression algorithms.

Despite the reconstruction-oriented performance metrics, task-oriented metrics have also been used to measure accuracy in semantic communication, given various underlying tasks. Question answering accuracy (QAA) is a widely used evaluation metric that measures the understanding performance of a model by calculating the ratio of correctly answered questions to the total number of questions. It is primarily used to evaluate the effectiveness of image or video understanding models in tasks such as visual-based reasoning, temporal event localization, and cross-modal comprehension. A higher QAA indicates better model performance in understanding visual contents. QAA can be calculated as

$$\mathrm{QAA} = \frac{p}{q}, \tag{2.34}$$

where $p$ is the number of questions answered correctly and $q$ is the total number of questions.

### 2.2.2 Timeliness-Based Metric

Most of the above performance evaluation indicators evaluate the system transmission performance by measuring the distortion between the reconstructed information and the original information. This distortion-based evaluation method is directly transferred from specific artificial intelligence tasks, focusing on the accuracy dimension of task completion and ignoring the timeliness dimension. This means that above evaluation methods cannot reflect the impact of information timeliness. The semantic communication system also pursues improving the transmission accuracy while ensuring the timeliness of information, rather than approaching the upper limit of accuracy regardless of cost. Since the distortion-based performance evaluation method cannot reflect the dimension of information timeliness, semantic communication further introduces timeliness indicators to evaluate performance in real-time systems.

**Fig. 2.1** The performance of three metrics versus time

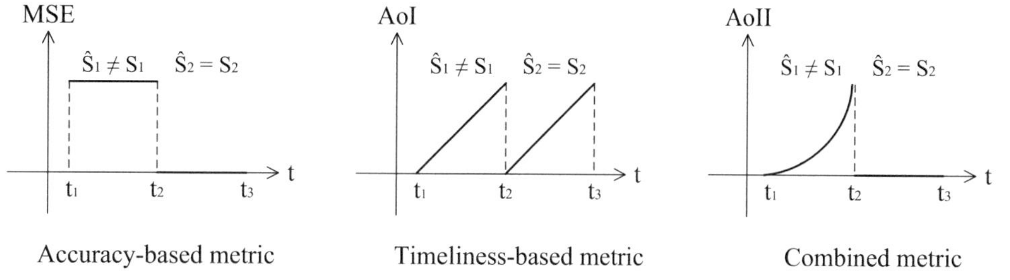


Delay is widely used to measure the quality of experience (QoE) from the time dimension in traditional communication systems, where each bit of information is treated independently and their value does not change over time. However, for real-time communication systems, data samples are correlated and the fresh sample should contain more semantics than the stale one. In this regard, the age of information (AoI) is proposed to measure one semantic property known as the freshness of information from the receiver's perspective [23]. AoI at a given time $t$ is defined as

$$\mathrm{AoI(t)} = t - u(t), \tag{2.35}$$

where $u(t)$ is the generation time of the most recently received data sample, which includes the end-to-end delay of a single packet from source to destination and the waiting time for the arrival next packet at the destination. AoI has successfully worked as an optimization tool in the design of optimal resource allocation strategies in various real-time status update and control systems.

Despite its contributions, the AoI can only capture the linear performance degradation from the time dimension regardless of the correlation of source modalities. In the example of unmanned aerial vehicle (UAV) tracking networks, the position of a moving drone changes rapidly with time while the position of a hovering drone changes slowly. For slowly changing sources, data packets are highly related thus old data may still be valuable. Therefore, the value of information (VoI) concept has been proposed to measure both freshness and correlation properties for usefulness. The VoI can be generalised as a non-decreasing (or non-increasing) and non-linear function of AoI, representing the dissatisfaction with stale information (or the utility of fresh information) [24]. One simple way to design the VoI function is using exponential AoI to model the usefulness of the quickly changing source; while using logarithmic AoI to model the slowly changing source. The other comprehensive method is using communication theory to design VoI. In particular, auto-correlation in probability theory, squared error in estimation theory, and mutual information in information theory have been adopted as a VoI metric.

The advantage of these metrics is that both AoI and VoI are general and simple, thus they can be applied to various scenarios and measure the performance among different data sources. However, neither relates to the real value or realisation of transmitted information. Therefore, they are not appropriate to evaluate the data semantics. To deal with the shortcomings of the aforementioned metrics, combined metrics are proposed to evaluate the performance by jointly considering accuracy and usefulness concerning the underlying transmission task. The age of incorrect information (AoII) has been introduced in [25] as an enabler for semantic-empowered communication systems. AoII is defined as the product of the timeliness-based function $f(\cdot)$ and accuracy-based function $g(\cdot)$, which is given as

$$\mathrm{AoII(t)} = f(\mathrm{AoI(t)}) \cdot g(X_t, \hat{X}_t). \tag{2.36}$$

Depending on the transmission task, they can take any of the available functions. Figure 2.1 gives an example to compare the performance of three types of semantic metrics. Although the combined metrics are comprehensive, they are not general and need to be redesigned based on specific tasks. The adoption of suitable functions is a great challenge to apply AoII in the real world.

Different from traditional communications, performance metrics in semantic communications are specified to the type of data sources and the underlying transmission tasks. If we transmit data with content-based semantics (e.g., a sentence or an image), accuracy-based metrics can be employed. If we transmit a sequence of real-time semantics (e.g., a sequence of real-time sensing data), timeliness-based metrics should be taken into account.

## 2.3 Benchmark Dataset

This book utilizes a diverse set of image datasets ranging from low-resolution benchmarks to large-scale, high-resolution collections, enabling a comprehensive evaluation of neural image compression and representation learning models across a wide range of computer vision tasks.

Among the widely used low-resolution datasets, CIFAR-10 consists of 60,000 32×32 color images across 10 object classes and serves as a standard benchmark for small-scale image classification. Similarly, MNIST contains 70,000 28

× 28 grayscale images of handwritten digits, offering a lightweight yet effective dataset for evaluating baseline models. The Kodak dataset, comprising 24 high-quality color images typically at 768×512 resolution, is widely adopted in image compression, restoration, and quality assessment as a benchmark for perceptual and quantitative evaluation.

In the domain of facial analysis, this book involve several specialized datasets: Celeba, which includes over 200,000 celebrity face images annotated with 40 labels, supports tasks such as facial prediction and generative modeling; UTK Face, with approximately 26,000 images covering various age groups, genders, and ethnicities, is used for demographic estimation; SCUT-FBP5500 has 5,500 images labeled with seven emotion categories, facilitates research on emotion recognition and facial expression analysis; and CelebAmasked-HQ, a high-quality masked face dataset, provides a challenging setting for tasks like face reconstruction and inpainting under occlusion.

To evaluate the capability and semantic representation quality of large-scale models, the book also incorporates the ImageNet dataset, containing over 14 million annotated images across more than 20,000 categories, which has become a benchmark in large-scale visual recognition. The LSUN-Churches and LSUN-Bedrooms datasets, each providing high-resolution images of indoor and architectural scenes, are employed to evaluate generative models and neural compression methods in complex, structured environments. Furthermore, the TGIF-QA dataset is a prominent benchmark in the field of VideoQA, specifically designed to facilitate research on spatio-temporal reasoning in visual content. It comprises 165000 question-answer (QA) pairs.

By leveraging this comprehensive suite of datasets, the book systematically investigates the effectiveness, robustness, and generalizability of various neural architectures and algorithms, covering both constrained and unconstrained image processing scenarios.

## 2.4 Summary

This chapter gives an overview of semantic communications. It discusses and highlights the basic fundamental mathematical theory of semantic communications and the role of the performance evaluation metrics in the design and implementation of semantic communication systems. In the next chapter, we are going to move to the technical details of semantic communications and the comparison with conventional communications.

## Problems

2.1 Which of the following is a key difference between Shannon's information theory and semantic communication?

A. Shannon's theory focuses on meaning, while semantic communication focuses on error rates
B. Shannon's theory quantifies information, while semantic communication focuses on transmitted meaning
C. Shannon's theory applies only to wired communication
D. Semantic communication ignores signal noise

2.2 In Shannon's theory, the fundamental limit of information transmission is given by:

A. Channel Capacity
B. Nyquist Rate
C. Entropy Function
D. Modulation Index

2.3 Which of the following is NOT a key metric in evaluating semantic communication performance?

A. Semantic Similarity
B. BLEU
C. BER
D. PSNR

2.4 What is the role of semantic coding in communication?

A. Ensuring all transmitted bits are error-free
B. Extracting and transmitting only meaningful content
C. Increasing the signal power for better reception
D. Minimizing latency in digital systems

2.5 Which of the following techniques is commonly used in semantic communication?

A. Huffman Coding
B. Parity Bit Checking
C. Frequency Division Multiplexing
D. Deep Learning

2.6 Why is conventional bit-level error correction insufficient for semantic communication?

A. It does not account for transmission delays
B. It only corrects syntax errors, not meaning distortion
C. It increases bandwidth usage
D. It requires additional hardware

2.7 Which of the following statements about Shannon entropy is true?

A. Shannon entropy measures the amount of information in a source
B. Lower entropy indicates higher unpredictability in data
C. Higher entropy means more redundancy in data
D. Shannon entropy is irrelevant in semantic communications

2.8 Which concept in Shannon's theory measures shared information between two random variables?

A. Entropy
B. Mutual information
C. KL divergence
D. Channel capacity

2.9 KL divergence is

A. Symmetric and measures channel noise
B. Non-symmetric and compares probability distributions
C. Used to calculate PSNR
D. A timeliness metric

2.10 The capacity of transmission data may exceed Shannon channel capacity when:

A. KL divergence is maximized
B. AoI is infinite
C. Receiver resolves semantic ambiguity
D. Entropy is zero

2.11 Which metric is applicable for speech signal transmission in semantic communications?

A. MSE
B. BLEU
C. PSNR
D. SDR

2.12 Frechet inception distance (FID) measures:

A. Speech quality
B. Similarity between image sets
C. Text translation accuracy
D. Packet delay

2.13 Age of information (AoI) quantifies:

A. Bit error rate over time
B. Transmission delay
C. Semantic similarity
D. Data freshness

2.14 A source transmits four possible messages with probabilities $\{0.5, 0.25, 0.15, 0.10\}$. Calculate the entropy using Shannon's formula.
**Answer:** $H(X) = 1.85$ bits/symbol

$$H(X) = -\sum p_i \log_2 p_i = 1.85$$

2.15 Given a channel with bandwidth $B = 1$ MHz and $SNR = 10$, calculate the maximum capacity.

## References


1. R. Carnap, Y. Bar-Hillel et al., *An Outline of a Theory of Semantic Information*. Research Laboratory of Electronics (Massachusetts Institute of Technology, 1952)
2. L. Floridi, Outline of a theory of strongly semantic information, *Minds and Machines*, vol. 14, pp. 197–221, May 2004
3. S. D'Alfonso, On quantifying semantic information. Information **2**(1), 61–101 (2011)
4. J. Bao, P. Basu, M. Dean, C. Partridge, A. Swami, W. Leland, J.A. Hendler, Towards a theory of semantic communication, in *Proceedings of the IEEE Network Science Workshop*, pp. 110–117, June 2011
5. K. Niu, P. Zhang, *A Mathematical Theory of Semantic Communication* (2024)
6. I.D. Melamed, *Measuring Semantic Entropy* (1997)
7. X. Liu, W. Jia, W. Liu, W. Pedrycz, AFSSE: an interpretable classifier with axiomatic fuzzy set and semantic entropy. IEEE Trans. Fuzzy Syst. **28**(11), 2825–2840 (2020)
8. A. Chattopadhyay, B.D. Haeffele, D. Geman, R. Vidal, *Quantifying Task Complexity Through Generalized Information Measures* (2021)
9. J. Choi, S.W. Loke, J. Park, A unified view on semantic information and communication: a probabilistic logic approach, in *IEEE International Conference on Communications Workshops (ICC Workshops)*, pp. 705–710 (2022)
10. G. Xin, P. Fan, EXK-SC: a semantic communication model based on information framework expansion and knowledge collision. Entropy **24**, 1842 (2022). (Dec.)
11. A. Kolchinsky, D.H. Wolpert, Semantic information, autonomous agency and non-equilibrium statistical physics, in *Interface Focus*, vol. 8, 6 Dec 2018
12. N.J. Venhuizen, M.W. Crocker, H. Brouwer, Semantic entropy in language comprehension. *Entropy*, vol. 21, no. 12 (2019)
13. C. Lu, From bayesian inference to logical bayesian inference, in *Intelligence Science II*, pp. 11–23 (Springer International Publishing, 2018)
14. J. Liu, W. Zhang, H.V. Poor, A rate-distortion framework for characterizing semantic information, in *Proceedings of the IEEE ISIT*, pp. 2894–2899, July 2021
15. T. Guo, Y. Wang, J. Han, H. Wu, B. Bai, W. Han, *Semantic Compression with Side Information: A Rate-distortion Perspective* (2022)
16. T. Okamoto, *A Unified Paradigm of Organized Complexity and Semantic Information Theory* (2016)
17. H. Xie, Z. Qin, G.Y. Li, B.-H. Juang, Deep learning enabled semantic communication systems. IEEE Trans. Signal Proc. **69**, 2663–2675 (2021)
18. Z. Weng, Z. Qin, Semantic communication systems for speech transmission. IEEE J. Sel. Areas Commun. **39**(8), 2434–2444 (2021)

19. H. Zhang, S. Shao, M. Tao, X. Bi, K.B. Letaief, Deep learning-enabled semantic communication systems with task-unaware transmitter and dynamic data. IEEE J. Sel. Areas Commun. **41**(1), 170–185 (2023)
20. Z. Wang, E.P. Simoncelli, A.C. Bovik, Multiscale structural similarity for image quality assessment, in *Proceedings of the Thirty-Seventh Asilomar Conference on Signals, System Computer*, vol. 2, pp. 1398–1402 (2003)
21. R. Zhang, P. Isola, A.A. Efros, E. Shechtman, O. Wang, The unreasonable effectiveness of deep features as a perceptual metric, in *Proceedings of the IEEE conference on computer vision and pattern recognition*, pp. 586–595 (2018)
22. Z. Li, K. Swanson, C. Bampis, L. Krasula, A. Aaron, Toward a better quality metric for the video community. *Netflix Technology Blog*, p. 2 (2020)
23. R.D. Yates, Y. Sun, D.R. Brown, S.K. Kaul, E. Modiano, S. Ulukus, Age of information: an introduction and survey. IEEE J. Sel. Areas Commun. **39**(5), 1183–1210 (2021)
24. A. Kosta, N. Pappas, A. Ephremides, V. Angelakis, The cost of delay in status updates and their value: non-linear ageing. IEEE Trans. Commun. **68**(8), 4905–4918 (2020)
25. A. Maatouk, M. Assaad, A. Ephremides, The age of incorrect information: an enabler of semantics-empowered communication. IEEE Trans. Wirel. Commun. **22**(4), 2621–2635 (2023)

# 3 Conventional Communication Versus Semantic Communication

Abstract

This chapter builds upon the concepts introduced in Chapters 1 and 2, further exploring traditional communication and semantic communication from an architectural perspective. In the section on traditional communication frameworks, we introduce the fundamental modules of digital communication systems. By explaining the process flow of bits and signal processing methods, readers can gain a necessary understanding of the basic operations in communication systems required in semantic communication research. While discussing traditional digital communication, we also emphasize the roles and research status of corresponding modules in semantic communication. In the section on semantic communication frameworks, we start with an overview of the architecture. Based on this, we introduce discussions on joint semantic-channel coding, semantic information transmission, as well as computational resources and semantic knowledge bases. Through this section, we aim to highlight the key concepts and design principles of semantic communication and inform readers that many related research efforts are ongoing.

Keywords

Communication framework • Semantic sampling • Joint semantic-channel coding • Semantic information transmission • Computing resource • Semantic knowledge base

## 3.1 Conventional Communication Framework

Communication refers to the exchange of information across different locations or times. In this context, *information* is broadly defined to include familiar formats such as text, audio, image, video, and others. Based on the way in which the signal parameters that represent the information are characterized, the signals can be classified as digital or analog. As for digital signals, the value of parameters is defined on finite sets. Binary digits are the most common digital form, while analog signal parameters take on a continuous value. Communication systems can then be categorized into analog or digital systems depending on the type of signal transmitted. Over the past several decades, digital communication has become the dominant paradigm of communication due to numerous advantages, including the regenerative nature of digital signals, the ease of applying channel coding and encryption algorithms, the rapid development of digital integrated circuits, and so on. Therefore, in this section, we will focus solely on introducing the framework of traditional digital communication systems.

Figure 3.1 illustrates the basic blocks of a general digital communication system. The various modules in this figure show a clear correspondence with each other. The upper part of the diagram represents the transmitter, while the lower part represents the receiver. At the transmitter, the digital communication system gradually converts the source information into digital bits, which are then transformed into a transmission waveform and sent into the channel for propagation. At the receiver, the waveform is progressively restored to bits, ultimately reconstructing the information. It is important to note that in the process illustrated in this figure, we have omitted key components of modern digital communication systems, such as multiplexing, synchronization, and multiple access. This simplification is intentional, as our goal is to help readers understand the flow of information in a digital communication system, laying the foundation for the subsequent introduction of the semantic communication framework, rather than delving into the specifics of digital communication. Next, we will further introduce the digital communication system framework from the following six aspects.

**Information Source and Information Sink** The information source is the origin of the information. The information to be transmitted is represented in various formats, including text, audio, images, video, and so on. The information sink is where the information is reproduced. It is worth noting that


W. Chen and Z. Qin, *Semantic Communications*, Textbooks in Telecommunication Engineering,
https://doi.org/10.1007/978-3-032-11005-3_3

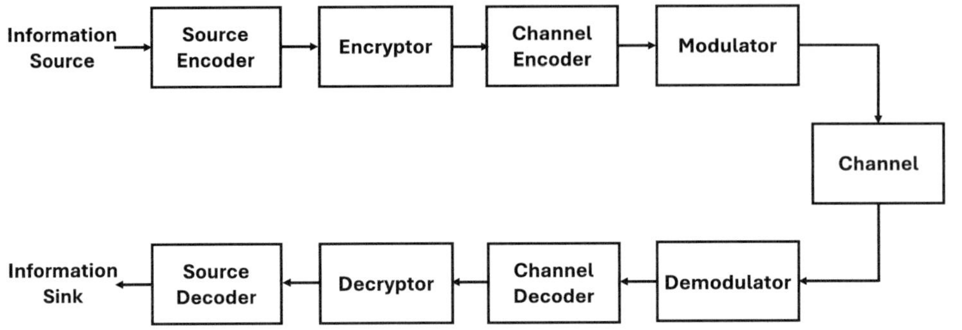


**Fig. 3.1** Framework of a typical digital communication system

in digital communication, we often consider the receiver as a device, such as a speaker, display, etc. This viewpoint is undoubtedly correct, as it highlights the bit-oriented nature of digital communication, focusing on whether the bit information is correctly reproduced at the receiver. In semantic communication, however, greater emphasis is placed on the concept of the information consumer at the receiver. Because semantic communication considers the tasks at the receiver end and aims to transmit semantic information effectively and reliably, the focus shifts to the receiver's perception or task performance. This is a defining feature of semantic communication.

**Source Coding and Decoding** The primary role of source coding is to reduce the redundancy in the transmitted information, thereby improving the efficiency of the communication system. By performing data compression, it lowers the required transmission symbol rate, which in turn reduces the bandwidth demand. Before implementing source coding algorithms, the source coding module typically completes steps such as signal conversion, sampling, and quantization, which will not be further elaborated here. Source decoding is the inverse process of source coding. Modern source coding and decoding methods have achieved remarkable performance. For instance, video compression algorithms like H.264/Advanced Video Coding (AVC) and H.265/High Efficiency Video Coding (HEVC) have saved substantial bandwidth resources for video transmission in today's digital society. However, these source coding and decoding methods are also facing developmental bottlenecks. The limitations imposed by bit-oriented design will be overcome in semantic encoding, further enhancing the performance of coding and decoding.

**Channel Coding and Decoding** Channel coding introduces redundancy according to specific rules to combat errors caused by factors such as noise, fading, and interference. By designing the code, channel coding achieves coding gain to improve the reliability of the communication system. Similarly, channel decoding is the inverse process of channel coding. The channel decoder at the receiver applies certain criteria and decoding algorithms to detect or correct errors during the decoding process. Modern channel coding has also achieved excellent performance. For example, low-density parity-check (LDPC) codes and Polar codes are widely used in 5G mobile communication systems. However, existing methods face challenges, such as the requirements for code length to meet the capacity, decoding delay, and the difficulty of joint optimization with source coding, which highlight the bottlenecks in the development of channel coding. Semantic communication leverages deep learning (DL) models by introducing impaired conditions during training and designing specific models to create a new encoding and decoding design paradigm.

**Encryption and Decryption** In scenarios where secure communication is required, the transmitted sequence undergoes an encryption process to ensure the safety of the information. This process typically uses a key and a specific encryption algorithm to generate ciphertext. At the receiver end, the corresponding decryption process is carried out. Depending on the type of encryption, such as symmetric or asymmetric encryption, the receiver will use different kinds of keys for decryption. Research on encryption and decryption in semantic communication is still relatively limited.

**Modulation and Demodulation** Modulation is a key step in digital communication systems, where the bitstream is converted into a waveform. The core function of modulation is to shift the spectrum of the baseband signal to a frequency band suitable for transmission over the channel. Common digital modulation schemes include amplitude shift keying (ASK), frequency shift keying (FSK), phase shift keying (PSK), and quadrature amplitude modulation (QAM), among others. The demodulation process involves shifting the signal spectrum back to baseband, performing out-of-band filtering, equalization, and other operations, and finally performing sampling and decision-making. It is worth noting that one step often overlooked in the modulation module is pulse shaping (baseband modulation), which is the actual process of converting the bits into a digital baseband waveform. The subsequent frequency shifting process is called bandpass modulation, resulting in a digital bandpass waveform. In semantic communication, research on the transmission module is relatively less developed compared to work on coding. However, some studies have explored the modulation module in semantic communication, which essentially requires an efficient method to map the output of the semantic communication encoding to a finite set of channel symbols.

**Channel** The channel is the physical medium through which signals are transmitted. Different channels have vastly different characteristics, and a good/accurate channel model is crucial for the design of digital communication systems. Apart from the thermal noise at the receiver, impairment caused by the channel is the main reason for signal distortion. Channels can typically be classified into wired and wireless channels. A wired channel can generally be modeled as a linear time-invariant (LTI) system, and the transfer function can be obtained through feedback from the receiver at the link establishment stage. Wireless channels, however, are much more complex. In addition to free-space fading, factors such as multipath effects caused by multiple transmission paths, shadowing effects due to obstructions in the transmission environment, and Doppler effects due to the movement of the transmitter, receiver, and objects in the environment all have a significant impact on wireless channels. Semantic communication must also pass through a physical channel. The key consideration is how to design semantic waveforms that enable high-quality transmission over a lossy channel while ensuring semantic fidelity. Additionally, semantic communication also considers an abstract semantic channel, which is established to characterize the flow of semantic information, primarily aimed at measuring the end-to-end performance of semantic communication.

Through the six aspects discussed above, we have gained a certain level of understanding of the traditional digital communication system framework. We can now quickly move on to the study of semantic communication. For those who wish to gain a more systematic and in-depth understanding of digital communication, there are numerous textbooks of high-quality available on the market, such as [1–9], among which [5, 7, 8] present digital communication from a more theoretical perspective.

## 3.2 Semantic Communication Framework

Research on semantic communication is still in its early stages. Currently, thanks to the powerful nonlinear representation capabilities of DL, the key modules in semantic communication systems are implemented through DL models. This section introduces the general framework of semantic communication systems, focusing on DL-enabled semantic communication. First, the architectures of the semantic communication system will be presented, followed by detailed explanations of some important components in the system: joint semantic-channel coding, semantic information transmission, and computing resources and knowledge bases in semantic communication systems.

### 3.2.1 Architecture of Semantic Communication System

Recent research highlights the significant potential of semantic communication to surpass traditional communication systems in several aspects. However, a universally accepted semantic communication architecture that matches the well-established paradigm of separate source and channel coding in traditional communications is still lacking. The design of such architectures is closely tied to foundational communication theories. For instance, the separate source and channel coding approach in digital communications is based on Shannon's source-channel separation theorem. The complexities and unresolved issues within semantic theories make it challenging to establish optimal semantic communication architectures. Nevertheless, the emerging DL based semantic communication models, incorporating various semantic processing blocks, resemble traditional communication systems in structure. Research into these models has demonstrated promising results in diverse applications, potentially offering new insights into the development of semantic communication theories.

A typical semantic communication system today consists of a semantic encoder and a channel encoder at the transmitter, along with corresponding decoders at the receiver. To handle diverse types of information and task-specific requirements, these systems employ a variety of neural network modules for the encoders and decoders. Sampling also plays a crucial role, as it can be optimized alongside encoders to work with the semantic information being processed. Semantic-aware sampling techniques are tailored for particular tasks, aiming to minimize uninformative samples. For example, we can incorporate a semantic sampling (or sensing) module into a semantic communication system, achieving spatially variable compression ratios for video transmissions. A detailed introduction on semantic sampling and joint semantic-channel coding is put in Sect. 3.2.2.

In practical deployments, some semantic communication system designs retain the core elements of semantic encoders and decoders while incorporating traditional channel coding and transmission techniques from digital communications. For instance, a semantic transmitter is proposed in [10], where source data is processed to extract semantic features, which are then transmitted over the physical channel using traditional methods, such as LDPC codes. The process of signal transmission and adaptation to the channel follows well-established communication principles that have been extensively studied over the past several decades. Overall, combining semantic encoding and decoding processes with existing communication technologies offers a practical approach for building initial semantic communication prototypes. Moving forward, it is crucial to develop semantic-aware transmission

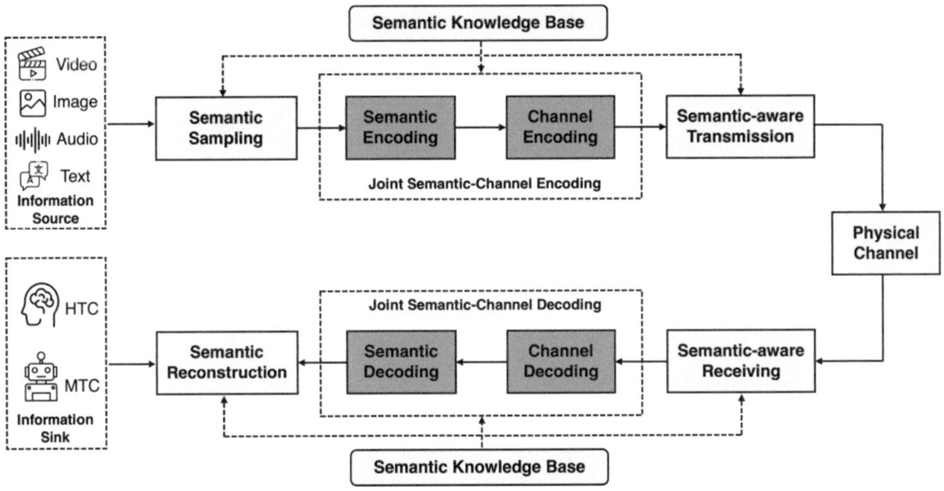


**Fig. 3.2** The general architecture of a semantic communication system

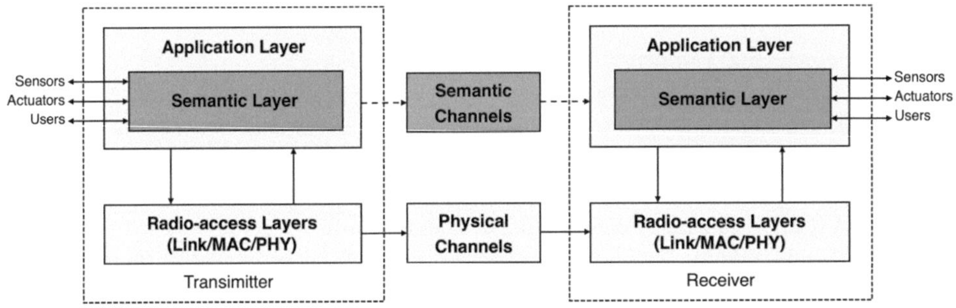


**Fig. 3.3** A semantic communication system with layered network architecture

technologies that convey semantic information more effectively, along with the corresponding semantic-aware receiving technologies. This will be further discussed in Sect. 3.2.3.

The general architecture of a semantic communication system, incorporating sampling, transmission, and task adaptation, is illustrated in Fig. 3.2. In this framework, the reconstructed data at the receiver can vary depending on the type of communication scenario: human-type communication (HTC) and machine-type communication (MTC). HTC focuses on human perceptual systems, while MTC is more concerned with extracting features for downstream tasks like classification, retrieval, and recognition.

Alternative semantic communication architectures also exist, diverging from the models described above. In [11], a novel data adaptation model is proposed, positioned before the semantic encoder, to transform the input source data into a form that is more suitable for semantic encoding. Drawing inspiration from human cognitive abilities, particularly system 2 contextual reasoning, [12] designed a semantic communication system with a virtual listener/receiver at both the transmitter and receiver ends. While these systems differ in structure, the central concept remains the same: leveraging semantic information in the data. This idea lies at the heart of all semantic communication system designs.

In addition to point-to-point communication, the multi-user scenario in semantic communication has also been a subject of exploration in semantic communication studies [13, 14]. To establish a more comprehensive communication framework, a semantic communication system with a layered network structure was proposed in [15]. A new semantic open system interconnection (OSI) model, as shown in Fig. 3.3, has been introduced to reflect this approach. In such a system, multiple users can be present, and user or environmental data is adaptively captured through semantic-aware sensors. The captured data undergoes semantic encoding and transmission, with an actuator at the receiving end completing the final communication task based on the transmitted semantic features and user-specific requirements. In the radio-access layers, the Link, MAC, and PHY layers correspond to the logical link control layer, medium access control layer, and physical layer, respectively. Similarly, a semantic communication system consisting of three protocol layers: the semantic-empowered physical-bearing layer, the semantic-empowered network protocol layer, and the semantic-empowered application-intent layer was introduced in [16]. This framework is underpinned by a new semantic information representation, called the semantic base. Additionally, in [17], a hybrid automatic repeat request (HARQ) scheme for semantic communication was proposed, which could be integrated into either the data link layer of the semantic OSI model or the S-NP layer mentioned above. Recently, large language models (LLMs) have exhibited exceptional capabilities in understanding and generating task-specific information. As a direct application, LLM can be a cross-layer resource manager. By leveraging semantic comprehension, domain-specific knowledge, and logical reasoning, these large models can play a critical role in enhancing the performance of semantic communication systems.

It is also noteworthy that many researchers have realized the importance of background knowledge in semantic communications, incorporating knowledge bases into system architectures to enhance the efficiency of semantic information extraction, compression, transmission, and reconstruction. And computing power has become more and more important in DL-enabled semantic communication systems. An effective semantic communication architecture must also support the sensing, measurement, and scheduling of computational resources. This is because the computational capacity directly influences the performance of DL-based models in semantic communication systems. Sufficient computational power and well-planned resource scheduling strategies are essential for enabling semantic communication systems to efficiently and sustainably complete a variety of tasks. The concept of a computing network, which merges computational resources with communication networks, has emerged as a promising solution. These two aspects will be explored further in Sect. 3.2.4.

All in all, as previously mentioned, semantic communications do not yet have clearly defined, universally recognized architectures like those in traditional digital communications. Developing a high-performance, practically applicable architecture for semantic communication is a formidable challenge, requiring breakthroughs in both semantic theory and focused research on aspects such as sampling, coding, and transmission.

### 3.2.2 Semantic Sampling and Joint Semantic-Channel Coding

Traditional image or video sampling systems, adhering to the Nyquist-Shannon sampling theorem, require the sampling rate to be at least twice the bandwidth of the original signal. Nowadays, acquiring data may come with a significant cost, both in terms of expensive hardware and high per-acquisition sensing expenses. For instance, holographic video acquisition bears a significant cost. Moreover, continuous full sampling can shorten the equipment's lifespan. Selectively targeting the most relevant information for transmission through appropriate downsampling can significantly enhance the accuracy of subsequent coded feature extraction and transmission performance. A semantic communication system with a semantic sampling module can maintain end-to-end system integrity. It will ensure that the information with high semantic fidelity is more accurately extracted and encoded by the subsequent Joint Source-Channel Coding encoder. To adapt to the dynamic channel environment, further optimization can be performed to utilize channel characteristics in the design of sampling and coding modules.

Two intuitive examples of semantic sampling are shown in Fig. 3.4. The examples in the upper part show how different blocks of an image are assigned varying levels of semantic sampling importance. The higher the value on a block, the greater its semantic significance. For example, the eyes in the image are given higher semantic importance compared to the

**Fig. 3.4** Examples of semantic sampling

background, and thus, more sampling resources are allocated to them. The lower part of the example demonstrates the application of semantic sampling in point cloud downsampling. From left to right, the results of downsampling at different rates are shown. It can be observed that the sampled points roughly outline the shape of the airplane, indicating that these sampled points have relatively higher semantic importance.

We take a semantic communication system for video transmission as a detailed example. In the semantic sampling phase, each video frame is divided into non-overlapping blocks, with each block assigned a specific sampling matrix during the sampling process. Once the overall sampling rate for the source media is determined, different sampling rates are applied to each block, depending on both the channel conditions and the semantic significance of the source. This allocation is carried out through the semantic scanning network, which incorporates a correction algorithm. The semantic scanning network consists of an attention-based channel adaptive module and other neural network layers. The semantic scanning network assesses the semantic importance of each region in the frame, adjusts it accordingly, and generates the corresponding sampling matrix for each block. This results in a semantic sampling ratio allocation map, which is shared with the receiver. The map indicates how many samples are allocated to each block.

In the reconstruction phase, the receiver employs both the semantic information and the channel conditions to reconstruct video frames. The inputs of the reconstruction module are the outputs from the decoder and the semantic sampling map shared by the transmitter. During each iteration of the reconstruction process, two key steps are carried out: (i) performing proximal gradient descent with a learnable step size and passing the result to the next stage for joint recovery, and (ii) integrating the semantic sampling map into the semantic extraction network with an attention-based channel adaptive module to retrieve accurate features. These features are then concatenated with the temporal results from the previous step to guide the reconstruction. The reconstruction process is finalized using additional deep neural networks, based on the proximal mapping. High-fidelity recovery is achieved after a set number of iterations.

As for JSCC, it plays a central role in the design of semantic communication systems. As discussed in Chap. 2, semantic theory is still evolving, and as a result, the methods for measuring and representing information in semantic communication systems are not as clearly defined as they are within Shannon's framework of information theory. Despite this, the progress made in the field of semantic communication has been substantial. DL has been crucial in advancing JSCC, enabling the development of semantic communication systems with outstanding performance.

The unique ability of DL models to extract features and represent information non-linearly has proven useful in semantic coding. For example, a DL-based image semantic coding approach can employ a convolutional encoder to extract semantic features and a generative decoder to reconstruct high-quality images even at low bit rates. Besides extracting semantic features directly, another approach focuses on discovering semantic links between different modalities of information. For instance, in a talking-head video, the most important semantic information is typically the face and lip movements. Therefore, after transmitting a frame where the person is directly facing the camera, subsequent frames can be represented as a text transcript. This method reduces the amount of data to be transmitted by leveraging semantic information, all while utilizing DL techniques for encoding and decoding. To take a step further, multimodal semantic coding can be developed. Multimodal semantic coding explores the semantic redundancy that exists between different modalities and achieves compression at the semantic level. A schematic diagram of multimodal semantic coding is shown in Fig. 3.5. For example, in a horse racing video with commentary, instead of separately compressing the video, text, and audio as traditionally done, we can explore the relationships between the visual, auditory, and

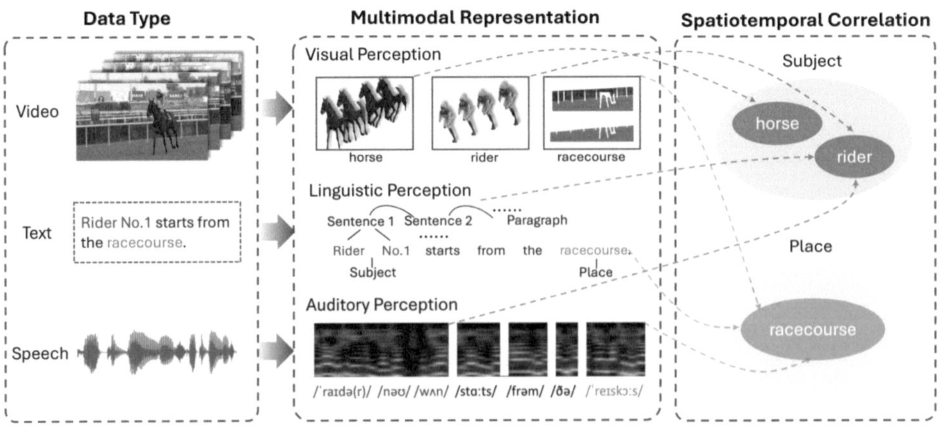


**Fig. 3.5** The concept of multimodal semantic coding

textual modalities and perform unified coding in the semantic space.

However, while relying solely on semantic encoders and decoders powered by DL models can provide promising results, network and channel conditions often degrade performance. This limitation has led to the exploration of channel encoding and decoding schemes that are tailored to the specific needs of semantic codecs, making JSCC a key focus in modern semantic communication research. JSCC shares some similarities with deep joint source-channel coding, where deep neural networks are used to jointly design the source and channel codecs under certain code length constraints. The main distinction of JSCC is its emphasis on the extraction of semantic information during the semantic encoding phase. This enables channel coding to be designed with specific semantic targets in mind, coupled with performance metrics and tailored training procedures. Joint source-channel coding has long been a classical topic in information theory, with some communication systems benefiting from its integrated design, which suggests that JSCC holds considerable promise for future semantic communication systems.

Recent studies have demonstrated that JSCC, which incorporates advanced semantic encoders and decoders, outperforms traditional deep joint source-channel coding approaches. For instance, in [18], a Transformer-based semantic encoder was proposed to extract semantic features from text. The performance metrics and loss functions used in this method incorporate sentence similarity within a semantic space. The corresponding channel coding is implemented through fully connected layers, with the semantic and channel coding modules jointly optimized during training. The results presented in [18] show that the proposed JSCC method outperforms both deep joint source-channel coding and the traditional separation scheme, which uses Huffman coding and Turbo coding. The significant gap in sentence similarity indicates that the underlying semantic information has been successfully extracted and reconstructed.

The extraction of semantic features from different types of information sources can be achieved using various DL models. As a result, channel coding modules need to be designed to match the specific characteristics of the extracted semantic features. For example, the channel coding module of a speech semantic encoder using a squeeze-and-excitation network might be a convolutional neural network to match the semantic feature maps. In video conferencing scenarios, facial semantics were extracted using an hourglass network, followed by channel coding through three dense layers. In some situations, even the packet-level channel coding should be considered. For instance, in [19], the synchronization between video and audio was addressed, with facial semantics being more effectively guided by speech-related semantic information. This study introduced packet-level channel coding to improve the compatibility of the semantic communication system with existing network transmission environments. Another design deviates slightly from the conventional JSCC approach, incorporating a semantic analysis transform, followed by semantic-guided JSCC and corresponding decoding modules. The semantic analysis transform extracts and organizes semantic information, and its functionality, along with part of the neural network layers in the semantic-guided JSCC, can be considered as part of the semantic encoding process, maintaining alignment with the joint semantic-channel coding framework.

In summary, despite the use of various DL models and some differences in current semantic communication system design, all these methods adhere to the joint semantic-channel coding structure. This unified approach holds great potential for enhancing communication system performance, underscoring the critical role that JSCC plays in advancing modern semantic communication systems.

### 3.2.3 Semantic Information Transmission

In recent studies, numerous end-to-end semantic communication systems have been proposed that utilize the JSCC framework. However, it is important to note that JSCC, in its current form, is a simplification of real-world communication systems. For a truly comprehensive semantic communication system, it is necessary to consider more than just the encoding and decoding processes. Effective transmission technologies that account for real-world channel conditions must be incorporated. Traditional communication systems, based on channel state information (CSI), typically focus on optimizing bit sequence transmission, assuming each bit carries equal weight. In contrast, semantic communication systems focus on transmitting semantically relevant features, where the importance of the features varies, affecting how they are transmitted over the channel. The performance of a semantic communication system is determined jointly by channel conditions and the significance of the semantic features. By leveraging CSI along with the semantic information generated by the joint encoder, various transmission strategies such as power control, modulation selection, precoding, and waveform design can be explored to enhance performance. Additionally, investigating adaptive transmission schemes based on both instantaneous and statistical CSI could improve the system's efficiency across different time scales. This highlights the remaining opportunities in the transmission technologies of semantic communication systems.

Digital modulation plays a pivotal role in today's communication systems. However, when directly applied to semantic communication systems, traditional digital modulation cannot be jointly trained with the DL-based encoding modules due to its non-differentiable nature. This separation results in suboptimal performance for the system as a whole. Actually, in semantic communication systems utilizing DL-based JSCC, the output typically consists of floating-point numbers. These numbers are often paired to form a constellation point, where one represents the in-phase component and the other represents the quadrature component. The precision of floating-point numbers is 32 bits or 64 bits, as usual. So the resulting channel symbols can be considered continuous, requiring the receiver to handle a constellation diagram with extremely high resolution. To address this issue, it is essential to map the floating-point outputs of JSCC to a finite set of symbols, ensuring compatibility with existing communication standards. A comparison between the direct output of a JSCC and the output of a digital modulation is presented in Fig. 3.6. Simply quantizing the floating-point values before modulation is not a straightforward solution. Using too few bits for quantization can degrade system performance, while using too many bits unnecessarily increases the symbol count. For instance, if floating-point numbers are quantized to 32 bits, a pair of such numbers would require 64 bits to represent a constellation point. This would necessitate 8 points in a 256-QAM constellation to represent the paired points, thereby multiplying the number of transmission symbols by eight.

To address the modulation issue in semantic communications, recent works have proposed trainable modulation techniques. For instance, a fully connected layer-based bit quantizer was introduced in [17], which allows joint optimization during training, though the traditional bit-symbol mapping is still used for modulation. More advanced solutions, such as two trainable modulation modules–one with full-resolution constellations and another with quantized-resolution constellations–have been proposed in [20]. Additionally, in [21], a DL-based joint coding-modulation mechanism was designed, where the semantic encoder outputs the probability of constellation points rather than traditional symbol mappings, and a sampling method was utilized to overcome the non-differentiable problem. These joint coding-modulation designs have demonstrated superior performance compared to conventional methods. However, a major limitation is that these approaches rely entirely on end-to-end joint training, resulting in a lack of interpretability. Future work should focus on leveraging semantic features, established modulation theories, and more targeted modulation designs to improve semantic-aware modulation techniques.

The transmission of semantic information over MIMO systems has also been the subject of extensive research. Traditional MIMO transmission technologies typically use singular value decomposition (SVD) precoding, which decomposes the MIMO channel into several independent parallel single-input single-output (SISO) channels. In the context of semantic communication, improving system performance can be achieved by allocating more important semantic features to SISO channels with higher SNRs. For example, a vision transformer-based self-attention mechanism can be designed to link the semantic features of images with the corresponding channel conditions. Research explored joint semantic coding and transmission strategies for MIMO systems under different scenarios [22–24]. Moreover, theoretical work on the upper and lower bounds of the joint source-channel coding distortion exponent for semantic communication was presented in [25]. However, whether neural networks should entirely govern the exploitation of MIMO channel degrees of freedom remains an open question. Further exploration is required to develop both data-driven and model-driven approaches for semantic-aware MIMO transmission systems.

Multiple access techniques have also been investigated in semantic communication systems. Specifically, non-orthogonal multiple access (NOMA) has been applied to both heterogeneous semantic-bit transmission in the downlink [26] and NOMA-enhanced semantic communication in the uplink [27]. These contributions illustrate the potential of integrating modern communication techniques with semantic communication frameworks.

In conclusion, the development of semantic information transmission is closely intertwined with advancements in semantic theory and the integration of DL-based modules

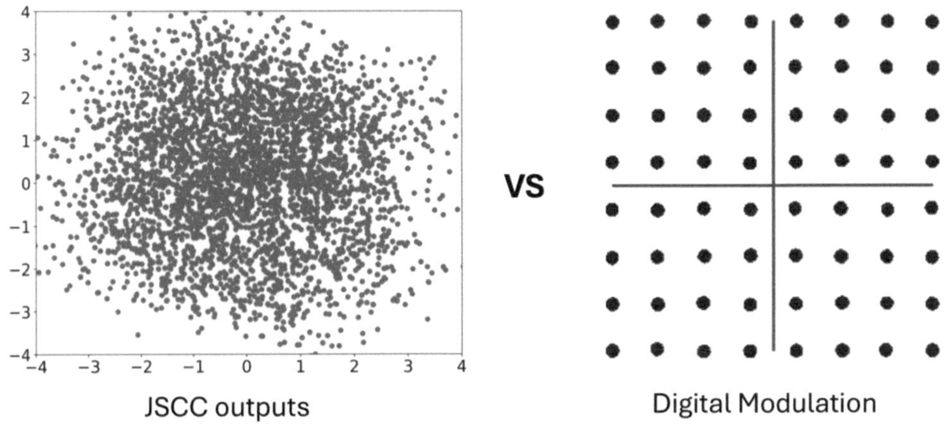


**Fig. 3.6** Constellation map for direct output of JSCC and digital modulation

in semantic communication systems. Unlike traditional wireless transmission, which focuses on bit-level communication, semantic transmission is concerned with conveying semantic features, which opens up additional avenues for optimization in aspects like constellation selection, power allocation, and waveform design. These optimizations are crucial because different semantic features have varying levels of importance and contribute differently to overall system performance. Semantic transmission technologies aim for global optimization by leveraging advanced transmission techniques such as MIMO and OFDM, rather than merely reaching the capacity limits dictated by bit-oriented communication. While considerable progress has been made in JSCC and other components of semantic communications, there is still much to be done in the development of effective and adaptive semantic-aware transmission technologies. The ultimate challenge remains to design reliable, efficient transmission schemes that can respond to both the characteristics of source semantic features and channel variations. To achieve an optimal end-to-end semantic communication system, further work is needed to integrate JSCC and other modules with advanced semantic-aware transmission mechanisms.

### 3.2.4 Computing Resource and Knowledge Base in Semantic Communication

Compared to areas like coding and transmission, there is relatively less research on computational resources and knowledge bases in semantic communication. However, computational resources and knowledge bases play a crucial role in developing a semantic communication system, and this section provides a brief introduction to these aspects.

For DL-enabled semantic communication systems, computational power is a crucial requirement for their proper functioning. To ensure that various devices within the semantic communication network have adequate computational resources, it is essential to establish a computing network.

The concept of a computing network revolves around interconnecting widely distributed cloud, edge, and end computing nodes. This network pools diverse computing, storage, and networking resources into a unified resource pool. It constantly monitors the availability and status of resources, dynamically allocating computing power based on user demands. The system also manages the partitioning and offloading of tasks to optimize resource usage. As a result, it ensures that application needs are met, improves resource utilization, and enhances the overall user experience. The integration of hierarchical computing frameworks with artificial intelligence can significantly accelerate the development of AI-driven applications. A typical computing network architecture is illustrated in Fig. 3.7.

The primary techniques employed in computing networks include computing power measurement, perception, and scheduling. Computing power measurement serves as the foundation for both perception and scheduling by establishing a universal model to evaluate the computing power demands of users and the status of available resources. These measurements guide the distribution and allocation of tasks across a heterogeneous computing infrastructure. Computing power perception involves the continuous real-time monitoring of resource availability and task demands, providing the necessary data for efficient power scheduling. The edges monitor their available communication and computation resources, which enables them to estimate end-to-end latency and achievable data rates. Furthermore, before accepting new connections, the system assesses the load to ensure that the overall demand does not exceed the edge nodes' processing capabilities. The process of computing power scheduling involves dividing tasks and offloading them to the appropriate computing nodes based on the requirements of each task. This ensures better user experiences and more efficient resource utilization. For tasks with extensive computational requirements, dividing the workload and distributing it across multiple computing nodes is necessary.

Despite these advancements, there are still several challenges facing the development of computing networks. One of the primary obstacles is managing resources in edge and end devices, which often feature heterogeneous hardware and software. This diversity complicates the unification of resource management. Another challenge is the real-time changes in the computing statuses of network nodes, making the real-time perception of computing power more difficult. Additionally, concerns about data sharing and security between computing nodes remain unresolved and require further investigation.

The semantic knowledge base plays a crucial role in semantic communication systems. Unlike traditional communication systems, which do not have an equivalent counterpart, the concept of a semantic knowledge base is still emerging in the field. In this context, we define the semantic knowledge base as a structured database that supports encoding, decoding, and transmission modules by providing relevant semantic descriptions and representations within a semantic communication framework.

Semantic knowledge bases in the realm of semantic communication can be broadly categorized into two types. The first is based on knowledge graphs from artificial intelligence, which assist in generating semantic representations during encoding by establishing links between entities, relationships, and attributes. These graphs can also be leveraged by the receiver for tasks such as matching and retrieving relevant semantic information, enabling multi-level downstream applications. The second type treats the parameters of DL models as a form of semantic knowledge base. This approach

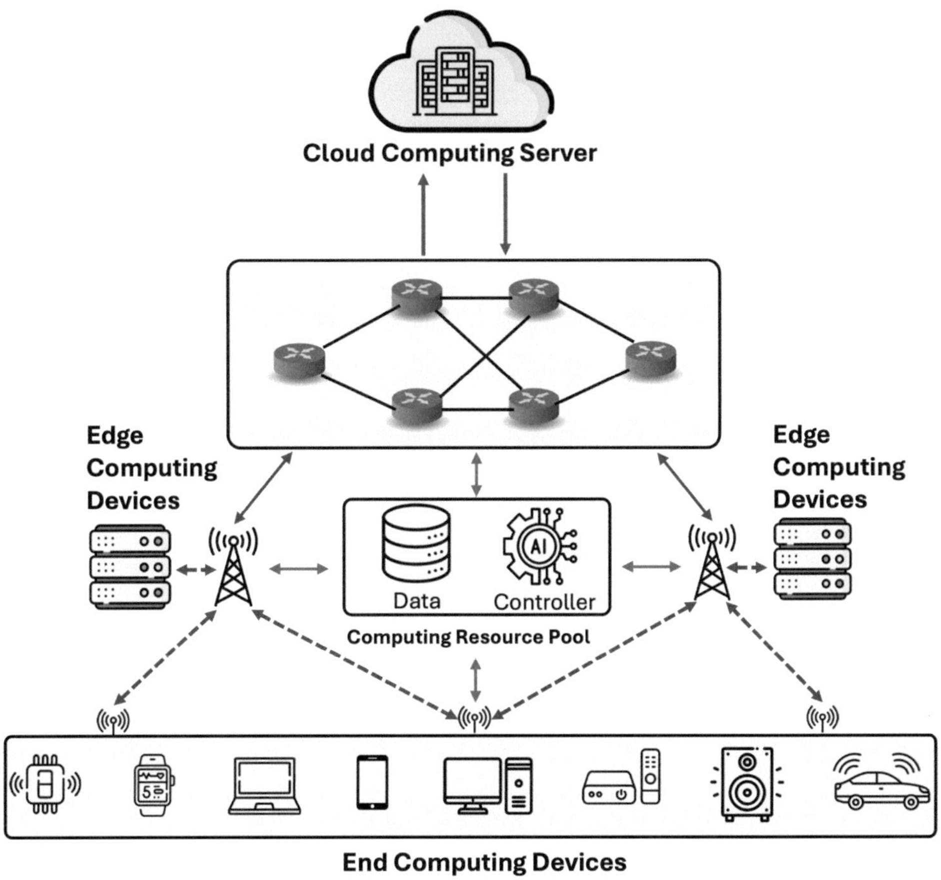


**Fig. 3.7** A general computing network architecture

is prevalent in many existing semantic communication systems that employ joint semantic-channel coding.

For instance, in [28], a knowledge graph model is used to extract semantic information from text in the form of triples, which serve as the basic unit for both the transmitter and receiver. Given that the structure of a knowledge graph aligns well with human cognitive logic, it allows for an intuitive calculation of semantic importance, thereby facilitating a more precise allocation of transmission resources in semantic communication systems. Similarly, [29] introduced a Text2KG aligner that converts text messages into semantic symbols, which are then decoded using a fine-tuned model based on the pre-trained model. An object-attribute-relation model, similar to a knowledge graph, has also been proposed for extracting semantic representations from visual data [30], and scene graphs have been employed in [31] for downstream tasks related to images in semantic communications. While knowledge graphs have certain advantages, such as lower computational and storage costs and faster response times, their ability to represent complex modalities like speech, images, and video is more limited compared to DL-based models. One of the challenges in deploying semantic knowledge bases is their integration into traditional communication systems, which do not have this module. Special attention must be given to how the knowledge base, in the form of a knowledge graph, can be stored and accessed on edge intelligent devices. How to optimize the caching of semantic knowledge bases for such deployments is of great importance.

DL models also serve as a form of semantic knowledge base, with the parameters of trained models constituting structured semantic representations. These parameters are learned during the training phase and help in mapping source information to the semantic space or transforming semantic features into formats suitable for downstream tasks. This approach is common in semantic communication systems that utilize joint encoding and decoding modules. In fact, the parameters of the semantic encoder and decoder can be treated as a knowledge base that supports semantic representation. Recent work also explores the use of extracted semantic features from various layers of DL models to build multi-level semantic knowledge bases, such as those used for zero-shot object recognition tasks [32].

Although knowledge graph-based or DL-based semantic knowledge bases are primarily used for semantic extraction at the transmitter or reconstruction at the receiver, their utility extends to the transmission environment as well. Two types of semantics are central to semantic communication systems: source semantics and channel environment semantics [33]. In

[34], images of real transmission environments are input into a semantic extraction network, which extracts features that can aid in tasks like mmWave beam prediction and blockage forecasting. By constructing semantic knowledge bases for the transmission environment, the system can achieve a quicker and more accurate understanding of the channel conditions.

Overall, research on semantic knowledge bases, when compared to other modules in semantic communications, is still in its early stages. The definition of a semantic knowledge base, as well as the determination of its specific implementation methods and functionalities, remains an open problem. Moreover, as a new module not found in traditional communication systems, its deployment, updating, and security challenges present significant hurdles.

## 3.3 Summary

This chapter builds upon the concepts introduced in Chaps. 1 and 2, further exploring traditional communication and semantic communication from an architectural perspective. In the section on traditional communication frameworks, the fundamental modules of digital communication systems are discussed. By explaining the process flow of bits and signal processing methods, an understanding of the basic operations required in a communication system is built. While discussing traditional digital communication, this section also emphasizes the roles and research status of corresponding modules in semantic communication. In the section on semantic communication frameworks, an overview of the architecture is introduced first. Based on this, discussions on joint semantic-channel coding, semantic information transmission, as well as computational resources and semantic knowledge bases are presented. This section highlights the key concepts and design principles of semantic communication and informs readers that many related research efforts are ongoing.

## Problems

3.1 Which module is **not** a core module of traditional digital communication systems?

A. Modulator
B. Channel Decoder
C. Knowledge Base
D. Encryptor

3.2 What is the main advantage of digital signals over analog signals in communication?

A. Less affected by fading
B. Support for multiplexing
C. Easy integration into analog channels
D. Regenerative nature enabling reliable transmission

3.3 Which of the following best distinguishes semantic communication from traditional digital communication at the receiver side?

A. It relies on digital demodulators
B. It reconstructs analog waveforms
C. It focuses on perceptual or task-based understanding
D. It reduces transmission power

3.4 Which module introduces redundancy to protect against channel noise?

A. Source Encoder
B. Channel Encoder
C. Modulator
D. Demodulator

3.5 Why is source coding considered a bottleneck in traditional communication systems?

A. It increases bandwidth usage
B. It lacks compatibility with encryption
C. It is limited by bit-oriented design
D. It cannot handle multimedia signals

3.6 Which of the following problems in traditional communication is directly addressed by Joint Semantic-Channel Coding?

A. Difficulty in synchronizing analog signals
B. Bit-oriented source and channel coding are difficult to jointly design
C. Lack of modulation diversity
D. Security vulnerabilities during encryption

3.7 Which of the following best explains why Joint Semantic-Channel Coding outperforms traditional separate source and channel coding in semantic communication systems?

A. It simplifies modulation and demodulation schemes
B. It enables error-free transmission over all channels
C. It aligns semantic feature extraction with channel characteristics during training
D. It eliminates the need for neural networks

3.8 What is the main function of a computing network in DL-enabled semantic communication systems?

A. Providing error correction during channel decoding
B. Scheduling and distributing computational resources among edge, cloud, and end nodes
C. Modulating semantic features for channel transmission
D. Encrypting semantic knowledge bases

3.9 What is the key advantage of multimodal semantic coding over traditional separate compression of modalities?

A. It uses less memory
B. It avoids the need for channel coding
C. It explores semantic redundancy across modalities for unified coding
D. It eliminates the need for decoding

3.10 What is a limitation of applying traditional modulation directly in semantic communication?

A. It supports only analog signals
B. It cannot handle long sequences
C. It is non-differentiable and incompatible with neural networks
D. It increases semantic noise

3.11 In semantic MIMO transmission, how are the more important semantic features treated?

A. Transmitted first regardless of channel
B. Mapped to channels with lower SNR
C. Ignored to reduce bandwidth
D. Mapped to SISO channels with higher SNR

3.12 What type of knowledge base allows mapping entities and relations for semantic transmission?

A. Speech dictionary
B. Semantic dictionary
C. Knowledge graph
D. Keyword filter

3.13 **What is Joint Semantic-Channel Coding, and why is it important**?

3.14 **Describe the function and benefits of semantic sampling in communication systems**.

3.15 **What are the two major types of semantic knowledge bases mentioned in the chapter**?

## References


1. I.M. Jacobs, J. Wozencraft, *Principles of Communication Engineering* (1965)
2. S.G. Wilson, *Digital Modulation and Coding* (Prentice-Hall, Inc., 1995)
3. J.G. Proakis, M. Salehi, *Digital Communications* (McGraw-Hill, 2008)
4. U. Madhow, *Fundamentals of Digital Communication* (Cambridge University Press, 2008)
5. R.G. Gallager et al., *Principles of Digital Communication*, vol. 1 (Cambridge University Press Cambridge, UK, 2008)
6. E.A. Lee, D.G. Messerschmitt, *Digital Communication*. Springer Science and Business Media (2012)
7. B. Rimoldi, *Principles of Digital Communication: A Top-Down Approach* (Cambridge University Press, 2016)
8. A. Lapidoth, *A Foundation in Digital Communication* (Cambridge University Press, 2017)
9. B. Sklar, *Digital Communications: Fundamentals and Applications* (Pearson, 2021)
10. K. Niu, J. Dai, S. Yao, S. Wang, Z. Si, X. Qin, P. Zhang, A paradigm shift toward semantic communications. IEEE Commun. Mag. **60**(11), 113–119 (2022)
11. H. Zhang, S. Shao, M. Tao, X. Bi, K.B. Letaief, Deep learning-enabled semantic communication systems with task-unaware transmitter and dynamic data. IEEE J. Sel. Areas Commun. **41**(1), 170–185 (2022)
12. H. Seo, J. Park, M. Bennis, M. Debbah, Semantics-native communication via contextual reasoning, in *IEEE Transactions on Cognitive Communications and Networking* (2023)
13. H. Xie, Z. Qin, X. Tao, K.B. Letaief, Task-oriented multi-user semantic communications. IEEE J. Sel. Areas Commun. **40**(9), 2584–2597 (2022)
14. W.F. Lo, N. Mital, H. Wu, D. Gündüz, Collaborative semantic communication for edge inference. IEEE Wirel. Commun. Lett. (2023)
15. Q. Lan, D. Wen, Z. Zhang, Q. Zeng, X. Chen, P. Popovski, K. Huang, What is semantic communication? A view on conveying meaning in the era of machine intelligence. J. Commun. Inf. Netw. **6**(4), 336–371 (2021)
16. P. Zhang, W. Xu, H. Gao, K. Niu, X. Xu, X. Qin, C. Yuan, Z. Qin, H. Zhao, J. Wei et al., Toward wisdom-evolutionary and primitive-concise 6G: a new paradigm of semantic communication networks. Engineering **8**, 60–73 (2022)
17. P. Jiang, C.-K. Wen, S. Jin, G.Y. Li, Deep source-channel coding for sentence semantic transmission with HARQ. IEEE Trans. Commun. **70**(8), 5225–5240 (2022)
18. H. Xie, Z. Qin, G.Y. Li, B.-H. Juang, Deep learning enabled semantic communication systems. IEEE Trans. Signal Process. **69**, 2663–2675 (2021)
19. Y. Tian, J. Ying, Z. Qin, Y. Jin, X. Tao, Synchronous multi-modal semantic communication system with packet-level coding (2024). arXiv:2408.04535
20. P. Jiang, C.-K. Wen, S. Jin, G.Y. Li, Wireless semantic communications for video conferencing. IEEE J. Sel. Areas Commun. **41**(1), 230–244 (2022)
21. Y. Bo, Y. Duan, S. Shao, M. Tao, Joint coding-modulation for digital semantic communications via variational autoencoder. IEEE Trans. Commun. **72**, 5626–5640 (2024). (Sept.)

22. Z. Weng, Z. Qin, X. Tao, Semantic-aware speech-to-text transmission over MIMO channels, in *2023 IEEE International Conference on Communications Workshops (ICC Workshops)* (IEEE, 2023), pp. 1362–1367
23. S. Yao, S. Wang, J. Dai, K. Niu, P. Zhang, Versatile semantic coded transmission over MIMO fading channels (2022). arXiv:2210.16741
24. C. Bian, Y. Shao, H. Wu, D. Gündüz, Space-time design for deep joint source channel coding of images over MIMO channels, in *2023 IEEE 24th International Workshop on Signal Processing Advances in Wireless Communications (SPAWC)* (IEEE, 2023), pp. 616–620
25. Y. Shi, S. Shao, Y. Wu, W. Zhang, X.-G. Xia, C. Xiao, Excess distortion exponent analysis for semantic-aware MIMO communication systems. IEEE Trans. Wirel. Commun. (2023)
26. X. Mu, Y. Liu, L. Guo, N. Al-Dhahir, Heterogeneous semantic and bit communications: a semi-NOMA scheme. IEEE J. Sel. Areas Commun. **41**(1), 155–169 (2022)
27. X. Mu, Y. Liu, Exploiting semantic communication for non-orthogonal multiple access. IEEE J. Sel. Areas Commun. (2023)
28. S. Jiang, Y. Liu, Y. Zhang, P. Luo, K. Cao, J. Xiong, H. Zhao, J. Wei, Reliable semantic communication system enabled by knowledge graph. Entropy **24**(6), 846 (2022)
29. F. Zhou, Y. Li, X. Zhang, Q. Wu, X. Lei, R.Q. Hu, Cognitive semantic communication systems driven by knowledge graph, in *ICC 2022-IEEE International Conference on Communications* (IEEE, 2022), pp. 4860–4865
30. Y. Duan, Q. Du, X. Fang, Z. Xie, Z. Qin, X. Tao, C. Pan, G. Liu, Multimedia semantic communications: representation, encoding and transmission. IEEE Netw. **37**(1), 44–50 (2023)
31. S. Sun, Z. Qin, H. Xie, X. Tao, Task-oriented explainable semantic communications based on structured scene graphs, in *GLOBECOM 2023-2023 IEEE Global Communications Conference* (IEEE, 2023), pp. 3222–3227
32. Y. Sun, H. Chen, X. Xu, P. Zhang, S. Cui, Zero-shot multi-level feature transmission policy powered by semantic knowledge base, in *GLOBECOM 2023-2023 IEEE Global Communications Conference* (IEEE, 2023), pp. 2069–2074
33. Z. Qin, F. Gao, B. Lin, X. Tao, G. Liu, C. Pan, A generalized semantic communication system: from sources to channels. IEEE Wirel. Commun. **30**(3), 18–26 (2023)
34. Y. Yang, F. Gao, X. Tao, G. Liu, C. Pan, Environment semantics aided wireless communications: a case study of mmwave beam prediction and blockage prediction. IEEE J. Sel. Areas Commun. (2023)

# Part II
# Joint Coding for Semantic Communication

# Joint Source and Channel Coding

# 4

**Abstract**

JSCC (Joint Source-Channel Coding) is a key technology commonly applied in current semantic communication research. By jointly designing the source coding and channel coding modules, JSCC enables communication systems to achieve higher bandwidth efficiency and a more graceful performance degradation curve under non-ideal conditions. In this chapter, we begin with a brief introduction to the principles of JSCC, exploring how the integration of source and channel coding provides a unified framework for efficient communication. Then we will give an overview of traditional JSCC design methods, focusing on some techniques and strategies that have been employed over the years.

**Keywords**

Joint source–channel coding • Source coding • Channel coding • Source-aided channel coding

## 4.1 Introduction

In this section, we will give a brief introduction to the joint source-channel coding. Let us begin with the concept of lossless communication. For a communication system to be lossless, the source encoder must represent a source with entropy $H$ (measured in bits per source sample) using a code with a rate $R_s$ (in bits per source sample) that satisfies the condition $R_s > H$.

Now, consider transmitting this encoded source over a channel that allows $n$ channel uses per source sample. According to Shannon's theory, to achieve transmission with an arbitrarily low probability of error over a channel with capacity $C$ (in bits per channel use), the channel code rate $R_c$ (in bits per channel use) must satisfy $R_c < C$.

By combining the two requirements—one ensuring lossless source representation and the other ensuring reliable transmission through a noisy channel—we obtain the following inequality:

$$H < R_s = nR_c < nC.$$

For simplicity, let us assume $n = 1$, reducing the condition to:

$$H < R < C,$$

where $R_s = R_c = R$ represents the overall communication rate.

This condition prompts two key questions:

1. Is $H < C$ a necessary and sufficient condition for the transmission a given source over a channel with an arbitrarily low probability of error?
2. How can we design a code with rate $R$ such that $H < R < C$, both lossless representation of the source and reliable transmission over the noisy channel?

These questions are fundamentally addressed by **Shannon's source-channel coding theorem**, which provides both the theoretical foundation and practical guidance for their resolution.

Consider the setup illustrated in Fig. 4.1. A discrete source defined over a finite alphabet $\mathcal{S}$ generates a random sequence of symbols, denoted as $\boldsymbol{s}^n$. This sequence is encoded into a codeword $\boldsymbol{x}(s^n)$ and transmitted over a discrete memoryless channel with capacity $C$. The channel is defined by the conditional probability distribution $P(\hat{\boldsymbol{x}}|\boldsymbol{x})$, where $\hat{\boldsymbol{x}}$ denotes the received codeword. Finally, the decoder reconstructs an estimate $\hat{\boldsymbol{s}}^n$ of the original symbol sequence from the received codeword.

Shannon's source-channel coding theorem states that for a stationary ergodic source can be transmitted with an arbitrarily small error probability if its entropy rate is less than the


W. Chen and Z. Qin, *Semantic Communications*, Textbooks in Telecommunication Engineering,
https://doi.org/10.1007/978-3-032-11005-3_4

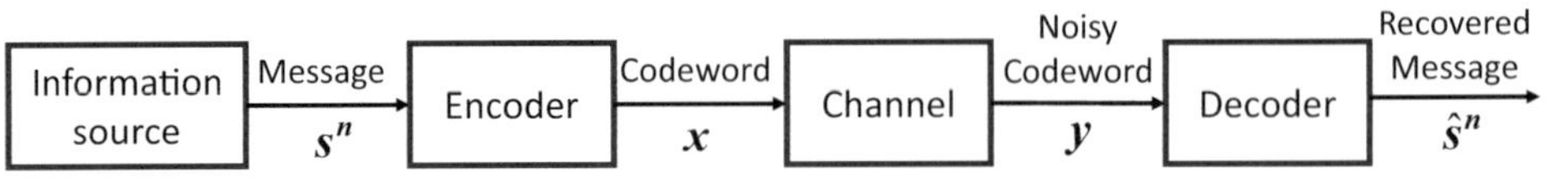


**Fig. 4.1** System model for joint-source-channel-coding

channel capacity. Assuming that the source sequence $s^n$ satisfies the asymptotic equipartition property (AEP). Under this assumption, there exists a typical set $A_\epsilon^{(n)}$ that contains at most $2^{n(H(\mathcal{S})+\epsilon)}$ distinct sequences of length-$n$ and has a probability $P[A_\epsilon^{(n)}] > 1-\epsilon$. Only the sequences in this typical set are encoded, while those outside it are discarded, ensuring that the encoding error does not exceed $\epsilon$. Each sequence in the typical set is then assigned a unique index, and lossless representation of these indices requires $n(H(\mathcal{S})+\epsilon)$ bits, given that there are no more than $2^{n(H(\mathcal{S})+\epsilon)}$ sequences.

Then, by applying channel coding principles, the effective source alphabet becomes the set of these indices. It can be demonstrated that transmitting the source message (represented by an index) is feasible with an error probability not exceeding $\epsilon$, provided that the channel code rate $R$ (in bits per channel use) is less than the channel capacity. Incorporating the bits required for the index representation, we have $H(\mathcal{S})+\epsilon = R < C$. As the sequence length $n$ approaches infinity, both the encoding error probability (bounded by $\epsilon$) and the transmission error probability converge to zero. Consequently, the source message can be transmitted with an arbitrarily small probability of error if $H(\mathcal{S}) < C$.

The separation of source and channel coding has traditionally been a prevalent strategy in the design of communication systems. This approach is primarily favored for its advantages in modularity, adaptability, and ease of implementation, rather than for the theoretical guarantee of no performance degradation implied by the source-channel separation theorem. In practice, it is well recognized that designing source and channel codes separately results in some performance loss. The absence of any performance penalty, as stated in the theorem, is contingent on the use of idealized codes that operate on infinitely long sequences—an assumption that incurs unacceptable complexity and delay in real-world applications.

Codes that are optimized for very long sequences may not achieve their theoretical performance under real-world conditions. The reason is that the statistical properties of a source can vary over time intervals that are shorter than the length of the input sequence processed by a source encoder, rendering a previously well-suited coding scheme less effective. For example, the statistical characteristics of a speech source might change within tens of milliseconds, which could involve fewer than two thousand bits and make a long-block-length design suboptimal. Similarly, a wireless channel may experience variations in its statistical behavior over periods shorter than those assumed by a channel encoder. In such cases, an initially efficient channel coding scheme becomes less optimal. Notably, the coherence time of a wireless channel depends on the relative speeds of the transmitter and receiver; it can be as short as a few milliseconds, although typical values are around one hundred milliseconds.

Furthermore, aiming for a source encoder that completely eliminates redundancy to achieve the most compact representation may not be the most effective design strategy. Highly compressed data are extremely sensitive to transmission errors because every bit becomes critical for accurately reconstructing the original source at the receiver. Retaining some residual redundancy in the encoded data stream can therefore enhance robustness against errors during decoding. Additionally, the conventional separate design of source and channel coding introduces an additional, less obvious performance drawback related to the error exponent. That is, the rate at which the probability of error of an optimal code decays with increasing block length. As demonstrated in [1], even for discrete memoryless sources and channels, joint source–channel coding schemes can achieve an error probability decay rate up to twice as fast as that achieved by separate designs, even when considering infinite-length sequences.

In the previous discussion, Shannon's source–channel coding theorem was derived by combining the implications of the AEP from the lossless source coding theorem with the joint AEP from the channel capacity theorem. Similarly, by integrating the joint AEP from the rate–distortion theorem with that from the channel capacity theorem, a lossy source–channel coding theorem can be established. In a system shown in Fig. 4.1, we assume that the encoding and decoding processes are characterized by a rate–distortion function $R(D)$. The average distortion between the source and the destination with the JSCC framework could be defined as:

$$D = \mathbb{E}[d(s^n, \hat{s}^n)] = \frac{1}{n}\sum_{i=1}^{n}\mathbb{E}[d(s_i, \hat{s}_i)]. \tag{4.1}$$

The lossy source–channel coding theorem asserts that this distortion $D$ is achievable if and only if $R(D) < C$. Naturally, as the coding rate increases, the distortion decreases, implying that the theorem characterizes the minimum distortion achievable for a channel with a given capacity. Together, the lossless and lossy source–channel coding theorems describe the conditions under which specific distortion levels can be attained in a communication channel. When the channel capacity exceeds the source entropy, zero distortion is attainable; conversely, if the channel capacity is less than the source entropy, the source message must undergo lossy compression, resulting in non-zero distortion.

## 4.2 Conventional Methods to Joint Source and Channel Coding

The development of effective JSCC schemes has long been a complex challenge in the field of information coding theory. Broadly, efforts to address this challenge fall into three main categories:

- Source-Aided Channel Coding: This part examines JSCC methods that integrate auxiliary side information at the decoders, beyond the primary source data. The work addresses challenges arising from the complexities of channel and source encoding, particularly under the constraints of non-stationarity and limited delay in the source. In practice, these complexities prevent source and channel codecs from operating as ideal, decoupled entities. Consequently, a mutual exchange of information between the source and channel decoders is not only possible but also beneficial. To mitigate the impact of these non-ideal behaviors, the method focuses on leveraging a priori source information, which encapsulates the statistical characteristics of the source, to enhance the error correction capabilities of the channel decoder. Additionally, it advocates for the utilization of a posteriori source information derived from the source decoding process, further improving the performance of channel decoding.
- Joint Optimization of Source Coding and Channel Coding: This approach is geared toward scenarios that the source coding (compression) and channel coding components are designed separately but then jointly optimized to enhance overall system performance. A key problem arises from the fact that independently designed components may not be optimally matched, leading to performance degradation when combined. The solution involves adjusting the parameters of both the compression algorithm and the channel code in jointly, thereby reducing the gap between theoretical limits and practical performance while ensuring robustness against the variability in source statistics and channel conditions.
- Joint Design of Source-Channel Coding: The last group moves toward a fully integrated design that unifies source and channel coding. It is motivated by the theoretical optimality of uncoded or linear transmission for Gaussian sources over Gaussian channels. However, practical scenarios—such as the transmission of images or videos—often involve non-Gaussian, non-linear data with intricate structures. The challenge here lies in the fact that pure linear or uncoded schemes fail to capture these complexities, leading to suboptimal performance. To overcome this, JSCC designs incorporate linear encoding strategies augmented with manually engineered non-linear elements.

Building on these foundational insights, researchers have proposed various methods to capitalize on source redundancy across a range of source and channel distributions. These designs aim to integrate redundancy into the coding process, enhancing the robustness of the transmitted signal against noise and other channel impairments. In the following subsections, we will provide serval examples of the conventional JSCC design methods.

### 4.2.1 Source-Aided Channel Coding

Currently, numerous joint source–channel coding (JSCC) designs are implemented by integrating source coding with established channel coding techniques, such as Turbo codes, low-density parity-check (LDPC) codes, and polar codes. These research strategies can be broadly categorized into source-controlled channel decoding approaches and channel coding-based JSCC methods. In the following discussion, several examples are outlined to facilitate a better understanding.

#### 4.2.1.1 Source-Controlled Channel Decoding

As early as 1948, Shannon noted in his seminal paper that "... any redundancy in the source will usually help if it is utilized at the receiving point. In particular, if the source already has redundancy and no attempt is made to eliminate it in matching to the channel, this redundancy will help combat noise." He also demonstrated a simple case where transmitting a non-uniform binary source without encoding over a binary erasure channel could achieve the optimal Shannon limit for separate source and channel coding (SSCC), provided the block length is infinite. These indicates that the source redundancy information can be leveraged to enhance noise resilience. One effective approach is to exploit the inherent redundancy in the source information to assist the channel decoder [2, 3]. Decoding processes can be improved by treating source and channel coding as interdependent processes rather than as separated functions. Specifically, incorporating source prior information, which reflects the statistical properties of the data, can enhance the channel decoder's error-correction capabilities. Additionally, the channel decoder can provide reliability metrics indicating the confidence level of each decoded bit, thereby guiding the source decoder in mitigating errors more effectively. Conversely, the source decoder can supply posterior information about the reconstructed data to further refine the channel decoding process. This bidirectional exchange of information fosters a cooperative decoding process that significantly enhances system efficiency and resilience in real-world scenarios.

In related researches, the information supplied to a decoding unit is quantified as the likelihood, confidence level, or reliability associated with each symbol. For a priori source information, the intuitive notion is that if a source generates bits with equal probabilities for 0 and 1, the a priori information value should be 0, indicating no bias toward either bit. As the probability of a bit being 1 increases, the a priori information assumes a larger positive value. Conversely, if the likelihood of a bit being 0 rises, the a priori value becomes increasingly negative. This information is commonly expressed as the log-likelihood ratio $L(x)$ for a binary random variable $x$, defined as $L(x) = \ln\left(\frac{P(x=+1)}{P(x=-1)}\right)$, where $P(x = +1)$ denotes the probability that $x$ equals 1. Convolutional channel codes employing Viterbi decoding are particularly well-suited to this log-likelihood representation.

Consider the system setup as Fig. 4.2. The source symbols are denoted as $\mathbf{s}$. After inputting to the source encoder, a bit sequence $\boldsymbol{u}$ is obtained. By inputting the $i$th bit into the convolutional channel encoder, a codeword $\boldsymbol{x} = \{x_1, x_2, \ldots, x_n\}$ consisting of $n$ coded words is produced. This codeword passes through a noisy channel, and after demodulated by the receive will result in the sequence $\boldsymbol{y}$. From $\boldsymbol{y}$, the estimated $\hat{\boldsymbol{u}}$ and $\hat{s}$ can be obtained by the channel decoder and source decoder, respectively.

With Viterbi decoding, it calculates all possible sequences and finds the one that maximizes the probability distribution $p(\boldsymbol{y}|\hat{\boldsymbol{u}})$ in order to complete the decoding process. The method involves mapping the estimated bit sequence to the estimated channel codeword state sequence $\boldsymbol{Q} = (q_0, q_1, ..., q_k)$, where $k$ represents the length of the bit sequence $\boldsymbol{u}$. These states can then be visualized as paths in the code's trellis diagram. Decoding is achieved by finding the path that maximizes the posterior probability $P({\boldsymbol{Q}_k}^l|\boldsymbol{y})$:

$$\begin{aligned}\hat{\boldsymbol{Q}}_k &= \arg\max_i P(\boldsymbol{Q}_k^l|\boldsymbol{y}),\\ &= \arg\max_i \frac{P(\boldsymbol{Q}_k^l, \boldsymbol{y})}{p(\boldsymbol{y})},\\ &\approx \arg\max_i P(\boldsymbol{Q}_k^l, \boldsymbol{y}),\\ &= \arg\max_i p(\boldsymbol{y}|\boldsymbol{Q}_k^l)P(\boldsymbol{Q}_k^l),\end{aligned} \tag{4.2}$$

where $l$ denotes the index of the trellis path. Assuming that the $k$-th bit $u_k$ influences the final Viterbi state sequence, $P(\boldsymbol{Q}_k^l)$ can be written as

$$P(\boldsymbol{Q}_k^l) = P(\boldsymbol{Q}_{k-1}^l)P(u_k). \tag{4.3}$$

In this case, the prior information from the source is used to guide the channel decoding [2, 3].

In addition to the Viterbi decoding mentioned above, the source-controlled channel coding method can also be implemented using iterative decoding methods. Since the introduction of Turbo codes in 1993, their capacity-approaching performance has been achieved through iterative decoding techniques that utilizing log-likelihood ratios (LLRs). The initial implementation of the Turbo encoder is a parallel concatenation of convolutional encoders, where interleavers are deployed to ensure the independence of the codewords. Meanwhile, it's also possible to implement Turbo codes with serially concatenated error correcting codes as shown in Fig. 4.3, which is of more interest for Joint Source-Channel Decoding (JSCD).

For concatenated channel codes, the Turbo principle is introduced as an iterative decoding approach. One channel decoder is deployed corresponding to each channel encoder, arranged in the same parallel or serial configuration used during encoding. Each decoder estimates the transmitted information bit sequence by selecting the sequence that maximizes the posterior probability distribution. The incorporation of log-likelihood ratios (LLRs) is fundamental to the implementation of the Turbo principle. Each bit is assigned an LLR value, which includes both a sign and a magnitude, representing the estimated source-encoded bit and the confidence level of the estimation, respectively. Each LLR can be expressed as the sum of the three values:

$$L(\hat{u}_i) = L_{int}(u_i) + L(u_i) + L_e(u_i). \tag{4.4}$$

The first term, referred to as "intrinsic information," represents the component of $L(\hat{u}_i)$ derived from the noisy symbol received at time $i$, denoted as $y_i$. The second term $L(u_i)$ corresponds to the a priori information associated with the bit $u_i$. The third term is "extrinsic information", which includes all indirect information about $u_i$ that the decoder can infer from

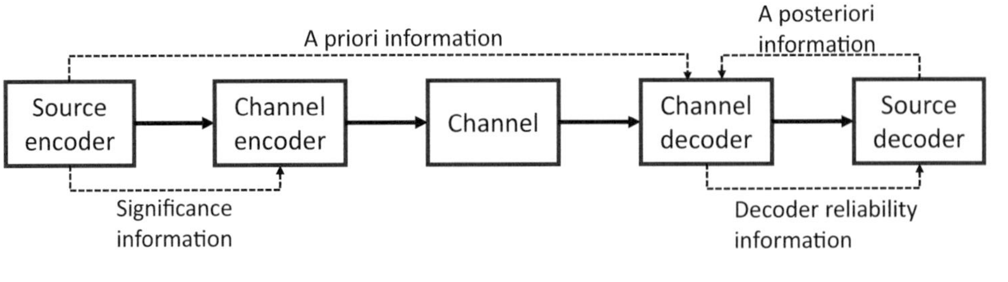


**Fig. 4.2** System model for source-aided channel coding

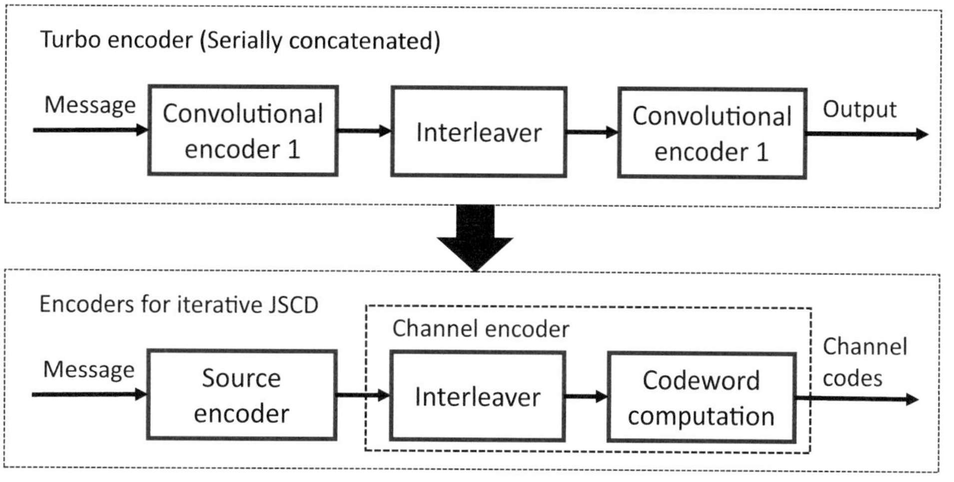


**Fig. 4.3** Serially concatenation of encoders under Turbo principle

the system. Since the decision on $u_i$ determined by the sign of $L(\hat{u}_i)$, these three LLR components contribute additively to the overall estimation.

Since the extrinsic information is derived from both the source symbols $\boldsymbol{s}$ and channel-coded $\boldsymbol{x}$, the likelihood information $L_e(u_i)$ is obtained by determining the most probable path in the decoding trellis through the decoding algorithm. However, processing the entire sequence of transmitted bits would incur substantial computational and storage demands. Therefore, to balance computational complexity and decoding performance, the $l$th bit is estimated using information from a few decoders both before and after $l$th decoder. Considering the joint utilization of extrinsic information by both decoders, the LLR value output from the first decoder at the onset of the decoding process can be expressed as

$$L(\hat{u}_i) = L_{int}(u_i) + L(u_i) + L_e^l(u_i), \tag{4.5}$$

where $l = 1$ when considering the extrinsic redundancy obtained from the first encoder. Assuming $P(u_i = +1) = P(u_i = -1)$, then a priori term $L(u_i) = 0$.

The extrinsic information $L_e^1(u_i)$ represents the indirect data inferred about encoder 1 from its structural design and serves as auxiliary information for decoder 2. Consequently, the extrinsic output from decoder 1 can be provided to decoder 2 as an additional input, thereby enhancing the LLR values derived from the noisy symbols. In this context, the output of decoder 2 can be expressed as

$$L(\hat{u}_i) = L_{int}(u_i) + L(u_i) + L_e^1(u_i) + L_e^2(u_i). \tag{4.6}$$

Based on this equation, it is evident that the a priori likelihood ratio is augmented by the extrinsic information provided by the previous decoders. Once computed, the same procedure can be employed to derive the output of decoder 1, which can be written as

$$L_2(\hat{u}_1) = L_{int}(u_1) + L(u_1) + L_e^1(u_1) + L_e^2(u_1). \tag{4.7}$$

Due to the iterative decoding process, the extrinsic message $L_e^2(u_1)$ is interdependent with $L_{int}(u_1)$ and $L_e^1(u_1)$, meaning that their combination is not entirely optimal.However, after completing the decoding iterations and applying the interleaver, the correlation among these messages does not significantly affect the overall decoding performance.

The above method has also been incorporated into JSCD schemes. In this context, the convolutional encoder is replaced with a concatenation of source and channel encoders as shown in Fig. 4.3. During the iterative decoding process, the output of the channel decoder serves as the extrinsic information for the serially concatenated source decoder. The source decoder's output is then fed back to the channel decoder in the next iteration to enhance the channel decoding performance. This iterative process continues until the reliability of the decoding results converges.

Consider the source encoder converts the $i$-th source symbol $s_i$ into multiple parameters $u_i^1, u_i^2, ..., u_i^k$. In conventional SSCC schemes, these parameters are typically quantized into $K$ bits, which are then fed into the channel encoder. To mitigate the effects of continuous noise, an interleaver is usually employed before channel encoding. The channel encoder generates the codeword $x_i$, which is transmitted over a noisy channel and received as the vector $y_i$. Assuming binary phase-shift keying (BPSK) modulation, the $l$th bit of the codeword of length $n$ as $x_i^l$. The impact of the channel can be described by the joint conditional probability density function of $\boldsymbol{y}_i$ given $\boldsymbol{u}_i$. In the case of an additive white Gaussian noise (AWGN) channel, we can write the pdf as

$$p(\boldsymbol{y}_i|\boldsymbol{u}_i) = \prod_{l=1}^{n} p(y_i^l|u_i^l). \tag{4.8}$$

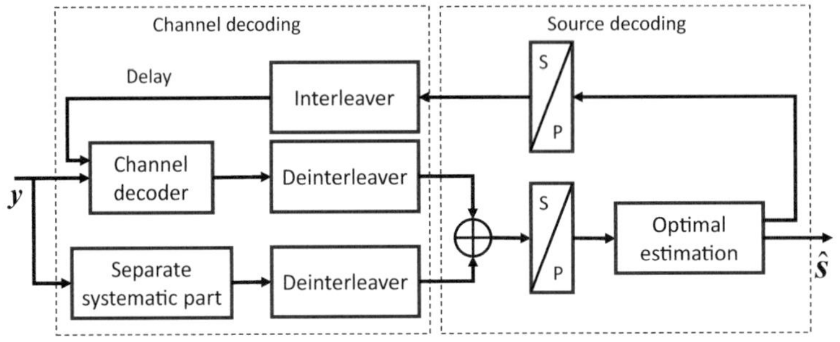


**Fig. 4.4** Block diagram of an iterative JSCD

The joint source-channel decoding algorithm in Fig. 4.4 leverages the residual redundancy remaining from the source coding stage, which arises due to practical implementation constraints. This redundancy is reflected in the dependency between quantization indices at consecutive time instances $u_i^k$ and $u_{i-1}^k$. To capture this dependency, it is modeled as a first-order stationary Markov process with index transition probabilities.

$$P(u_i^k = \upsilon^k | u_{i-1}^k = \psi^k), \tag{4.9}$$

where $\upsilon^k$ and $\psi^k$ take values 0, 1, ..., $2^n - 1$ and $k = 1, ..., K$.

At the receiver, source and channel decoding should be treated as an integrated process. Mathematically, this JSCD operation is formulated as the objective of obtaining the optimal estimate of $s_i$, which corresponds to minimizing the mean-squared error of the recovered bits:

$$min E[\sum_{k=1}^{K} |\hat{u}_i^k - u_i^k|^2 | y_i], \tag{4.10}$$

where the expectation is conditioned on the set of noisy vector $\boldsymbol{y}_i$. Considering all the realization $\boldsymbol{\upsilon}$ for the vector $\boldsymbol{x}_i$, the expected end-to-end mean-squared error (4.10) can be written as

$$\sum_{\forall \upsilon \in \Upsilon} \sum_{k=1}^{K} |\hat{u}_i^k - u_i^k|^2 P(\boldsymbol{u}_i = \boldsymbol{\upsilon} | \boldsymbol{y}_i). \tag{4.11}$$

Accordingly, the solution to the JSCD hinges on computing the a posteriori probabilities of all $\boldsymbol{\upsilon}$:

$$\hat{x}_i^k = \sum_{\forall \upsilon^k \in \Upsilon^k \upsilon^k} P(u_i^k = \upsilon^k | \boldsymbol{y}_i). \tag{4.12}$$

#### 4.2.1.2 Channel Coding-Based JSCC Methods

Various channel coding techniques, including LDPC codes, Polar codes, and rateless codes such as Raptor codes, have been investigated for integration into JSCC schemes.A prominent example is the use of Double-LDPC [4] and Double-Polar codes [5], where the source compression process is substituted with the corresponding coding encoder, followed by the addition of redundancy to facilitate error correction.

In some communication systems, when source information is not pre-compressed to retain redundancy before channel coding, JSCC can be employed to simultaneously achieve compression and noise robustness. Double-LDPC codes are designed based on this concept. Initially, an LDPC-based code is used to compress the source data, resulting in a compressed sequence $\boldsymbol{u} = \mathbf{H}_{sc}\boldsymbol{s}$, where $\mathbf{H}_{sc}$ is a sparse $l \times n$ parity-check matrix with a with a compression rate of $R_{sc} = l/n < 1$. This compressed sequence is then encoded using another LDPC code with a rate of $R_{cc} = l/m < 1$, producing the final codeword $\mathbf{x} = \mathbf{G}_{cc}\mathbf{u}$.

In the Tanner graph shown in Fig. 4.5, the compressed source information is connected to the channel code's variable nodes via the the check nodes of the source code. Initially, the variable nodes transmit their log-likelihood ratios (LLRs) to the check nodes, which then impose constraints on these LLRs. At the receiver, a separate belief propagation decoder performs the decoding. For more details, refer to [4]. Furthermore, subsequent studies have examined the use of Protograph-based LDPC and Spatially Coupled LDPC codes in JSCC applications. Similarly, Double Polar codes [5] adopt an similar strategy, where the source information is first compressed with a Polar-based code and subsequently protected by a second Polar code.

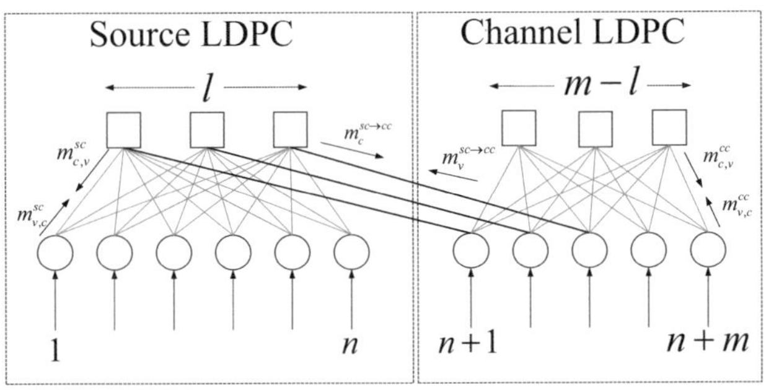


**Fig. 4.5** Tanner graph of Double-LDPC codes

## 4.2.2 Joint Optimization of Source Coding and Channel Coding

In the preceding subsection, most research on JSCC architecture design assumed that the source is either uncompressed or that its information is only used to assist the channel decoding process. However, practical sources typically exhibit significant, structured redundancy, and efficient, source-specific compression techniques are available for various types of data.

This subsection examines JSCC techniques that combine source-specific compression methods—designed for applications such as speech, image, and video compression—with channel coding. Typically, these approaches treat source compression and channel coding as separate processes while jointly optimizing their parameters.

Here are some specific examples of JSCC approaches. In [6], the authors integrated two-dimensional (2D) differential pulse coded modulation (DPCM) for image coding with a convolutional code, a method later extended to incorporate a 2D discrete cosine transform (DCT) within the same study. In contrast, Comstock and Gibson in [7] combined 2D-DCT with Hamming codes. During this period, substantial attention was also given to JSCC for speech signals. For example, Goodman and Sundberg in [8] examined the combined impact of quantization and transmission errors on the performance of embedded DPCM. Subsequently, their findings were applied to embedded DPCM speech encoding and punctured convolutional codes in [9]. Additionally, [10] presented an approach that integrated DPCM speech encoding with self-orthogonal convolutional codes.

Next, rather than focusing on a specific coding technique, we turn to explaining the joint design of two separate codewords and the associated rate allocation problem.

### 4.2.2.1 Bit Allocation in Concatenated Joint Source-Channel Coding

A straightforward method to implementing JSCC is to configure the source and channel coding blocks as two separate components arranged in a serial manner. In this configuration, the source encoder first processes the signal samples, and its output is subsequently fed into the channel encoder. Although these encoders are typically designed independently, concatenated JSCC achieves integrated coding by aligning their operating parameters and employing code designs that facilitate straightforward adjustment and interfacing. In the following, we will examine the structure and design principles of concatenated JSCC systems.

Figure 4.6 illustrates a conventional architecture for concatenated JSCC. In this configuration, the transmitter processes a block of source samples through a source encoder to produce encoded data, which is then passed to a channel encoder that generates the channel-encoded output for transmission. At the receiver, the data is first decoded by the channel decoder and then processed by the source decoder to reconstruct an approximation of the original source block. Although the source and channel codecs are typically designed as separate entities, their primary configuration parameter—the coding rates—can be adjusted to ensure compatibility. This adaptability is often achieved through codec designs that allow for easy rate modification, as exemplified by embedded source encoders and rate-compatible punctured channel encoders.

In concatenated JSCC, the main design goal is to optimize the allocation of source and channel coding rates. For simplicity, we focus on an AWGN channel. The objective is to jointly determine the optimal source and channel coding rates for a given channel signal-to-noise ratio (SNR). Moreover, it is assumed that the channel's transmission bit rate is fixed, so that the total number of bits transmitted over an entire encode-transmit-decode cycle remains constant at $N$ bits. Although more advanced communication systems can dynamically adjust the transmission bit rate (for example, via adaptive modulation), such mechanisms are beyond the scope of the concatenated JSCC bit rate allocation problem discussed here.

To allocate the coding rates, we need to balance the distortion originating from source coding, which arises from quantization and compression at the encoder, and the distortion introduced by the channel transmission errors. The goal is to minimize the overall impact of these distortions, often termed end-to-end distortion, which is typically approximated by summing the individual distortions from source coding and transmission errors. For example, if a source signal $s(t)$ is ssampled, quantized, and transmitted, the reconstructed signal $\hat{s}(t)$ at the receiver exhibits an end-to-end distortion that is measured using the mean squared error (MSE), which is given by

$$D = \int_T E[(s(t) - \hat{s}(t)^2]dt, \tag{4.13}$$

where $T$ denotes the duration during which the source signal is sampled and the distortion is evaluated. Since $s(t)$ is sampled, it can be reconstructed from its samples $s_i$, with $i = 1, ..., K$. By employing an orthonormal basis set $\phi_i$, the signal is transformed into interpolated samples as follows:

$$s(t) = \sum_{i=1}^{K} s_i \phi_i(t). \tag{4.14}$$

A common set used in Nyquist-Shannon sampling theorem can be written as

$$\phi_i(t) = \frac{\sin(\frac{\pi}{T_s}(t - kT_s))}{\frac{\pi}{T_s}(t - kT_s)}, \tag{4.15}$$

where $T_s$ is the sampling period.

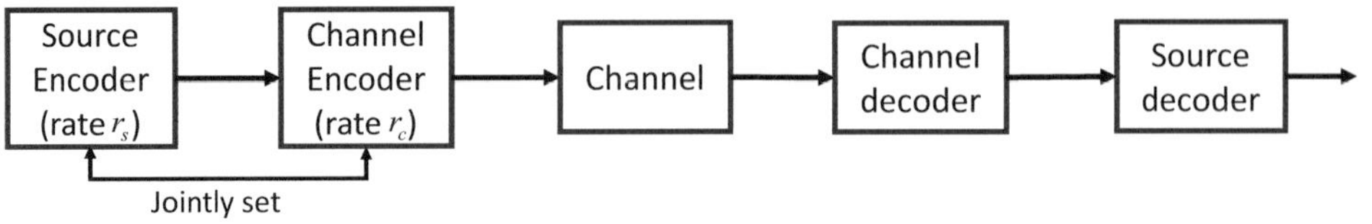


**Fig. 4.6** Block diagram of a concatenated joint source-channel coding scheme

After quantization, the samples $s_i$ are transmitted through a channel that may introduce errors. Although FEC codes are employed to rectify these errors, some error patterns might remain uncorrected. Typically, the channel coding is a concatenated scheme that combines an FEC code with an error detection code capable of reliably identifying errors missed by the FEC. So the source decoder inputs packets of received samples that are marked as being error free or containing an error. As a final safeguard, a robust source decoder is designed to manage the unlikely situation in which a packet is incorrectly marked as error-free despite containing an error. For packets identified as containing an error, the decoder utilizes an error concealment mechanism—such as interpolating from neighboring samples or using an average value—to compensate for the lost information. Since error concealment is not perfect, the residual information loss introduces distortion in the reconstructed source, corresponding to the transmission errors uncorrected by the FEC.

Let the quantized sample of $s_i$ be denoted as $\bar{s}_i$, the distortion in (4.13) can be written as

$$\begin{aligned} D &= E\left[\sum_{i=1}^{K}(s_i - \hat{s}_i)^2\right] \\ &= E\left[\sum_{i=1}^{K}(s_i - \bar{s}_i + \bar{s}_i - \hat{s}_i)^2\right] \\ &= E\left[\sum_{i=1}^{K}(s_i - \bar{s}_i)^2] + E[\sum_{i=1}^{K}(\bar{s}_i - \hat{s}_i)^2\right] \\ &\quad + 2E\left[\sum_{i=1}^{K}(s_i - \bar{s}_i)(\bar{s}_i - \hat{s}_i)^2\right] \\ &= D_S + D_C + 2E[\sum_{i=1}^{K}(s_i - \bar{s}_i)(\bar{s}_i - \hat{s}_i)^2\Big]. \end{aligned} \tag{4.16}$$

In this context, $D_S$ denotes the distortion resulting from source encoding, while $D_C$ represents the distortion due to transmission errors, quantified as the mean squared error between the original and reconstructed signals. The magnitude of the final term is influenced by the interrelationships among samples at different locations within the system. As demonstrated in [11], if the quantizer is optimized to minimize the average quantization distortion, this final term becomes zero.

This result is especially relevant in situations where error concealment is achieved by substituting lost samples with their expected value. Specifically, if we assume that the source has a zero mean, any received samples $\hat{s}_i$ that is lost due to uncorrectable transmission errors is replaced by $E(s_i) = 0$. The distortion incurred by this error concealment operation for a single sample is given by

$$D_C = E[(\bar{s} - \hat{s})^2] = E[\bar{s}^2]. \tag{4.17}$$

Now, by expanding the expression for source encoding distortion we get:

$$D_S = E[(\bar{s} - s)^2]. \tag{4.18}$$

As shown in [4.1] that $E[s\bar{s}] = E[\bar{s}^2]$, we can obtain:

$$\begin{aligned} D_S &= E[s^2] - E[\bar{s}^2] \\ &= \sigma_s^2 - E[\bar{s}^2] \\ &= \sigma_s^2 - D_C. \end{aligned} \tag{4.19}$$

Based on this analysis, when uncorrected errors persist after FEC decoding and error concealment is performed by replacing lost samples with their expected value, the overall end-to-end distortion is given by $D = D_S + \sigma_s^2 - D_S = \sigma_s^2$.

In summary, a quantized source sample transmitted can be received without errors—yielding an end-to-end distortion of $D = D_S$—or it can be affected by channel errors that the decoder fails to correct, in which case the end-to-end distortion becomes $D_F = D_S + D_C = \sigma_s^2$ (for a zero mean source). Let $P_r(\gamma)$ represent the probability of encountering uncorrectable errors, where $\gamma$ and $r$ denotes channel SNR and channel coding rate, respectively. The overall average end-to-end distortion is then given by

$$D(x, r, \gamma) = D_C P_r(\gamma) + D_S(x)(1 - P_r(\gamma)), \tag{4.20}$$

where $x$ denotes the source encoding rate in bits per sample. (4.20) indicates that when transmission occurs without errors, the overall end-to-end distortion is simply the source encoding distortion $D_S(x)$. Conversely, if the channel decoder fails to correct all errors, some or all of the source samples are discarded and substituted through an estimation or error concealment process, the resulting in a distortion of $D_C$ for a zero-mean source.

The value of $\sigma_s^2$ is influenced by the source statistics, the compression level, and the error concealment algorithm employed by the source decoder to address missing data. Typically, it is assumed to be substantially greater than $D_S(x)$. Furthermore, when the source samples follow a normalized Gaussian distribution and the lost samples are replaced with their expected value $E[s] = 0$, it follows that $\sigma_s^2 = 1$.

Since users evaluate communication quality based on the reconstructed source samples, the overall end-to-end distortion captures all transmission effects. Consequently, in a concatenated JSCC system, the joint optimization of source and channel coding rates is directed at minimizing this distortion.

In concatenated JSCC, the source and channel coding rates are selected jointly as a function of the channel state, e.g., SNR. In practical scenarios, these rates cannot be chosen from a continuous spectrum; even with codecs that allow for fine-grained adaptation, only a finite set of discrete options is available. Let $\mathcal{X} = \{x_1, x_2, ...\}$ and $\mathcal{R} = \{r_1, r_2, ...\}$ as the sets of available source and channel encoding rates, respectively. Each pairing of a source and channel coding rate defines an operating mode, denoted by $\Lambda_j$. These rates are related to the transmission bit rate—or more specifically, the number of bits sent per transmission period $C$—by the relation $C \geq Kx/r$, where $K$ represents the number of source samples processed during one transmission period. Once the parameters $r$, $C$, $K$, and the channel SNR are fixed, thereby determining $P_r(\gamma)$, choosing the largest allowed $x$ will minimizes the distortion. As a result, the optimal operating mode is identified by selecting an appropriate source coding rate from $\mathcal{X}$ or channel coding rate from $\mathcal{R}$. For any operating mode $\Lambda_j$, the distortion can be expressed as

$$d_{\Lambda_i}(\gamma) = D_C P_{\Lambda_i}(\gamma) + D_S(\Lambda_i)(1 - P_{\Lambda_i}(\gamma)). \quad (4.21)$$

The optimal operating mode minimizes the end-to-end distortion for a given channel SNR. When evaluated over a range of SNR values, these operating modes form a function that relates the minimum end-to-end distortion to the channel SNR. This approach exemplifies the practical implementation of concatenated JSCC: the transmitter stores a table that maps each segment of the D-SNR curve, defined by a minimum distortion level and its associated channel SNR boundaries, to a specific operating mode. The operating mode is then selected based on channel estimation outputs, in a manner akin to adaptive modulation schemes in modern communication systems.

#### 4.2.2.2 JSCC Designs for Unequal Error Protections

In practical systems, source encoders often integrate multiple techniques to achieve high compression efficiency. Consequently, the output bit stream can consist of several components produced by distinct compression methods. For example, a video bit stream may include motion vector data obtained via motion compensation, texture information that has been differentially and entropy coded, and headers carrying crucial metadata. Since these components vary in their significance for accurate source reconstruction, errors in the communication channel will affect each part differently. An error in one segment might introduce minimal distortion, whereas a similar error in another could severely degrade the reconstructed signal. In such scenarios, strictly following Shannon's source–channel coding theorem may incur a performance penalty. A more effective strategy is to apply varying levels of error protection to different sections of the bit stream according to their impact on the overall distortion. This method, known as unequal error protection (UEP) within JSCC techniques, is the focus of this discussion.

The impacts of channel errors on reconstruction quality are influenced by the importance and sensitivity of the affected portions. For instance, headers are critical since they typically include the information necessary for decoder configuration; if these are corrupted, large portions of the transmitted data may become undecodable. Similarly, certain parts of the bit stream—such as those encoded using variable-length entropy coding—are more vulnerable, because an error in an early bit can force the discarding of all subsequent bits. In differential encoding, an error in one sample can propagate to several following samples due to their interdependence.

To mitigate these issues, forward error correction code is commonly implemented before transmission. While adding redundant bits can help detect and correct some errors, it also increases the bit rate. Therefore, it is essential to carefully balance the amount of redundancy, since a uniform level may be excessive for less critical parts and insufficient for more crucial segments. The optimal compromise between distortion performance and error control overhead is achieved by assigning different levels of error protection to various parts of the encoded bit stream based on the potential impact of channel errors. This targeted approach is referred to as UEP.

The design of an error correction subsystem depends on the nature of the errors and the chosen method for handling those that cannot be entirely corrected. When a packet, frame, or block is detected as erroneous, a common approach is to remove it from the receiver's processing stream. Efficient error detection codes—such as cyclic redundancy check codes—can identify errors over large blocks with minimal extra bits, although they do not provide correction; hence, corrupted packets are typically discarded. For later decoding stages, this removal process effectively emulates a packet loss channel. This behavior is similar to that of the Internet, where routers or other intermediate nodes may drop packets during congestion, leading to packet loss statistics that are influenced by a variety of factors.

A number of techniques that provide UEP when the channel is modeled as a packet loss process. Many UEP methods derive from the concept of priority encoding transmission (PET) [12]. In PET, a message is divided into segments of varying priority. Each segment is encapsulated in one or more packets and is assigned FEC with strength proportional to its importance. This FEC generates additional redundancy packets that, combined with data packets, form segments through which the original message is transmitted. In this framework, the priority of a segment is defined by the fraction of its packets that must be successfully received to recover that part of the message; thus, high-priority segments are encoded to be decodable even if only a small portion of their packets is received.

A main task in designing a PET scheme is determining the priorities for each message segment, which in turn dictates the channel coding strength for each source packet. In [13], PET is applied to protect images encoded with an embedded source encoder. Embedded encoders produce a progressive bit stream, where the quality of the reconstructed source improves as more bits are received. This progressive property naturally informs the assignment of priorities, with the earliest bytes being the most critical—not only because they provide a baseline description that significantly reduces distortion compared to receiving no data, but also because any data received after the first uncorrectable error becomes unusable. In the context, PET approach was implemented for an image encoded with an embedded coder, resulting in a message of $M$ bytes.The progressive nature of the encoder means that the first byte is the most important and the last the least. The transmission used $N$ packetss, and the message is divided into $L$ segments, each with $N$ bytes, for a total of $L \times N = M$ transmitted bytes. When FEC is applied, each segment $i$ contains $m_i$ source-coded bytes along with $N - m_i$ redundancy bytes. Figure 4.7 illustrates this arrangement with $M = 16$, $N = 4$, and $L = 4$. The most important byte in the message is $B_1$ and the least important is $B_{30}$. In this example, the message is truncated at byte $B_{10}$, meaning that bytes $B_1$ through $B_{10}$ (the 10 transmitted bytes) are used, while bytes $B_{11}$ through $B_{30}$ are omitted. In the figure, $B_F$ denotes the redundancy bytes added for FEC.

The figure illustrates how the PET-based scheme allocates varying levels of packet loss protection across segments. The

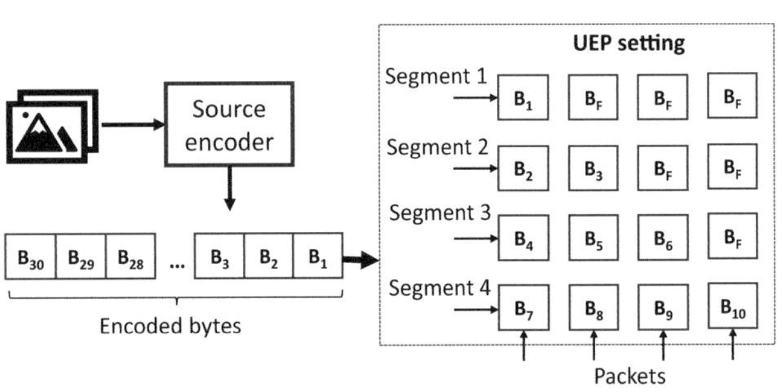


**Fig. 4.7** Unequal error protection setting for transmission of an encoded image using 4 segments and 4 packets

FEC subsystem employs a ReedÃ¢â,¬âŁœSolomon $(n, k_i)$ coder, with $n$ fixed as $N$ (the total number of packets) and $k_i$ (equals to $m_i$) determined during the design phase according to the criteria outlined below. For instance, in the scenario depicted in Fig. 4.7, if the third packet is lost during transmission, the data bytes $B_1$, $B_2$, $B_3$, and $B_4$ in the first packet are recovered successfully, since for segments where $k_i = i$, recovery is possible as long as at least $k_i$ bytes are received. In contrast, for segment 4, where $k_4 = 4$, reception of all four packets is required to recover the source encoded bytes. Therefore, the loss of byte $B_9$ renders that segment unrecoverable. Due to the progressive structure of the source encoded stream, any bytes following $B_9$ must be discarded, even if they are received without error.

This example illustrates how applying PET to an embedded-coded image confers differentiated levels of protection against packet losses. Selecting the values $m_i$ entails balancing the trade-off between the source encoding rate and the overall channel coding rate. When the source encoder produces a number of bytes equal to the channel's maximum capacity $M$, the decision to include FEC redundancy both sets the priority for each source encoded byte and determines the number of transmitted source bytes, given by $M - \sum_{i=1}^{L}(N - m_i)$. Let $f_i = N - m_i$ and let $p_d(f_i)$ denote the probability that segment $i$ is successfully decoded (i.e., that $f_i$ or fewer packets are lost, as determined by the transmitter–channel–receiver characteristics), the design objective becomes finding a vector $\mathbf{f} = (f_1, f_2, ..., f_L)$ that specifies the FEC redundancy allocation for each segment. To quantify the benefit of receiving each segment, we define the incremental quality improvement (measured in PSNR for the image transmission example) obtained when segment $i$ is received, assuming all preceding segments have been decoded correctly, as

$$g_i(\mathbf{f}) = Q[F(i, \mathbf{f})] - Q[F(i-1, \mathbf{f})], \tag{4.22}$$

where $F(i, \mathbf{f})$ represents the sequence of bytes from the source encoded stream from the first byte through the end of segment $i$,and $Q[F(i, \mathbf{f})]$ is the reconstructed source quality achieved using only that fragment. Thus, $g_i(\mathbf{f})$ quantifies the quality gain when the $i$-th segment is successfully decoded. It is now possible to state the design problem as

$$\begin{aligned} \max_{\mathbf{f}G(\mathbf{f})} &= \sum_{i=1}^{L} g_i(\mathbf{f}) p_d(f_i) \\ s.t.\ m_i + f_i &= N, i = 1, 2, ..., L \\ f_i &\geq f_{i+1}, i = 1, 2, ..., L-1. \end{aligned} \tag{4.23}$$

The expression $G(\mathbf{f})$ represents the expected reconstructed quality of the source. This formulation not only enforces the intrinsic segment priority order of PET but also reflects the progressive nature of the source encoded stream—ensuring that if segment $i$ is decodable, all prior segments are as well. This optimization problem can be tackled using a low-complexity, suboptimal heuristic algorithm that iteratively adjusts the allocation by adding or subtracting between 1 and Q FEC bytes bytes per segment while maintaining the condition $f_i \geq f_{i+1}$. Under this design, the segments deemed most critical (i.e., those with lower indices) are allocated a greater number of FEC bytes, making them decodable even when a smaller fraction of packets is received. As shown in [13], this unequal loss protection strategy enables the reconstructed image quality to degrade gracefully as packet loss increases, exhibiting a smooth decline and outperforming a comparable equal loss protection scheme over most packet loss rate ranges.

### 4.2.3 Joint Design of Source-Channel Coding

Although the designs discussed earlier are framed within the JSCC context, they fundamentally depend on separate source and channel codes that are either jointly decoded or optimized in tandem. Ultimately, a communication process produces an analog signal—continuous in both time and amplitude—that propagates through a physical medium. Since the essence of the joint source–channel coding problem is to map a message onto such an analog signal for transmission, it is logical to consider JSCC schemes that perform coding in a single integrated step, directly converting the source output into an analog waveform. In this chapter, we provide an example of this approach, illustrating how both analog and digital source signals can be directly mapped onto a signal suitable for physical transmission.

Furthermore, inspired by the theoretical optimality of uncoded transmission under certain source and channel conditions, practical JSCC methods have been proposed for image and video transmission. This concept was later popularized as SoftCast in [14], where uncoded transmission is applied directly after JPEG compression. SoftCast utilizes a discrete cosine transform on the input image and transmits the DCT coefficients over the channel using a dense modulation scheme. Numerous variations and modifications of SoftCast have been developed; interested readers can refer to the literature for additional details. Here, we present a straightforward example to elucidate the idea.

#### 4.2.3.1 Analog Joint Source–Channel Coding in Vector Spaces

In the is example, we examine communication systems in which analog sources are transmitted directly as continuous signals without distinct source and channel coding, a method known as analog source-channel coding. Here, the transmitter adapts the analog signal produced at the source so that it meets the specific characteristics and requirements of the channel.

The foundational concepts behind analog JSCC techniques date back to Claude Shannon's work in "Communication in the Presence of Noise," where he elegantly integrated the sampling theorem with the notion of channel capacity. Shannon's key insight was to represent signals, messages, channel-induced distortion, as well as the roles of the transmitter and receiver, as elements within vector spaces, thereby capturing the fundamental components of a communication process in geometric terms.

Assume that the source communicates by transmitting modulated signals of duration $T$, which are constrained by a channel bandwidth $W$. According to the sampling theorem, each transmitted signal can ideally be represented without any loss of information by a sequence of samples taken $1/(2W)$ seconds apart.Thus, a signal lasting $T$ seconds can be represented without loss of information by a set of $2WT$ samples.These samples serve as the coordinates of a vector in a $2WT$-dimensional space, known as the signal space.

Shannon's geometric representation permits key aspects of signal and communications theory to be interpreted as geometric entities. For instance, if the $2WT$ sample coordinates are referenced to mutually orthogonal axes, the distance from the origin to a point with coordinates $(x_1, x_2, ..., x_{2WT})$ (or equivalently, the length of the vector associated with a signal with samples $\{x_n\}_1^{2WT}$) is as follows:

$$d = \sqrt{\sum_{n=1}^{2WT} x_n^2}. \tag{4.24}$$

The sampling theorem tells us that a signal $f(t)$ can be reconstructed from its $2WT$ samples through the addition of time-shifted sinc functions:

$$f(t) = \sum_{n=1}^{2WT} x_n \frac{sin(2W\pi t - n\pi)}{\pi(2Wt - n)}. \tag{4.25}$$

The signal energy $E$ is as follows:

$$E = \int_{-\infty}^{+\infty} f(t)^2 dt = \frac{d^2}{2W}. \tag{4.26}$$

Thus, the square of the distance from the origin is $2W$ times the energy of the signal, i.e., $d^2 = 2WE = 2WTP$, where $P$ represents the average power. This geometric viewpoint also implies that the distance between two points in signal space equals $\sqrt{2WT}$ times the RMS difference between the corresponding signals, and that all signals with an average power less than $P$ lie within a sphere of radius $r = \sqrt{2WTP}$. Additionally, the impact of additive noise can be interpreted as displacing the signal's representative point by an amount proportional to the noise's RMS value.

Following the same thinking, messages produced by a source can be viewed as points in a separate message space. For example, consider a speech source generating sounds of duration $T_m$ seconds and a maximum frequency $W_m$. Similar to the signal space, the message space will have $2W_m T_m$ dimensions.

With both a signal space and a message space defined, it is natural to conceptualize the transmitter and receiver operations geometrically. The transmitter performs a mapping from the message space to the signal space, while the receiver inversely maps the signal space back into the message space.

In a discrete-time system, this mapping can be formalized by assuming the source produces $M$ samples and the transmitter outputs $K$ channel symbols. Since the dimensions of the source and channel spaces need not be identical, the mapping may either maintain, expand, or reduce dimensionality. Assuming the durations of the signals and messages are equal ($T = T_m$), a change in dimensionality reflects either bandwidth expansion (with $2W_m T_m < 2WT$ or $K > M$) or bandwidth compression (with $2W_m T_m > 2WT$ or $K < M$) during the operation of the transmitter.

The mapping function not only provides a framework for converting an $M$-dimensional source sample space into a $K$-dimensional channel symbol space but also enables the system either to accommodate more users via bandwidth compression or to enhance noise resilience through bandwidth expansion.

To understand the relationship between this mapping and bandwidth adjustments, consider matching the source information rate to the channel capacity. In the simplest case—where the channel is discrete-time, memoryless, and affected by additive white Gaussian noise (AWGN) with SNR $\gamma_c$, Shannon's formula gives the channel capacity in bits per channel use as

$$C = 1/2 \log_2(1 + \gamma_c). \tag{4.27}$$

Assuming sampling at the Nyquist rate, if the channel has a bandwidth $W$, it would be possible to transmit $2W$ symbols per second, making the channel capacity measured in bits per second be:

$$C = 2W \frac{1}{2} \log_2(1 + \gamma_c) = W \log_2(1 + \gamma_c). \tag{4.28}$$

Meanwhile, if the source generates i.i.d. samples from a zero-mean Gaussian distribution with variance $\sigma_X^2$, the source coding rate in bits per sample for a mean squared error (MSE) distortion $D$ is give by

$$R = \begin{cases} \frac{1}{2} \log_2\left(\frac{\sigma_x^2}{D}\right), & \text{for } 0 \le D \le \sigma_x^2 \\ 0, & \text{for } D \ge \sigma_x^2. \end{cases} \tag{4.29}$$

If the source signal occupies a bandwidth $B$, following the Nyquist–Shannon sampling theorem, the number of transmitted samples per second is $2B$. Consequently, for a mean squared error distortion $D$, the source coding rate, now measured in bits per second, is as follows:

$$R = \begin{cases} 2B \cdot \frac{1}{2} \log_2\left(\frac{\sigma_x^2}{D}\right), & \text{for } 0 \le D \le \sigma_x^2 \\ 0, & \text{for } D \ge \sigma_x^2. \end{cases} \tag{4.30}$$

To minimize the source distortion, the source coding rate must be as high as possible while still being transmitted over the channel. Equating the source coding rate with the channel capacity yields

$$B \log_2\left(\frac{\sigma_x^2}{D}\right) = W \log_2(1 + \gamma_c). \tag{4.31}$$

Solving for the coded source SNR results in what is known as the optimal performance theoretically attainable (OPTA) curve:

$$\frac{\sigma_x^2}{D} = (1 + \gamma_c)^{W/B}. \tag{4.32}$$

This result is influenced by the ratio of the channel to source bandwidth ($W/B$). Since the derivation relies on the Nyquist–Shannon sampling theorem, a change in bandwidth from the source to the channel entails a corresponding change in sampling rate. Accordingly, the bandwidth ratio can be approximated by the ratio of the number of channel samples $K$ to the number of source samples $M$ that are mapped to form these $K$ samples, i.e., $W/B \approx K/M$. Notably, larger bandwidth expansion ratios yield higher source signal quality and implicitly improve resilience against channel errors, allowing the same source SNR to be achieved at a lower channel SNR when more channel bandwidth is employed.

## 4.3 Summary

In this chapter, we began with a brief introduction to the principles of Joint Source-Channel Coding (JSCC), highlighting how the integration of source and channel coding offers a unified framework for efficient communication. We then provided an overview of traditional JSCC design approaches, focusing on representative techniques and strategies developed over the years. Specifically, we reviewed three categories of conventional JSCC methods: source-aided channel coding, joint optimization of source and channel coding, and the design of source-channel coding schemes.

## Problems

4.1 By combining the requirements for source coding and channel coding (assuming each source sample corresponds to $n = 1$ channel use), what is the resulting constraint?

A. $H < R < C$
B. $H > R > C$
C. $H = R = C$
D. $R < H < C$

4.2 According to the passage, the separate design of source and channel coding has limitations in practical applications primarily because:

A. The separate design always achieves optimal performance.
B. The separate design is simpler to implement.
C. The idealized assumption of "infinitely long sequences" leads to unacceptable complexity in real-world applications.
D. The separate design offers a faster error exponent.

4.3 Which of the following measures does NOT enhance performance in Joint Source-Channel Decoding (JSCD) systems?

A. Modeling the residual source redundancy as a Markov process.
B. Treating the source and channel decoders as independent modules to optimize separately.
C. Iteratively exchanging soft information between decoders.
D. Estimating the minimum mean squared error (MSE) based on posterior probability.

4.4 In the Turbo principle for iterative decoding, the overall LLR for a bit $\hat{u}_i$ is expressed as a sum of several components. Which of the following correctly identifies these components?

A. Intrinsic information only.
B. Intrinsic information and channel noise.
C. Intrinsic information, a priori information, and extrinsic information.
D. A priori information and modulation parameters.

4.5 In a concatenated source–channel coding system, what is the main characteristic of the source encoder and the channel encoder?

A. They are designed completely independently, with no relationship between their parameters.
B. They are designed independently, but their coding rates are adjusted to achieve matching and interfacing.
C. They must be designed jointly and cannot be optimized separately.
D. Their designs are irrelevant as long as the transmission bit rate remains fixed.

4.6 During transmission—even when employing FEC and error detection mechanisms—if there remain uncorrected errors, what approach is typically used to handle these errors?

A. Retransmit the entire packet.
B. Use error concealment techniques to replace lost samples with neighboring samples or an average value.
C. Discard the erroneous packet without any compensation.
D. Increase the FEC redundancy and retry the transmission.

4.7 In the discussion on distortion decomposition, assuming that the source signal has zero mean and that the distortion caused by uncorrected transmission errors is denoted as $D_C$, how can the source coding (quantization) distortion $D_S$ be expressed?

A. $D_S = D_C$
B. $D_S = \sigma_s^2 - D_C$
C. $D_S = \sigma_s^2 + D_C$
D. $D_S = \sigma_s^2 \cdot D_C$

4.8 Which of the following scenarios best demonstrates the necessity of applying unequal error protection (UEP)?

A. All data is encoded using fixed-length codes.
B. A video bit stream that includes motion vector data, texture information, and header information, where the header is critical for decoding.
C. An audio signal in which all bits are uniformly distributed with no priority distinction.
D. Data packets that are never lost during transmission.

4.9 When applying a PET scheme to perform UEP for an embedded-coded image, as the packet loss rate increases, the reconstructed image quality:

A. Declines sharply.
B. Remains completely unchanged.
C. Gradually declines in a stable manner and outperforms an equal error protection scheme.
D. Varies randomly.

4.10 One major advantage of a single-step JSCC scheme that directly maps the source to an analog signal is:

A. It avoids the sampling process and directly transmits continuous signals.
B. It simplifies the encoding structure, reduces system latency, and directly adapts to the continuous nature of the physical channel.
C. It completely eliminates source redundancy.
D. It achieves ideal source compression.

## References


1. Y. Zhong, F. Alajaji, L. Campbell, On the joint source-channel coding error exponent for discrete memoryless systems. IEEE Trans. Inf. Theory **52**(4), 1450–1468 (2006)
2. J. Hagenauer, Source-controlled channel decoding. IEEE Trans. Commun. **43**(9), 2449–2457 (1995)
3. A. Murad, T. Fuja, Joint source-channel decoding of variable-length encoded sources, in *1998 Information Theory Workshop (Cat. No.98EX131)* (1998), pp. 94–95
4. M. Fresia, F. Perez-Cruz, H.V. Poor, Optimized concatenated ldpc codes for joint source-channel coding, in *2009 IEEE International Symposium on Information Theory* (2009), pp. 2131–2135
5. Y. Dong, K. Niu, J. Dai, S. Wang, Y. Yuan, Joint source and channel coding using double polar codes. IEEE Commun. Lett. **25**(9), 2810–2814 (2021)
6. J. Modestino, D. Daut, Combined source-channel coding of images. IEEE Trans. Commun. **27**(11), 1644–1659 (1979)
7. D. Comstock, J. Gibson, Hamming coding of dct-compressed images over noisy channels. IEEE Trans. Commun. **32**(7), 856–861 (1984)
8. D.J. Goodman, C.-E. Sundberg, Transmission errors and forward error correction in embedded differential pulse code modulation. Bell Syst. Tech. J. **62**(9), 2735–2764 (1983)
9. D.J. Goodman, C.-E. Sundberg, Combined source and channel coding for variable-bit-rate speech transmission. Bell Syst. Tech. J. **62**(7), 2017–2036 (1983)
10. C. Moore, J. Gibson, Self-orthogonal convolutional coding for the dpcm-aqb speech encoder. IEEE Trans. Commun. **32**(8), 980–982 (1984)
11. R. Totty, G. Clark, Reconstruction error in waveform transmission (corresp.). IEEE Trans. Inf. Theory **13**(2), 336–338 (1967)
12. A. Albanese, J. Blomer, J. Edmonds, M. Luby, M. Sudan, Priority encoding transmission. IEEE Trans. Inf. Theory **42**(6), 1737–1744 (1996)
13. A. Mohr, E. Riskin, R. Ladner, Unequal loss protection: graceful degradation of image quality over packet erasure channels through forward error correction. IEEE J. Sel. Areas Commun. **18**(6), 819–828 (2000)
14. S. Jakubczak, D. Katabi, Softcast: one-size-fits-all wireless video. SIGCOMM Comput. Commun. Rev. **40**, 449–450 (2010). (Aug.)

# Deep Joint Semantic-Channel Coding

# 5

**Abstract**

This chapter provides an overview of the joint semantic-channel coding framework that integrates both analog and digital semantic communication paradigms. Analog semantic communication directly maps semantic representations into continuous constellation points, which can achieve efficient transmission but also introduce considerable challenges for practical implementation, such as robustness to channel noise, hardware constraints, and difficulties in standardization. To address these limitations, digital semantic communication frameworks are introduced, in which semantic features are discretized and encoded in a digital manner. This digital extension enables seamless compatibility with existing digital communication infrastructures, while preserving the task-oriented characteristics of semantic information. By combining analog and digital schemes, the joint semantic-channel coding framework provides a flexible solution to balance efficiency, robustness, and practicality. Furthermore, this chapter discusses design considerations, implementation challenges, and potential applications of the joint framework, aiming to bridge the gap between theoretical advances in semantic communication and its deployment in real-world systems.

**Keywords**

Semantic communication • Analog communication • Digital communication • Semantic-channel coding • Compatibility • Practical deployment

## 5.1 Introduction

Traditional separate source and channel coding schemes may struggle to achieve high efficiency in dynamic communication environments, leading to the development of DJSCC. DJSCC integrates source and channel encoding into a unified process, enabling more efficient transmission in wireless communication. As shown in Fig. 5.1. For example, in the low SNR regime, more symbols are allocated for channel coding, and fewer symbols are allocated for source coding. This prioritization ensures that error correction is robust enough to handle the higher noise levels. Conversely, in the high SNR regime, fewer symbols are allocated for channel coding, and more symbols are allocated for source coding [1]. This allocation strategy optimizes the usage of available resources. In addition, by directly mapping source signals to modulated channel inputs, DJSCC eliminates the need for bit-level conversions, simplifying the design of the system and improving its overall performance. This approach also provides greater flexibility and adaptability.

This joint optimization enables higher communication efficiency, improved robustness in low-SNR scenarios, and graceful performance degradation, effectively overcoming the "cliff effect" observed in conventional schemes [2–5]. Beyond these advantages, DJSCC exhibits strong flexibility and scalability. It can be integrated with various neural network architectures—such as GNNs [6], GANs [7], Transformers [8], VAEs [9], and even Diffusion Models [10]—to further enhance its feature representation and generalization capabilities. These advancements position DJSCC as a promising foundation for next-generation communication paradigms. However, despite its notable progress, the primary design objective of DJSCC remains focused on signal reconstruction rather than on optimizing for task-specific performance. In contrast, this limitation has led to the emergence of semantic communication, a paradigm that shifts the communication goal from faithfully reconstructing the original signal to enabling the receiver to successfully perform the intended task using only the received semantic information.

Notably, semantic communication inherits and extends many of the key advantages of DJSCC. Similar to DJSCC, it leverages deep neural networks to enable end-to-end opti-


W. Chen and Z. Qin, *Semantic Communications*, Textbooks in Telecommunication Engineering,
https://doi.org/10.1007/978-3-032-11005-3_5

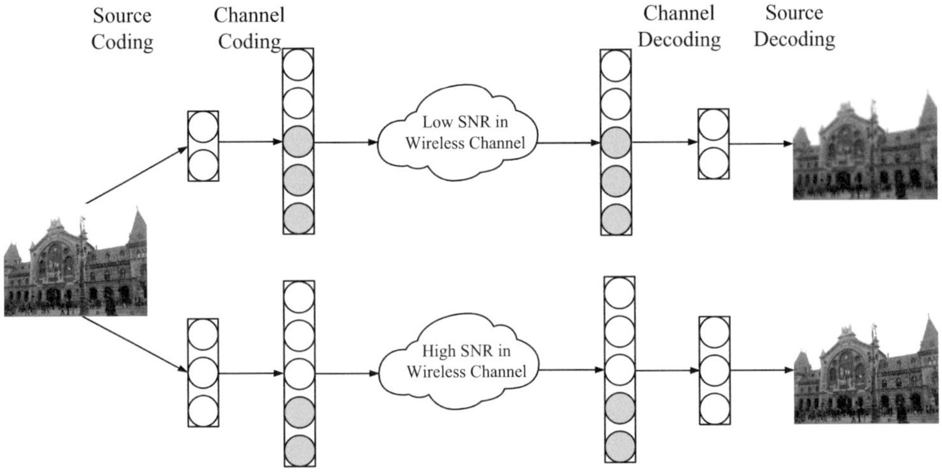


**Fig. 5.1** Illustration of DJSCC for dynamic SNR conditions

mization, ensuring robust and efficient information transmission even under noisy and bandwidth-limited conditions. By jointly learning semantic and channel representations, semantic communication eliminates the rigid separation between semantic and channel coding in traditional systems. This integrated design facilitates intelligent resource allocation, allowing the system to prioritize the transmission of task-relevant information while maintaining resilience to dynamic channel variations.

Moreover, semantic communication demonstrates notable advantages in three critical aspects: transmission efficiency, semantic-channel adaptability, and technical scalability.

- In terms of transmission efficiency, semantic communication significantly improves data quality under constrained bandwidth. For example, [11] demonstrates that deep neural network-based semantic schemes achieve lower end-to-end distortion than traditional digital compression methods [12, 13], under the same channel coding length, by prioritizing semantically meaningful content.
- In terms of semantic-channel adaptability, by jointly semantic and channel coding, semantic communication enables real-time adaptation. Studies such as [1, 14] proposed adaptive strategies based on bandwidth conditions and attention mechanisms, respectively, enhancing robustness across SNR variations. Feedback-aware methods [15, 16] further refine this adaptability. Additionally, like DJSCC, semantic communication benefits from graceful performance degradation—avoiding the "cliff effect" [4, 5] and ensuring resilience under adverse channel conditions [17, 18].
- In terms of technical scalability, semantic communication frameworks are highly extensible, supporting the integration of advanced architectures such as GNNs [6, 19, 20], GANs [7, 21, 22], Transformers [8, 23, 24], VAEs [9, 25, 26], and diffusion models [10, 27, 28]. These models enhance the expressiveness of semantic encoders and decoders across diverse modalities, promoting flexible and efficient task-oriented communication.

However, semantic communication goes a step further by aligning the communication process with task-specific objectives. In image-based scenarios, for instance, the system focuses on task-relevant semantic elements—such as object boundaries, class labels, or spatial relationships—while discarding non-essential background content [29, 30]. This selective attention not only improves task performance but also significantly reduces the amount of data that needs to be transmitted.

Moreover, semantic communication inherits DJSCC's advantage of graceful performance degradation. Unlike traditional separate coding schemes that suffer from the abrupt "cliff effect", both DJSCC and semantic communication maintain a gradual decline in performance under deteriorating channel conditions. This ensures that critical semantic information remains accessible, enabling the system to support task execution even in adverse environments.

In terms of scalability and flexibility, semantic communication also mirrors the strengths of DJSCC. Just as DJSCC benefits from the integration of diverse deep learning models—such as CNNs, VAEs, GANs, Transformers, and diffusion models—to boost representation and reconstruction performance, semantic communication can be adapted across various tasks and data modalities, including vision, language, and multi-modal fusion. This makes it particularly well-suited for deployment in bandwidth-constrained and latency-sensitive intelligent systems.

In summary, semantic communication not only builds upon the foundational strengths of DJSCC—such as end-to-end optimization, robustness, and adaptive feature prioritization—but also introduces a paradigm shift by embedding task awareness directly into the encoding and transmission pipeline. This transition from signal-level fidelity to task-level utility marks a fundamental evolution in communication systems, paving the way for intelligent, efficient, and goal-oriented information transmission.

## 5.2 Analog Semantic Channel Coding

Analog semantic communication represents a distinct paradigm within semantic communication systems. Its unique feature lies in its ability to directly generate and transmit constellation symbols without relying on predefined constellation maps. The core innovation of this approach is rooted in the encoding process, where the semantic encoder extracts meaningful features from the input data—such as images, text, or audio—and maps these features directly onto continuous constellation symbols. This method bypasses traditional bit-level representations, focusing instead on preserving the semantic essence of the transmitted information.

A defining characteristic of analog semantic communication is its use of continuous encoding spaces. Unlike conventional digital communication methods, which rely on discrete source coding and channel coding, analog semantic communication employs continuous mapping between the source signal and the transmitted symbols. This continuous mapping provides a more direct representation of the information, minimizing the loss typically associated with the separation of source and channel encoding processes. In the context of semantic communication or DJSCC, both source encoding and channel coding are jointly optimized, resulting in a more efficient and reliable communication system.

Early research on semantic communication or DJSCC has primarily focused on analog communication, where feature vectors are directly mapped to analog signals for transmission. While this approach has demonstrated significant potential, it faces considerable challenges in real-world deployment due to hardware limitations. Many commercial hardware systems, such as wireless transmitters, utilize hard-coded modulators optimized for specific modulation formats like 16QAM or 64QAM. These hardware constraints restrict the flexibility of semantic communication or DJSCC systems, as they must conform to predefined modulation schemes. This limitation poses a significant barrier to practical implementation, particularly in systems that demand scalability and adaptability. As a result, bridging the gap between the theoretical advantages of analog semantic communication and the practical requirements of existing hardware remains a critical challenge for advancing this technology.

The encoder in an analog semantic communication system typically produces complex-valued symbols that correspond to the semantic features extracted from the source data. As shown in Fig. 5.2, these symbols are not restricted by predefined constellations, allowing for more flexible and

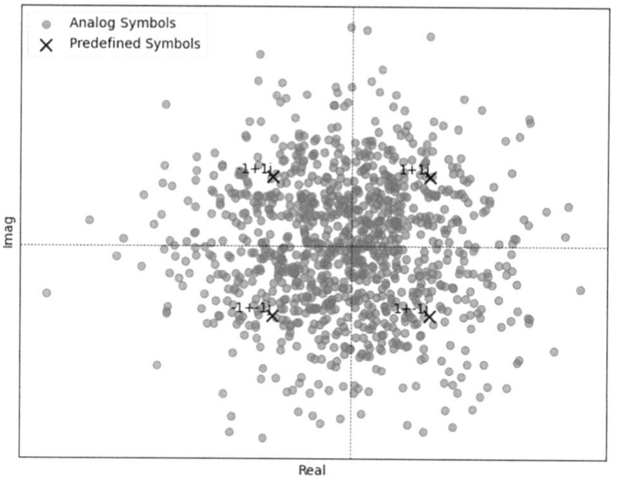


**Fig. 5.2** Illustration of arbitrary constellation points

continuous representations of the data. The process can be expressed mathematically as follows:

$$x = E(s; \alpha) \in \mathbb{R}^{zw_2h_2}, \tag{5.1}$$

where $x$ represents the vector reshaped from the semantic features. $E(\cdot; \alpha)$ represents the semantic encoder with its parameters $\alpha$. $s$ represents the input data. The semantic features extracted from the input data are typically divided into real and imaginary components, which are then used to form complex-valued symbols that will be transmitted over the communication channel. This can be mathematically expressed as

$$x_t = x_1 + jx_2 \in \mathbb{C}^{zw_2h_2/2}, \tag{5.2}$$

where $x_t$ denotes the transmitted complex symbols. $x_1$ is the real component of the transmitted complex symbols, and $x_2$ is the imaginary component. $j$ is the imaginary unit. The combined representation of the semantic features in the complex plane can be written as

$$x = [x_1, x_2], \tag{5.3}$$

indicating that both the real and imaginary components are derived from the output of the semantic encoder.

### 5.2.1 Source Coding for Different Source Data

In analog semantic communication, the encoder design is tailored to the specific type of input data and the categories of source coding are summarized in Table 5.1. Various data structures—such as images, text, audio, and non-structured data—require distinct encoding schemes to effectively capture and transmit the essential semantic information.

**Table 5.1** Categories of research related to source/semantic coding

| | Field | Categories |
|---|---|---|
| Source coding | Source data | Image |
| | | Audio |
| | | Video |
| | | Unstructured data |
| | | Multimodal |
| | Coding network | CNN |
| | | RNN |
| | | Transformer |
| | | VAE |
| | | GNN |

**Table 5.2** Categories of research related to channel coding

| | Field | Categories |
|---|---|---|
| Channel coding | Wireless channel | AWGN |
| | | Rayleigh |
| | | MIMO |
| | | BSC |
| | | BEC |
| | Adaption | SNR |
| | | Bandwidth |
| | | Channel |
| | | CSI |

For high-dimensional data such as images, CNNs are commonly used. CNNs excel at detecting spatial features and structures within image data, making them well-suited for image classification, object recognition, and image generation. Additionally, for video data, techniques combining CNNs with temporal models, like 3D CNNs or RNNs, are employed to capture both spatial and temporal dependencies.

For sequential data, such as text or speech, RNNs or Transformers are more appropriate. RNNs are effective for tasks like speech recognition, where temporal dependencies are crucial. In contrast, Transformers are commonly used in applications such as machine translation or text summarization, due to their ability to handle long-range dependencies and improve the efficiency of sequence modeling.

Moreover, with the increasing diversity of data types, multimodal encoding techniques have been developed. These techniques enable the simultaneous processing of multiple data forms, such as combining images with corresponding text for tasks like image captioning or multimodal translation. This is particularly beneficial in applications where various types of data need to be fused or transmitted together, such as autonomous systems, video conferences, or mixed-reality environments.

The flexibility of the encoder design in analog semantic communication allows for efficient adaptation to different data structures, thus improving both the source and channel encoding processes. This flexibility ensures that the encoding is optimized based on the specific characteristics of the data, improving overall transmission efficiency.

### 5.2.2 Channel Coding for Different Channel Conditions

Channel coding in analog semantic communication maintains reliable transmission in different channel conditions, such as AWGN, Rayleigh and MIMO. The goal is to ensure that the encoded symbols are accurately transmitted despite the presence of noise, interference, or other impairments in the communication channel.

Channel coding in analog semantic communication also needs to be adapted to different channel conditions. The goal of channel coding is to ensure reliable transmission of the encoded symbols even in the presence of noise and interference. Unlike traditional digital communication, which often relies on discrete symbol maps, analog semantic communication allows for more flexible encoding schemes that can be tailored to the specific characteristics of the communication channel. The categories of channel coding are summarized in Table 5.2.

In static channel conditions, where the noise and interference are relatively predictable, a simpler channel coding scheme may suffice. However, in dynamic or extreme channel conditions, where the transmission environment fluctuates significantly, more sophisticated adaptive channel coding techniques are required. These techniques can adjust the modulation and error-correction parameters in real-time, ensuring robust communication even in challenging environments.

In AWGN, simpler coding schemes with channel coding network like repetition codes or LDPC can be used, while in more dynamic environments like Rayleigh fading channels, advanced techniques such as Turbo Codes and Polar Codes are needed to handle signal variations. For MIMO channels, exploiting spatial diversity with Space-Time or Space-Frequency Codes helps improve capacity and mitigate fading effects to improve the communication performance.

In the context of CSI, the receiver can use feedback from the channel to optimize the decoding process. This allows for a more efficient transmission by adjusting the receiver's decoding strategy based on the current channel conditions, further improving the system's robustness and performance.

## 5.3 Digital Semantic Channel Coding

Analog semantic communication faces significant constraints due to limitations in hardware, protocols, and encryption. Commercially available hardware is typically designed to comply with specific communication standards, such as Bluetooth or Wi-Fi, which enforce predefined constellation orders and fixed configurations. These rigid specifications limit its suitability for diverse application scenarios and the rapidly evolving demands of modern communication systems.

To address these challenges, digital semantic communication has emerged as a promising alternative. By enabling the quantization and encoding of semantic information into discrete formats, digital semantic communication is more compatible with modern error correction and modulation techniques. This compatibility allows for improved performance in noisy environments, greater adaptability to varying channel conditions, and better integration with existing digital communication infrastructure.

Compared to analog communication, digital communication offers several advantages that make it well-suited for real-world deployment. These include stronger error correction capabilities, better resistance to interference, and enhanced security. These characteristics make digital systems more reliable in practical communication scenarios. As a result, digital communication has become a prominent focus in the field of semantic communication. Researchers are actively exploring ways to integrate digital semantic communication into practical systems. By combining the semantic compression benefits of semantic communication with the efficiency of digital channel coding and modulation, digital semantic communication provides a new pathway for large-scale deployment.

The transition from analog to digital semantic communication marks a significant advancement in the field. Early work, such as [31, 32], proposed a digital approach to semantic features, utilizing linear quantization functions to convert full-precision features into bit sequences. Later, [33] extended this concept to IoT scenarios by combining linear quantization with DNN compression techniques. This approach demonstrated significant improvements in data transmission efficiency, particularly in bandwidth-limited conditions. To enhance system flexibility, [34] introduced an innovative multi-bit rate selection method. By deploying a policy network at the transmitter, this method dynamically adjusts the bit length of semantic coding based on channel conditions, optimizing transmission while reducing the number of bits required for reliable communication.

However, research has revealed that feature maps often exhibit long-tailed distributions, which are non-uniform in nature [35–37]. Linear quantization, which assigns equal quantization levels to each interval, is not well-suited for these types of distributions. As a result, quantization precision decreases, limiting overall system performance. To address this issue, nonlinear quantization techniques have been developed. These methods dynamically adjust quantization intervals, enabling more accurate representation of the true distribution of semantic features, thereby enhancing both quantization precision and overall communication performance.

In [38], the DeepJSCCQ (DJSCCQ) method was introduced, where "Q" stands for quantization. This approach uses softmax functions to map analog constellation points to predefined constellation points, enabling nonlinear quantization of semantic symbols. Additionally, [39] introduced the JCM method, which incorporates VAEs to further explore the potential of nonlinear quantization in digital semantic communication. Both DJSCCQ and JCM use softmax-based nonlinear quantization directly on the constellation diagram, where the quantized constellation points are determined by the Euclidean distance between the analog outputs and the predefined constellation points. However, the non-uniform distribution of analog outputs on the constellation diagram still poses challenges for these softmax-based nonlinear quantization methods.

Digital semantic communication represents an innovative solution that combines the benefits of semantic communication with the advantages of digital communication. It seamlessly integrates semantic compression, resistance to interference, and encryption capabilities, all while ensuring compatibility with hardware constraints. This research area plays a pivotal role in enabling the large-scale deployment of semantic communication systems, especially in resource-constrained environments that demand high reliability. The potential of digital semantic communication is vast, offering a solid foundation for future advancements in communication technologies.

Current research in digital semantic communication primarily investigates two main approaches. The first approach employs neural networks to directly generate constellation points for analog communication. These generated points are then quantized into predefined constellation points using softmax functions, as illustrated in Fig. 5.3. While this method facilitates end-to-end generation of constellation points, its

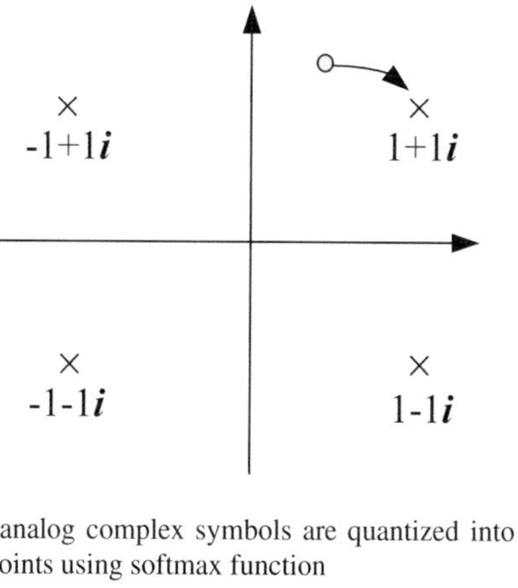


**Fig. 5.3** The analog complex symbols are quantized into predefined constellation points using softmax function

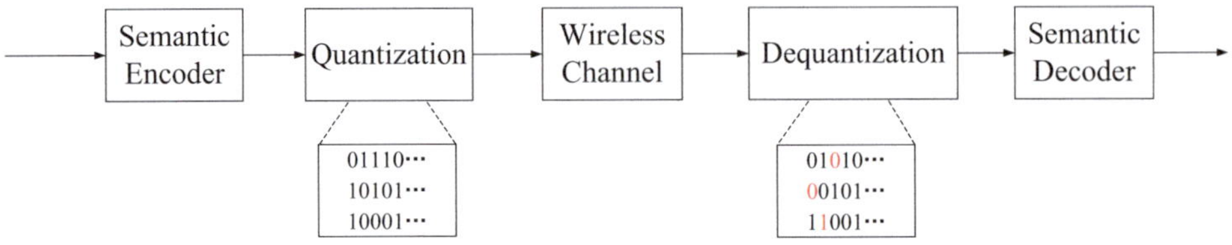


**Fig. 5.4** The semantic features are quantized into bit sequences and transmitted by the digital wireless communication system

flexibility is limited by the reliance on fixed constellation maps. Additionally, it faces challenges in effectively handling the non-uniform distributions commonly observed in simulated constellation points, which restricts its applicability to diverse modulation schemes.

The second approach involves direct quantization of feature values, where the quantized values are converted into bit sequences and subsequently mapped to constellation points through encoding, as shown in Fig. 5.4. This method provides precise control over the quantization process, enabling adaptation to various data distributions and supporting flexible bit mapping for different modulation schemes.

### 5.3.1 Symbol Quantization in Digital Semantic Communication

In this section, a digital approach is introduced that quantizes the analog constellation symbols generated by the network into predefined discrete constellation symbols. This method ensures that the encoder outputs channel symbols from a finite constellation, aligning the communication system with standard modulation techniques while maintaining semantic relevance.

To quantify the analog constellation symbols generated by the encoder into predefined discrete constellation symbols, the quantization process is applied to the complex-valued vector $\boldsymbol{x}_t$, mentioned in the Eq. 5.2. Assuming the predefined constellation map is $C$, The mapping of a complex-valued symbol $x_t^{(i)} \in \boldsymbol{x}_t$ to the nearest constellation point $c_j \in C$ can be expressed as

$$\bar{x}_t^{(i)} = \arg\min_{c_j \in C} \left| x_t^{(i)} - c_j \right|^2, \tag{5.4}$$

where $\left| x_t^{(i)} - c_j \right|^2$ is the squared Euclidean distance between the symbol $x_t^{(i)}$ and $c_j$. $\bar{x}_t^{(i)} \in \bar{\boldsymbol{x}}_t$ denotes the quantized complex-valued vector. The limitation of this operation is non-differentiability, which poses challenges for gradient propagation during network end-to-end training. To address these challenges, a soft quantization approach based on the softmax can be employed. This approach ensures differentiability, enabling the constellation mapping to be optimized during the training process. The soft quantization process can be expressed as

$$\bar{x}_t^{(i)} = \sum_{j=1} \frac{e^{-T d_{ij}}}{\sum_{n=1} e^{-T d_{in}}} c_j, \tag{5.5}$$

where $T$ is a temperature parameter that controls the sharpness of the softmax.

The quantized symbol is represented as a weighted sum of all predefined constellation points, where the weights are determined by the distances between the analog symbols and the constellation points. As $T$ increases, the soft quantization approaches hard quantization like Eq. 5.4, effectively assigning each analog symbols to its nearest constellation point. This smooth approximation enables the computation of gradients in an end-to-end training. The gradients in this quantization process are expressed as

$$\frac{\partial L}{\partial \boldsymbol{x}_t} = \frac{\partial L}{\partial \bar{\boldsymbol{x}}_t} \frac{\partial \bar{\boldsymbol{x}}_t}{\partial \boldsymbol{x}_t}, \tag{5.6}$$

$$\frac{\partial \bar{x}_t^{(i)}}{\partial x_t^{(i)}} = -2T \sum_{j=1} w_{ij} \left( x_t^{(i)} - c_j \right) c_j + 2T \sum_{j=1} w_{ij} c_j \sum_{n=1}^{M} w_{in} \left( x_t^{(i)} - c_n \right), \tag{5.7}$$

where $w_{ij} = \frac{e^{-T d_{ij}}}{\sum_{n=1} e^{-T d_{in}}}$ denotes the weights for each constellation point are determined by the softmax function. This gradient computation ensures that the parameters of the model are updated effectively during backpropagation. The soft quantization framework provides differentiability through the use of softmax weights $w_{ij}$, enabling smooth and efficient optimization of the parameters. The temperature parameter $T$ plays a crucial role in controlling the balance between soft and hard assignments, with larger $T$ values approaching hard quantization behavior.

### 5.3.2 Feature Quantization in Digital Semantic Communication

In this section, a different approach is introduced that contrasts with directly quantizing symbols. Instead, this method first quantizes the continuous input data into discrete values, and then maps these quantized values onto the constellation diagram. By separating the quantization and mapping processes, this approach offers greater flexibility in adapting to diverse data distributions and modulation schemes.

The digital semantic communication system designed for edge-device collaboration is illustrated in Fig. 5.5. This system comprises a digital semantic encoder at the transmitter and a digital semantic decoder at the receiver. Unlike analog semantic communication, which relies on arbitrary complex-valued constellation symbols, the digital semantic encoder transforms input data (e.g., images) into digital features represented as bit sequences. These bit sequences are transmitted through a wireless channel to the receiver. The digital semantic decoder reconstructs the semantic information from the received bit sequences. Taking image reconstruction as an example, the image data is represented as $\boldsymbol{s} \in \mathbb{R}^{3w_1h_1}$, where $3w_1h_1$ indicates the dimensions of the image. At the device, the digital semantic encoder encodes the image into digital semantic features, which is mathematically expressed as

$$\boldsymbol{D} = Q\left(E\left(\boldsymbol{s};\boldsymbol{\alpha}\right);\boldsymbol{\theta}\right) \in \mathbb{R}^{zw_2h_2}, \tag{5.8}$$

$$\boldsymbol{b} = f_{a\to d}\left(\boldsymbol{D}\right) \in \{0,1\}^{zqw_2h_2}, \tag{5.9}$$

where $\boldsymbol{D}$ represents the quantized features. $E(\cdot;\boldsymbol{\alpha})$ represents the semantic encoder with parameter $\boldsymbol{\alpha}$. $Q(\cdot;\boldsymbol{\theta})$ represents the quantization module with parameter $\boldsymbol{\theta}$. $\boldsymbol{b}$ denotes the bit sequence output by the digital semantic encoder. The quantized features have $2^q$ discrete quantization levels, with each level represented by a $q$-bit sequence.

We consider a binary symmetric channel (BSC) for transmission. The received bit sequence is denoted as $\hat{\boldsymbol{b}} \in \{0,1\}^{zqw_2h_2}$. The digital semantic decoder maps these bit sequences back to the corresponding $2^q$ quantization levels and reconstructs the semantic information. The reconstructed semantic information is used for inference or reconstruction tasks. For image reconstruction, this process is expressed as

$$\hat{\boldsymbol{D}} = f_{d\to a}\left(\hat{\boldsymbol{b}}\right) \in \mathbb{R}^{zw_2h_2}, \tag{5.10}$$

$$\hat{\boldsymbol{s}} = D\left(\hat{\boldsymbol{D}};\boldsymbol{\beta}\right), \tag{5.11}$$

where $\hat{\boldsymbol{D}}$ is the reconstructed digital features obtained through mapping and decoding. $\hat{\boldsymbol{s}}$ is the reconstructed image derived from the reconstructed features $\hat{\boldsymbol{D}}$. $D(\cdot;\boldsymbol{\beta})$ represents the semantic decoder with parameter $\boldsymbol{\beta}$.

### Detail of the Features Quantization

This chapter introduces the edge-device collaborative digital semantic communication system, as shown in Fig. 5.5. The digital semantic encoder network extracts semantic features and quantizes them into discrete levels for efficient transmission, while the digital semantic decoder network reconstructs and infers semantic information from the received bit sequence.

#### (1) Quantization Modules

To better accommodate diverse modulation schemes and data distributions, this section focuses on the approach of quantizing continuous input values and subsequently mapping the quantized values onto a constellation diagram. In this context, both uniform quantization and a nonlinear quantization function suitable for semantic communication are introduced.

The primary goal of the nonlinear quantization function is to address the inherent limitations of uniform quantization by allowing for flexible adjustments of quantization levels. This adaptability enables the system to better align with complex data distributions, ultimately enhancing its ability to handle diverse communication scenarios and improving overall performance under various modulation conditions.

**Uniform Quantization** is a commonly used method that divides input features into fixed intervals, making it well-suited for data distributions that are relatively uniform. The quantization function for uniform quantization is mathematically expressed as

$$Q(\boldsymbol{X}) = \min\left(\boldsymbol{X}\right) + \frac{1}{q_w} \cdot \operatorname{round}\left(q_w \cdot \left(\boldsymbol{X} - \min\left(\boldsymbol{X}\right)\right)\right), \tag{5.12}$$

where $q_w = \frac{2^q - 1}{\max(\boldsymbol{X}) - \min(\boldsymbol{X})}$ is a scaling factor that maps continuous values into discrete levels, which are represented as a $q$-bit sequence. round$(\cdot)$ is the rounding operation used to assign input values to the nearest quantization level. One significant limitation of uniform quantization is its discrete and non-differentiability, which makes gradient propagation during neural network training challenging. This non-differentiability limits the system's ability to fine-tune parameters effectively in end-to-end training. To overcome these challenges, the quantization function can be approximated using a combination of multiple unit step functions. Although unit step functions are inherently non-differentiable at their transition points, smoothing them with sigmoid functions introduces continuous transitions. This smoothing facilitates gradient propagation during backpropagation, mitigating vanishing gradient issues and improving the training performance of neural networks in quantization tasks. An ideal linear quantization function can

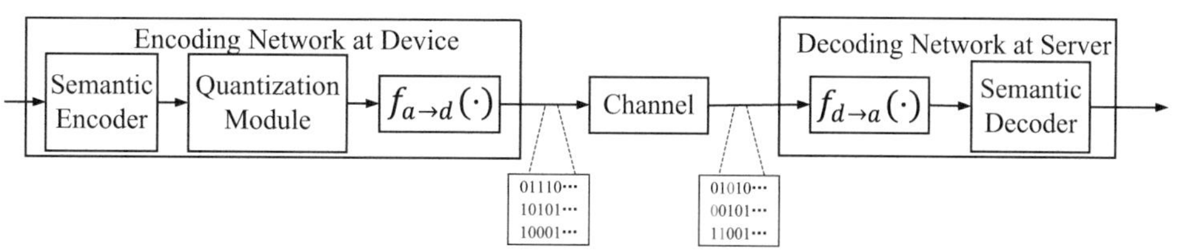


**Fig. 5.5** The digital semantic communication system for device-edge cooperative inference

be expressed as a linear combination of several unit step functions:

$$Q_{\text{ideal}}^{\text{linear}}(\boldsymbol{X}) = \min(\boldsymbol{X}) + \frac{1}{q_w} \sum_{i=1}^{2^q-1} \mu\left(\boldsymbol{X} - (\min(\boldsymbol{X}) + iq_w) \cdot \mathbf{1}\right), \tag{5.13}$$

where $\mu(\cdot)$ represents the unit step function. $\min(\boldsymbol{X}) + iq_w$ represents discrete quantization values. $\mathbf{1} \in \mathbb{R}^{zw_2h_2}$ is a matrix of ones matching the feature map's dimensions. This formulation provides a theoretical foundation for incorporating smoothed quantization functions into neural networks, making them more trainable and effective in handling the complexities of discrete operations. Uniform quantization assumes that the input data is uniformly distributed. However, this assumption often fails when dealing with complex and diverse data distributions. In such cases, the fixed intervals of uniform quantization fail to adapt to the distribution's characteristics, resulting in increased quantization errors. These errors, in turn, degrade the overall performance of the system. While uniform quantization is simple and efficient for uniform data, its lack of adaptability to complex data distributions highlights the need for more flexible quantization methods.

**Nonlinear Quantization** is used to replace fixed discrete values with trainable parameters to overcome the limitations of uniform quantization as shown in Fig. 5.6. This approach allows the quantization levels to adapt dynamically in a data-driven manner, enabling better representation and capture of complex data distributions. As a result, it significantly enhances the quantization performance and overall system efficiency. Let $\boldsymbol{\theta}$ denote the set of quantized discrete values. An ideal nonlinear quantization function with variable intervals can be expressed as

$$Q_{\text{ideal}}(\boldsymbol{X}; \boldsymbol{\theta}) = \boldsymbol{\theta}[0] \cdot \mathbf{1} + \sum_{i=1}^{2^q-1} (\boldsymbol{\theta}[i] - \boldsymbol{\theta}[i-1]) \cdot \mu\left(\boldsymbol{X} - \boldsymbol{\theta}[i] \cdot \mathbf{1}\right), \tag{5.14}$$

where $\boldsymbol{\theta} \in \mathbb{R}^{2^q}$. The terms $(\boldsymbol{\theta}[i] - \boldsymbol{\theta}[i-1])$ and $\boldsymbol{\theta}$ represent the scaling factor and offset of the $i$-th unit step function, respectively. $\boldsymbol{\theta}$ also corresponds to the $i$-th discrete quantization value. Since the unit step function is non-differentiable at its transition points, it leads to vanishing gradient problems during backpropagation. To address this issue, a nonlinear quantization function with a linear combination of multiple smoothed sigmoid functions is expressed as

$$Q(\boldsymbol{X}; \boldsymbol{\theta}, T) = \boldsymbol{\theta}[0] \cdot \mathbf{1} + \sum_{i=1}^{2^q-1} (\boldsymbol{\theta}[i] - \boldsymbol{\theta}[i-1]) \cdot \sigma_i(\boldsymbol{X}; T), \tag{5.15}$$

where $\sigma_i(x; T) = \frac{1}{1+\exp(-T(x-\boldsymbol{\theta}[i]))}$ is the sigmoid function with a tunable parameter $T$ that controls the steepness of the transition. This formulation serves as a differentiable approximation of the ideal nonlinear quantization function, allowing effective gradient propagation and mitigating vanishing gradient issues. By adjusting $T$, the sigmoid function can closely approximate the behavior of an ideal step function while maintaining smoothness, thereby facilitating efficient backpropagation during training.

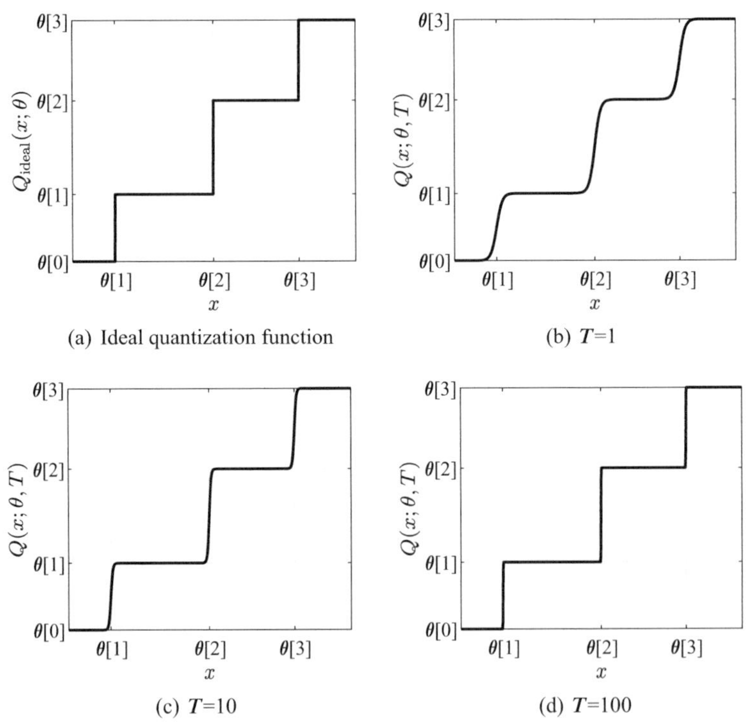


**Fig. 5.6** The ideal non-linear quantization function and the "soft" quantization function. During the training stage, the "soft" quantization function goes from a straight line to steps as the $T$ increases

During backpropagation, optimizing the nonlinear quantization function and its associated model parameters requires effective gradient propagation through the quantization process to the preceding layers of the model. Specifically, this involves calculating the gradients of the loss function with respect to both the input $\boldsymbol{X}$ and the quantization parameters $\boldsymbol{\theta}$. Accurately computing these gradients enables the model to fine-tune and optimize the quantization parameters effectively. The gradients in the quantization process are expressed as

$$\frac{\partial L}{\partial \boldsymbol{X}} = \frac{\partial L}{\partial Q(\boldsymbol{X};\boldsymbol{\theta},T)} \frac{\partial Q(\boldsymbol{X};\boldsymbol{\theta},T)}{\partial \boldsymbol{X}}, \tag{5.16}$$

$$\frac{\partial Q(\boldsymbol{X};\boldsymbol{\theta},T)}{\partial \boldsymbol{X}} = T\sum_{i=1}^{2^q-1}(\boldsymbol{\theta}[i]-\boldsymbol{\theta}[i-1])\cdot\sigma_i(\boldsymbol{X};T)\otimes(\mathbf{1}-\sigma_i(\boldsymbol{x};T)), \tag{5.17}$$

$$\frac{\partial L}{\partial \boldsymbol{\theta}[i]} = \frac{\partial L}{\partial Q(\boldsymbol{X};\boldsymbol{\theta},T)} \frac{\partial Q(\boldsymbol{X};\boldsymbol{\theta},T)}{\partial \boldsymbol{\theta}_{[i]}} \tag{5.18}$$

$$\frac{\partial Q(\boldsymbol{X};\boldsymbol{\theta},T)}{\partial \boldsymbol{\theta}[i]} = \begin{cases} \mathbf{1}-\sigma_1(\boldsymbol{X};T), & i=0 \\ -T(\boldsymbol{\theta}[i]-\boldsymbol{\theta}[i-1])\cdot\sigma_i(\boldsymbol{X};T)\otimes(\mathbf{1}-\sigma_i(\boldsymbol{X};T)) \\ \quad +\sigma_i(\boldsymbol{X};T)-\sigma_{i+1}(\boldsymbol{X};T), & 0<i<2^q-1 \\ -T\left(\boldsymbol{\theta}_s[2^q]-\boldsymbol{\theta}[2^q-1]\right)\cdot\sigma_{2^q}(\boldsymbol{X};T)\otimes(\mathbf{1}-\sigma_{2^q}(\boldsymbol{X};T)) \\ \quad +\sigma_{2^q}(\boldsymbol{X};T), & i=2^q-1. \end{cases} \tag{5.19}$$

where $\mathcal{L}$ represents the objective to be minimized. The gradient computation in the nonlinear quantization module ensures that, during backpropagation, both the model parameters and the nonlinear quantization intervals are effectively updated. This process allows the quantization parameters to adapt dynamically to different data distributions, ultimately optimizing the quantization performance and enhancing overall system efficiency.

**(2) Conversion from Quantized Values to Bit Sequences**

After applying nonlinear quantization on the neural network's feature maps, the resulting quantized values can be converted into bit sequences for digital representation and transmission. Specifically, the feature maps of the neural network are processed by the nonlinear quantization module and quantized into discrete values $\boldsymbol{\theta}$. These quantized values are then further converted into bit sequences $\boldsymbol{b} = f_{a\to d}(\boldsymbol{D}) \in \{0,1\}^{zqw_2h_2}$, where each quantization level corresponds to a $q$-bit sequence. This bit representation enables modulation, encoding, and transmission in digital communication. For example, in the case of 2-bit quantization ($q = 2$), the conversion function can be expressed as

$$f_{a\to d}(x;\boldsymbol{\theta}) = \begin{cases} 11 & x > \dfrac{\boldsymbol{\theta}[2]+\boldsymbol{\theta}[3]}{2} \\ 10 & \dfrac{\boldsymbol{\theta}[1]+\boldsymbol{\theta}[2]}{2} < x \le \dfrac{\boldsymbol{\theta}[2]+\boldsymbol{\theta}[3]}{2} \\ 01 & \dfrac{\boldsymbol{\theta}[0]+\boldsymbol{\theta}[1]}{2} < x \le \dfrac{\boldsymbol{\theta}[1]+\boldsymbol{\theta}[2]}{2} \\ 00 & x \le \dfrac{\boldsymbol{\theta}[0]+\boldsymbol{\theta}[1]}{2} \end{cases} \tag{5.20}$$

**Reconstruction of Digital Semantic Feature**

At the receiver side, the transmitted digital semantic features are decoded and reconstructed. This involves converting the received bit sequences into feature maps composed of discrete quantized values. The reconstructed feature map is expressed as $\hat{\boldsymbol{D}} = f_{d\to a}\left(\hat{\boldsymbol{b}}\right) \in \mathbb{R}^{zw_2h_2}$. These reconstructed features are used to complete subsequent inference tasks, such as classification or detection. For example, in the case of 2-bit quantization ($q = 2$), the decoding function is defined as

$$f_{d\to a}(x;\boldsymbol{\theta}) = \begin{cases} \boldsymbol{\theta}[3] & x=11 \\ \boldsymbol{\theta}[2] & x=10 \\ \boldsymbol{\theta}[1] & x=01 \\ \boldsymbol{\theta}[0] & x=00. \end{cases} \tag{5.21}$$

The reconstructed feature $\hat{\boldsymbol{D}}$ is then fed into the digital semantic decoder network on the edge server, where it undergoes further processing to complete tasks such as classification or detection.

To evaluate the performance of two distinct digital semantic communication systems— feature-quantized (FQ) digital semantic communication and softmax-quantized (SQ) digital semantic communication baseline comparison schemes were introduced: (1) a traditional separated digital coding method combining BPG for source coding and ideal capacity-based channel coding labeled "BPG+Capacity", and (2) an analog scheme [38] labeled "DJSCC(analog)". The traditional method employs BPG for source coding and assumes ideal channel coding, while the analog scheme transmits arbitrary complex constellation points without quantization.

The FQ digital semantic communication was compared with the SQ digital semantic communication under various modulation schemes, including 16QAM and 256QAM, where the results are shown in Fig. 5.7. Experiments were conducted on an Additive White Gaussian Noise (AWGN) channel using M-QAM modulation (where M denotes the modulation order). The "SQ (M QAM)" curve represents the softmax-based quantization method applied to classification tasks.

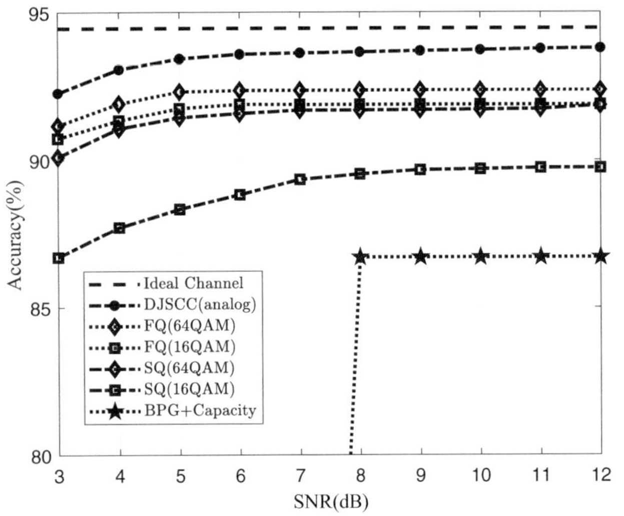


**Fig. 5.7** Performance of the digital semantic communication under diverse high-order modulations [40]

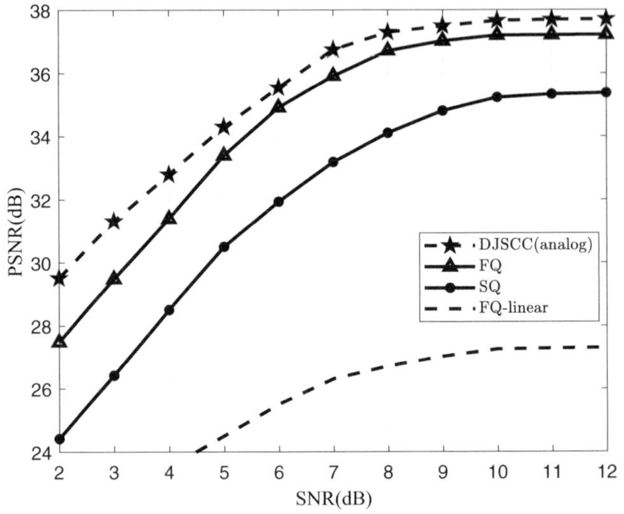


**Fig. 5.8** Comparison between the digital semantic communication in image reconstruction tasks under 16-QAM

The analog scheme "DJSCC (analog)" allows flexible transmission of continuous constellation points, whereas digital systems rely on discrete fixed constellations. While analog scheme achieves higher performance [38], it faces compatibility challenges with existing hardware and modulation standards. Experimental results show that under low-order modulation (e.g., 16QAM), the FQ outperforms SQ by 4% at 3dB SNR and 2% at 12dB SNR. As the modulation order increases, the performance gap narrows, with both digital methods approaching analog performance under high-order modulation (e.g., 256QAM). This highlights the scalability of the FQ approach, making it suitable for real-world deployment in resource-constrained environments.

The FQ was further evaluated on the CIFAR-10 dataset for image reconstruction tasks under 16QAM modulation, compared with SQ [38] and the FQ-Linear [33]. All methods shared identical network architectures have the same computational complexity and parameters. As shown in Fig. 5.8, the FQ achieved superior reconstruction quality, validating its effectiveness across diverse tasks. This versatility makes it ideal for applications requiring precise semantic recovery, such as medical imaging and remote sensing.

## 5.4 Summary

This chapter presents a joint semantic-channel coding framework that integrates both analog and digital semantic communication approaches. While analog semantic communication generates continuous constellation points, it faces practical implementation challenges. To address these limitations, digital semantic communication frameworks are proposed, offering compatibility with existing digital hardware while preserving task-oriented semantic features. The advantages and trade-offs of both methods are discussed, highlighting their roles in advancing efficient and deployable semantic communication systems.

## Problems

5.1 The core advantage of DJSCC is:

A. Increase redundant information to enhance error correction capability
B. Separate design reduces system complexity
C. Joint optimization achieves end-to-end efficiency maximization

5.2 The primary role of deep learning in semantic communication is:

A. Replace all traditional encoding modules
B. Automatically extract task-relevant semantic features
C. Increase signal transmission power

5.3 The core manifestation of robustness in semantic communication systems is:

A. Achieving theoretical limits under ideal channel conditions
B. Maintaining performance despite mismatched training-deployment channel conditions
C. Completely eliminating channel noise effects

5.4 The key advantage of the gradual degradation characteristic in semantic communication is:

A. Completely avoiding transmission errors
B. Preserving basic semantic integrity during channel deterioration
C. Reducing encoder/decoder computational load

5.5 The fundamental difference between semantic communication and traditional communication lies in:

A. Only optimizing physical layer transmission rate
B. Using semantic information rather than bit streams as transmission units
C. Completely relying on traditional channel coding schemes

5.6 The fundamental distinction between digital and analog semantic communication lies in:

A. Transmission power requirements
B. Constellation point representation (discrete vs continuous)
C. Channel coding complexity

5.7 Key advantages of digital semantic communication include:

A. Complete elimination of channel noise
B. Compatibility with existing digital communication hardware/protocols
C. Improved transmission power efficiency

5.8 The essence of constellation point quantization is:

A. Mapping continuous features to standardized modulation constellations
B. Increasing feature dimensions for higher capacity
C. Eliminating channel coding requirements

5.9 Primary benefit of feature map quantization:

A. Perfect reconstruction of analog signals
B. Bitstream compatibility with multiple digital modulations
C. Simplified channel estimation

5.10 The strategic advantage of nonlinear quantization over linear is:

A. Easier implementation with uniform intervals
B. Allocating more quantization levels to critical feature regions
C. Reducing total quantization bits

## References


1. J. Xu, B. Ai, W. Chen, A. Yang, P. Sun, M. Rodrigues, Wireless image transmission using deep source channel coding with attention modules. IEEE Trans. Circuits Syst. Video Technol. **32**(4), 2315–2328 (2022)
2. J. Xu, T.-Y. Tung, B. Ai, W. Chen, Y. Sun, D. Gunduz, Deep joint source-channel coding for semantic communications. IEEE Commun. Mag. **61**(11), 42–48 (2023)
3. J. Yang, S. Hong, L. Wang, A task-guided normalized min-sum decoding network for ldpc codes-based djscc. IEEE Commun. Lett. **27**(8), 1934–1938 (2023)
4. Y. Shao, D. Gunduz, Semantic communications with discrete-time analog transmission: a paper perspective. IEEE Wirel. Commun. Lett. **12**(3), 510–514 (2023)
5. M. Yang, C. Bian, H.-S. Kim, Deep joint source channel coding for wireless image transmission with ofdm, in *ICC 2021—IEEE International Conference on Communications* (2021), pp. 1–6
6. Z. Wu, S. Pan, F. Chen, G. Long, C. Zhang, P.S. Yu, A comprehensive survey on graph neural networks. IEEE Trans. Neural Netw. Learn. Syst. **32**(1), 4–24 (2021)
7. J. Mao, K. Xiong, M. Liu, Z. Qin, W. Chen, P. Fan, K.B. Letaief, A gan-based semantic communication for text without csi. IEEE Trans. Wirel. Commun. **23**(10), 14498–14514 (2024)
8. B. Wang, R. Li, J. Zhu, Z. Zhao, H. Zhang, Knowledge enhanced semantic communication receiver. IEEE Commun. Lett. **27**(7), 1794–1798 (2023)
9. Q. Hu, G. Zhang, Z. Qin, Y. Cai, G. Yu, G.Y. Li, Robust semantic communications with masked vq-vae enabled codebook. IEEE Trans. Wirel. Commun. **22**(12), 8707–8722 (2023)
10. J. Huang, C. Liu, D. Liu, Semantic segmentation-based low-rate image communication with diffusion models, in *2024 16th International Conference on Wireless Communications and Signal Processing (WCSP)* (2024), pp. 1412–1417
11. E. Bourtsoulatze, D. Burth Kurka, D. Gunduz, Deep joint source-channel coding for wireless image transmission. IEEE Trans. Cogn. Commun. Netw. **5**(3), 567–579 (2019)
12. J. Dai, S. Wang, K. Tan, Z. Si, X. Qin, K. Niu, P. Zhang, m. IEEE J. Sel. Areas Commun. **40**(8), 2300–2316 (2022)
13. Y.M. Saidutta, A. Abdi, F. Fekri, Joint source-channel coding over additive noise analog channels using mixture of variational autoencoders. IEEE J. Sel. Areas Commun. **39**(7), 2000–2013 (2021)
14. D.B. Kurka, D. Gunduz, Bandwidth-agile image transmission with deep joint source-channel coding. IEEE Trans. Wirel. Commun. **20**(12), 8081–8095 (2021)
15. Y. Guo, W. Chen, J. Xu, L. Li, B. Ai, Deep joint CSI feedback and multiuser precoding for MIMO OFDM systems. IEEE Trans. Veh. Technol. 1–6 (2024)
16. Y. Guo, W. Chen, F. Sun, J. Cheng, M. Matthaiou, B. Ai, Deep learning for csi feedback: one-sided model and joint multi-module learning perspectives. IEEE Commun. Mag. 1–8 (2024)
17. M. Liu, W. Chen, J. Xu, B. Ai, A comprehensive study of papr reduction techniques for deep joint source channel coding in ofdm systems, in *2023 IEEE 23rd International Conference on Communication Technology (ICCT)* (2023), pp. 1395–1399

18. M. Liu, W. Chen, J. Xu, B. Ai, Real-time implementation and evaluation of sdr-based deep joint source-channel coding, in *2022 IEEE 96th Vehicular Technology Conference (VTC2022-Fall)* (2022), pp. 1–5
19. J. Guo, C. Yang, Learning precoding for semantic communications, in *2022 IEEE International Conference on Communications Workshops (ICC Workshops)* (2022), pp. 163–168
20. C.K. Thomas, E.C. Strinati, W. Saad, Reasoning with the theory of mind for pragmatic semantic communication, in *2024 IEEE 21st Consumer Communications and Networking Conference (CCNC)* (2024), pp. 662–668
21. Y. Miao, J. Yan, Y. Wang, Z. Li, D. Hu, A semantic communication system based on vector quantization and generative model, in *2024 6th International Conference on Communications, Information System and Computer Engineering (CISCE)* (2024), pp. 46–50
22. R. Tang, D. Gao, M. Yang, T. Guo, H. Wu, G. Shi, Gan-inspired intelligent jamming and anti-jamming strategy for semantic communication systems, in *2023 IEEE International Conference on Communications Workshops (ICC Workshops)* (2023), pp. 1623–1628
23. L.X. Nguyen, Y.L. Tun, Y.K. Tun, M.N.H. Nguyen, C. Zhang, Z. Han, C. Seon Hong, Swin transformer-based dynamic semantic communication for multi-user with different computing capacity. IEEE Trans. Veh. Technol. **73**(6), 8957–8972 (2024)
24. Y. Wang, S. Ma, D. Gao, G. Shi, Swin kansformer-based semantic communication systems for wireless image transmission, in *2024 IEEE/CIC International Conference on Communications in China (ICCC Workshops)* (2024), pp. 265–270
25. S. Ma, W. Qiao, Y. Wu, H. Li, G. Shi, D. Gao, Y. Shi, S. Li, N. Al-Dhahir, Task-oriented explainable semantic communications. IEEE Trans. Wirel. Commun. **22**(12), 9248–9262 (2023)
26. D. Chen, W. Hua, Hierarchical vae based semantic communications for pomdp tasks, in *ICASSP 2024—2024 IEEE International Conference on Acoustics, Speech and Signal Processing (ICASSP)* (2024), pp. 5540–5544
27. H. Zhang, Z. Bao, H. Liang, Y. Liu, C. Dong, L. Li, Diffusion-based wireless semantic communication for vr image, in *2024 IEEE/CIC International Conference on Communications in China (ICCC Workshops)* (2024), pp. 639–644
28. Y. Zeng, X. He, X. Chen, H. Tong, Z. Yang, Y. Guo, J. Hao, Dmce: diffusion model channel enhancer for multi-user semantic communication systems, in *ICC 2024—IEEE International Conference on Communications* (2024), pp. 855–860
29. X. Song, L. Yuan, Z. Qu, F. Zhou, Q. Wu, T.Q.S. Quek, R.Q. Hu, Knowledge graph driven uav cognitive semantic communication systems for efficient object detection, in *ICC 2024—IEEE International Conference on Communications* (2024), pp. 1685–1690
30. C. Liu, C. Guo, S. Wang, Y. Li, D. Hu, Task-oriented semantic communication based on semantic triplets, in *2023 IEEE Wireless Communications and Networking Conference (WCNC)* (2023), pp. 1–6
31. H. Li, C. Hu, J. Jiang, W. Zhi, W. Zhu, Jalad: joint accuracy-and latency-aware deep structure decoupling for edge-cloud execution, in *IEEE 24th International Conference on Parallel and Distributed Systems (ICPADS)* (2018), pp. 671–678
32. A.E. Eshratifar, A. Esmaili, M. Pedram, Bottlenet: a deep learning architecture for intelligent mobile cloud computing services, in *IEEE/ACM International Symposium on Low Power Electronics and Design (ISLPED)* (2019), pp. 1–6
33. H. Xie, Z. Qin, A lite distributed semantic communication system for internet of things. IEEE J. Sel. Areas Commun. **39**(1), 142–153 (2021)
34. Q. Zhou, R. Li, Z. Zhao, Y. Xiao, H. Zhang, Adaptive bit rate control in semantic communication with incremental knowledge-based harq. IEEE Open J. Commun. Soc. **3**, 1076–1089 (2022)
35. J. Fang, A. Shafiee, H. Abdel-Aziz, D. Thorsley, G. Georgiadis, J. Hassoun, Post-training piecewise linear quantization for deep neural networks, in *European Conference on Computer Vision (ECCV)* (2020)
36. E. Park, S. Yoo, P. Vajda, Value-aware quantization for training and inference ofneural networks, in *European Conference on Computer Vision (ECCV)* (2018)
37. S. Jain, S. Venkataramani, V. Srinivasan, J. Choi, K. Gopalakrishnan, L. Chang, Biscaled-dnn: Quantizing long-tailed datastructures with two scale factors for deep neural networks, in *ACM/IEEE Design Automation Conference (DAC)* (2019), pp. 1–6
38. T.-Y. Tung, D.B. Kurka, M. Jankowski, D. Gündüz, DeepJSCC-Q: constellation constrained deep joint source-channel coding. IEEE J. Sel. Areas Inf. Theory **3**(4), 720–731 (2022)
39. Y. Bo, Y. Duan, S. Shao, M. Tao, Joint coding-modulation for digital semantic communications via variational autoencoder. IEEE Trans. Commun. **72**(9), 5626–5640 (2024)
40. L. Guo, W. Chen, Y. Sun, B. Ai, Digital-SC: Digital semantic communication with adaptive network split and learned non-linear quantization. IEEE Trans. Cogn. Commun. Netw. 1–13 (2024)

# 6 Channel Adaptive Semantic Communication


**Abstract**

With the rapid evolution of communication technologies and the increasing complexity of wireless environments, semantic communication systems face significant challenges in maintaining performance under dynamic channel variations. Most existing deep joint source-channel coding (DJSCC) methods are designed under the assumption of stable transmission conditions, particularly a fixed or consistent signalto-noise ratio (SNR) between training and deployment. However, real-world wireless channels are highly dynamic due to multipath fading, interference, and environmental changes, leading to severe performance degradation when such assumptions are violated. To address this issue, this chapter introduces two novel channel adaptation mechanisms for semantic communication. First, an attention-based channel adaptation strategy is developed to dynamically adjust feature importance according to fluctuating SNR levels, improving the robustness of semantic representations. Second, a diffusion-based adaptive scheme is proposed, where channel noise is modeled as the forward process of a diffusion model, and reverse denoising is performed at the receiver to recover high-fidelity semantic features. Together, these mechanisms enhance the adaptability and reliability of semantic communication systems, bridging the gap between theoretical models and practical wireless deployments.



**Keywords**

Semantic communication • Channel adaptation • Attention mechanism • Diffusion model • Wireless robustness • Multipath fading • Interference resilience • Noise denoising


## 6.1 Introduction

The swift advancement of communication technologies and the growing intricacy of wireless settings have rendered the ability to adapt to various channels and scenarios essential for contemporary communication systems. Typically, deep learning approaches to semantic communication presuppose fixed or predetermined channel conditions throughout training and deployment. Yet, in practice, communication systems are naturally dynamic. Channel conditions, including SNR, fading behavior, and interference, fluctuate considerably over time and across diverse environments.

For example, in Internet of Things (IoT) applications, devices typically function in varied and unpredictable settings. For instance, in cities, significant interference and multipath effects can cause swift fluctuations in the signal-to-noise ratio (SNR), whereas in rural regions, weak signals and extended transmission ranges may lead to conditions with low SNR. Furthermore, in IoT scenarios, devices may shift between indoor environments to outdoor environments characterized by extensive scattering and Rayleigh fading channels. Without effective adaptation mechanisms, communication systems would experience performance degradation, resulting in unreliable data transmission, increased latency, and reduced energy efficiency.

Another example is autonomous driving systems, where vehicles must communicate in highly dynamic scenarios. In urban areas, the presence of buildings, pedestrians, and other vehicles can cause rapid changes in channel characteristics. In addition, the transition between different scenarios, such as moving from an open area to a tunnel, can introduce sudden and drastic changes in channel conditions.

Semantic communication systems often face complex and dynamic channel environments, which pose significant



W. Chen and Z. Qin, *Semantic Communications*, Textbooks in Telecommunication Engineering,
https://doi.org/10.1007/978-3-032-11005-3_6

challenges to existing DNN-based JSCC methods. These methods typically assume that the channel conditions, particularly the SNR, during the optimization phase are consistent with those in the actual deployment phase. However, studies have shown [1] that mismatches in channel conditions between optimization and deployment stages can lead to substantial performance degradation, potentially offset the advantages of deep learning-based JSCC. Thus, communication systems have to incorporate adaptive mechanisms that can dynamically adjust to both channel conditions and scenario characteristics.

**1. Channel Quality Adaptation**

- **SNR Adaptation**: The systems should be capable of adapting to varying SNR levels, ensuring reliable performance in high-SNR and low-SNR conditions. This can be achieved through techniques such as dynamic modulation and coding schemes, as well as real-time SNR estimation and feedback.
- **Channel-Type Adaptation**: Systems must adapt to different channel types, such as AWGN, Rayleigh fading, and Rician fading. This requires models that can generalize across diverse channel characteristics and dynamically adjust their encoding and decoding strategies.

**2. Channel Scenario Adaptation**

- **Multi-Scenario Adaptation**: Systems should be designed to handle transitions between different scenarios, such as urban to rural areas, indoor to outdoor environments, or highway to urban driving. This involves training models on diverse datasets that capture the unique characteristics of each scenario and developing mechanisms to detect and adapt to scenario changes in real time.
- **Environment-Specific Optimization**: For specific scenarios, such as industrial IoT or autonomous driving, systems should incorporate environment-specific optimizations to maximize performance. For example, in IoT, systems may prioritize robustness to interference, while in autonomous driving, low latency, and high reliability are critical.

To address the challenge in dynamic channel environments, traditional solutions often rely on training multiple networks, each tailored to specific SNR conditions, and selecting the most suitable network during transmission or reception based on real-time channel conditions. However, this approach significantly increases the computational and storage overhead of transceivers, making it impractical for deployment in resource-constrained scenarios. In addition, existing methods focus primarily on SNR variations and often overlook the diversity of channel types, such as Rayleigh fading, Rician fading, and MIMO channels. This narrow focus limits their applicability in complex and heterogeneous wireless environments, where channel conditions can vary widely in both SNR and fading characteristics.

Therefore, developing channel/scenario adaptation methods capable of dynamically adapting to varying channel conditions has become an urgent challenge. Such methods must not only exhibit robustness across a wide range of SNRs but also adapt to multiple channel types, ensuring reliable and efficient communication in diverse and dynamic scenarios.

## 6.2 Channel Quality Adaptive Semantic Communication

In this section, we introduce two channel adaptation methods: attention-based adaptation and diffusion-based adaptation. Both methods offer distinct advantages in the semantic communication systems.

The attention-based adaptation method dynamically adjusts transmission signals to adapt to varying channel conditions. By taking the SNR as input, this method generates weights for the transmission signals based on the specific channel state. This approach is highly flexible and scalable, functioning as a plug-and-play module that can be integrated into various semantic communication tasks. Its versatility and ease of deployment make it an excellent choice for applications across diverse communication scenarios.

The diffusion-based adaptation method employs advanced techniques of diffusion models. It maps the transmission process to a forward process and applies the reverse diffusion process to eliminate noise. This approach demonstrates superior performance in reconstruction tasks compared to the attention-based method. However, its applicability is limited by its reliance on diffusion-based architectures. Consequently, the diffusion-based adaptation method is suited as a plug-and-play module for semantic communication systems based on diffusion.

These two methods complement each other, allowing for their application in scenarios that align with their respective strengths, providing flexible and effective solutions for channel adaptation in semantic communication.

### 6.2.1 Attention-Based Channel Adaptation

The design inspiration is derived from resource allocation strategies in cascaded source-channel coding, where the source and channel coding rates can be adjusted to adapt to varying channel conditions. Specifically, under the constraint of fixed transmission resources.

When transmitting an image under a fixed bandwidth constraint, the strategy varies with the SNR level. In low SNR conditions, the source is compressed more heavily to save bandwidth, while the channel coding rate is increased to protect against noise-induced errors. Conversely, in high SNR conditions, less aggressive compression is used to preserve image quality, and the channel coding rate is reduced, requiring less redundancy. This strategy enables efficient transmission.

This resource allocation strategy enables good transmission performance under constant channel bandwidth. Unfortunately, existing deep learning-based joint source-channel coding methods often lack a flexible network structure capable of dynamically adapting to varying channel conditions.

#### 6.2.1.1 Attention-Based Network Design

The neural network encoder and decoder of ADJSCC are built using two types of modules: Feature Learning (FL) Module and Attention Feature (AF) Module. As illustrated in the upper part of Fig. 6.1, FL and AF modules are alternately connected to form the ADJSCC encoder and decoder. The FL module extracts features from its input, while the AF module takes the channel condition (SNR) and the output of the FL module as input to generate a sequence of scaling factors. The product of the scaling factors from the AF module and the output of the FL module serves as a filter for the importance of the learned features.

The architecture of the Attention Feature (AF) module based on channel-wise soft attention is illustrated in the lower part of Fig. 6.1. Let $\mathbf{F}^G = \left[F_1^G, F_2^G, \ldots, F_c^G\right] \in \mathbb{R}^{h\times w\times c}$ denote the feature map extracted by the Feature Learning (FL) module, where $c$ represents the number of features and $h \times w$ represents the size of each feature. Similarly, let $\mathbf{F}^A = \left[F_1^A, F_2^A, \ldots, F_c^A\right] \in \mathbb{R}^{h\times w\times c}$ denote the scaled feature map produced by the AF module.

In summary, the AF module processes $\mathbf{F}^G$ using a global average pooling function. The results from global average pooling are concatenated with the channel's signal-to-noise ratio (SNR) to form the context information, which is then fed into a fully connected (FC) neural network to predict a scaling factor. The scaled features $\mathbf{F}^A$ are obtained by element-wise multiplication of $\mathbf{F}^G$ and the scaling factor. This process dynamically adjusts $\mathbf{F}^A$. The AF module comprises three key components: Context Extraction, Factor Prediction and Feature Recalibration.

**Context Extraction**

The context information consists of the channel SNR and the feature information, i.e., $\mathbf{I}^G$. Image features extracted by convolutional kernels typically cover only local receptive fields and may lack a global perspective, especially when smaller kernels are used. To address this limitation, global average pooling $G(\cdot)$ aggregates global information by averaging all elements $u_{jk}$ of $F_i^G$ as follows:

$$I_i^G = G\left(F_i^G\right) = \frac{1}{h\cdot w}\sum_{j=1}^{h}\sum_{k=1}^{w} u_{jk} \in \mathbb{R}. \tag{6.1}$$

These global information features $I_i^G$ are concatenated with the channel SNR to produce the context information $\mathbf{I}$:

$$\mathbf{I} = \left(\mathrm{SNR}, I_1^G, I_2^G, \ldots, I_c^G\right) \in \mathbb{R}^{c+1}. \tag{6.2}$$

**Factor Prediction**

The scaling factor $\mathbf{S}$ is predicted by a neural network $P_w(\cdot)$, which uses the context information $\mathbf{I}$ as input. To maintain

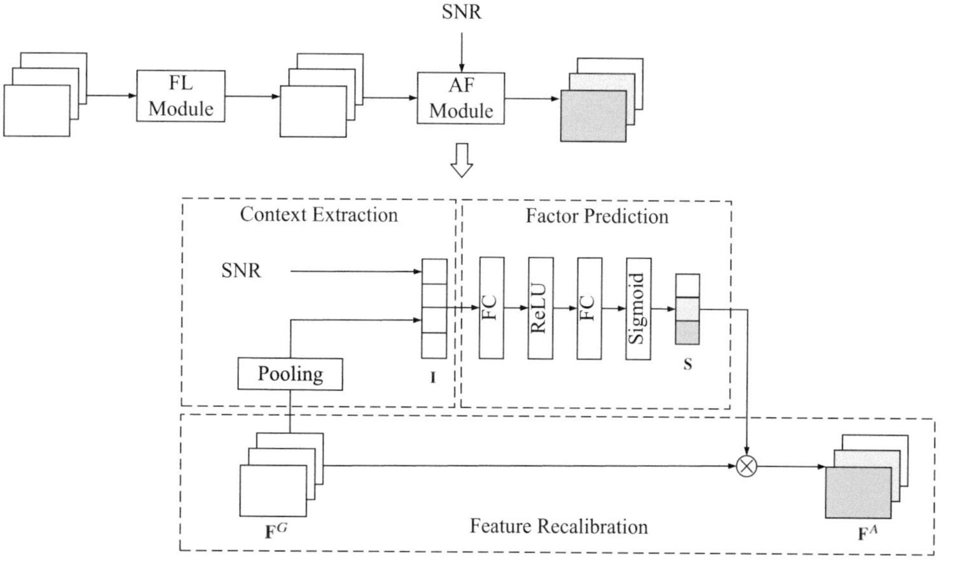


**Fig. 6.1** The relationship between FL modules and AF modules in ADJSCC [2]. AF module contains three parts: **a** Context Extraction, **b** Factor Prediction, **c** Feature Recalibration

computational efficiency, $P_w(\cdot)$ is a simple neural network comprising two fully connected layers. The first FC layer uses the ReLU activation function, while the second FC layer employs a sigmoid activation function to constrain the output to the range (0, 1). The scaling factor is computed as

$$\mathbf{S} = P_w(\mathbf{I}) = \sigma\left(\mathbf{W}_2\delta\left(\mathbf{W}_1\mathbf{I} + \mathbf{b}_1\right) + \mathbf{b}_2\right) \in \mathbb{R}^c, \qquad (6.3)$$

where $\mathbf{W}_1$ and $\mathbf{b}_1$ are the weights and biases of the first FC layer, $\mathbf{W}_2$ and $\mathbf{b}_2$ are the weights and biases of the second FC layer, and $\delta$ and $\sigma$ represent the ReLU and sigmoid activation functions, respectively. The parameter set of the factor prediction network is $w = (\mathbf{W}_1, \mathbf{b}_1, \mathbf{W}_2, \mathbf{b}_2)$.

**Feature Recalibration**

The feature recalibration step adjusts the feature map $\mathbf{F}^A$ based on $\mathbf{F}^G$ and the scaling factor $\mathbf{S}$. Each feature $F_i^A$ in $\mathbf{F}^A$ is obtained as follows:

$$F_i^A = R\left(F_i^G, S_i\right) = S_i \cdot F_i^G, i = 1, 2, \ldots, c, \qquad (6.4)$$

where $F_i^A$, $F_i^G$, and $S_i$ are the $i$-th elements of $\mathbf{F}^A$, $\mathbf{F}^G$ and $\mathbf{S}$, respectively.

Figure 6.2 compares the performance of the attention-based channel adaptation methods (called as ADJSCC) and the methods without channel adaptation (called asd BDJSCC) at bandwidth ratios $R = 1/12$ and $R = 1/6$.

In Fig. 6.2a, where $R = 1/12$, the ADJSCC model consistently outperforms the BDJSCC models trained at specific $\text{SNR}_{\text{train}}$. As $\text{SNR}_{\text{test}}$ increases, the ADJSCC model achieves higher performance, exceeding the BDJSCC model ($\text{SNR}_{\text{train}} = 1dB$) by up to 6 dB. Even as $\text{SNR}_{\text{test}}$ decreases, the ADJSCC model maintains superior performance compared to BDJSCC models trained at higher $\text{SNR}_{\text{train}}$ values, such as $\text{SNR}_{\text{train}} = 13dB$ or $\text{SNR}_{\text{train}} = 19dB$. Remarkably, the ADJSCC model outperforms BDJSCC models even when $\text{SNR}_{\text{train}} = \text{SNR}_{\text{test}}$. This highlights the robustness of ADJSCC, as it is trained over a specific SNR range yet adapts effectively to varying channel conditions. The performance advantage of ADJSCC becomes even more evident as $\text{SNR}_{\text{test}}$ deviates from $\text{SNR}_{\text{train}}$.

Figure 6.2b, with $R = 1/6$, reveals similar trends as Sect. 6.2.1.1. However, with the increased bandwidth ratio, the performance gap between the ADJSCC model and BDJSCC models ($\text{SNR}_{\text{train}} = 13dB$ or $\text{SNR}_{\text{train}} = 19dB$) diminishes in high $\text{SNR}_{\text{test}}$ regimes. This suggests that ADJSCC delivers more substantial performance gains in low-bandwidth scenarios.

Overall, these results demonstrate that ADJSCC provides a more efficient and practical solution compared to using multiple BDJSCC models trained at different $\text{SNR}_{\text{train}}$ levels. Implementing multiple BDJSCC models would require complex transmitter and receiver designs without offering any significant performance improvements over the ADJSCC strategy. In contrast, ADJSCC achieves superior performance while maintaining simplicity and adaptability across varying channel conditions.

### 6.2.2 Diffusion-Based Channel Adaptation

The diffusion-based generative semantic communication system, illustrated in Fig. 6.3, represents an advanced framework for channel-adaptive environments. This system integrates a pre-trained diffusion model with a semantic encoder to extract high-level semantic features from input images. The semantic features are transmitted through a wireless channel to the receiver. By incorporating the SNR and the reverse denoising process of the diffusion model, the system effectively eliminates channel noise, further restoring the semantic features. Finally, the denoised features are reconstructed into a high-quality image that closely approximates the original input using the semantic decoder.

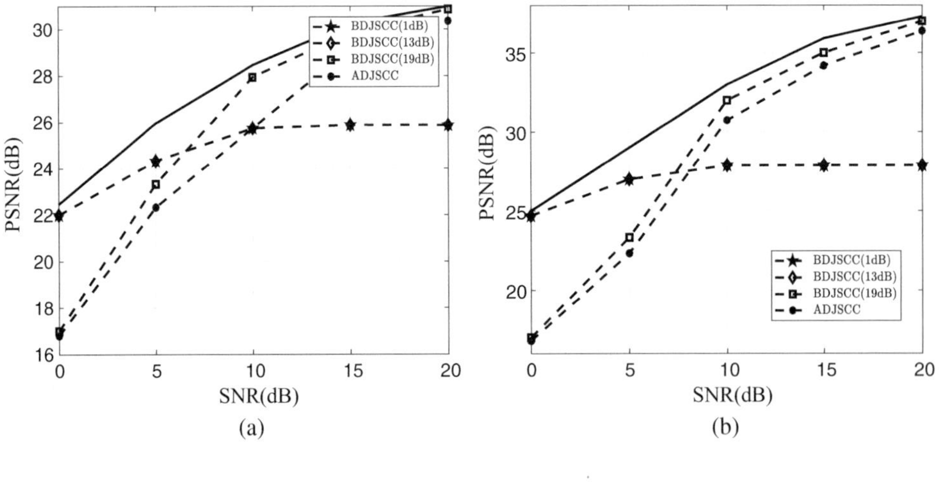


**Fig. 6.2** Performance of ADJSCC and BDJSCC on CIFAR-10 test images. **a** R = 1/12 and **b** R = 1/6. The curve of ADJSCC is trained under the uniform distribution of SNR from 0dB to 20dB. Each curve of BDJSCC is trained at a specific SNR

**Fig. 6.3** The framework of the semantic encoder and decoder based on diffusion [3]

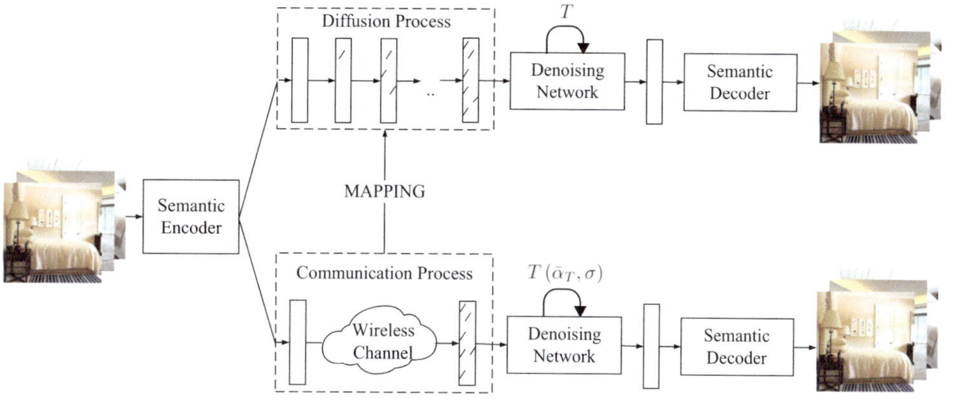


The input image is reshaped into a vector $\boldsymbol{s} \in \mathbb{C}^{3WH}$, where $W$ and $H$ represent the width and height of the image, respectively. The semantic encoder transforms the input into semantic features, which can be represented as

$$\boldsymbol{x} = E(\boldsymbol{s}; \boldsymbol{\theta}), \tag{6.5}$$

where $\boldsymbol{x} \in \mathbb{C}^{whc}$ is the vector reshaped from the semantic feature. $E(\cdot; \boldsymbol{\theta})$ denotes the semantic encoder with the parameter $\boldsymbol{\theta}$. The semantic feature is transmitted over the channel, which is expressed as

$$\boldsymbol{y} = \boldsymbol{x} + \boldsymbol{n}_0, \tag{6.6}$$

where $\boldsymbol{y}$ is the vector reshaped from received feature with channel noise following Gaussian distribution with variance $\sigma^2$, i.e., $\boldsymbol{n}_0 \sim \mathcal{N}\left(\mathbf{0}, \sigma^2 \cdot \mathbf{1}\right)$. $\mathbf{0}$ and $\mathbf{1}$ are vectors where each element equals zero and one, respectively. The transmission process can be represented as a conditional probability distribution, which follows a Gaussian distribution, i.e., $p_c(\boldsymbol{y} \mid \boldsymbol{x}) \sim \mathcal{N}\left(\boldsymbol{x}, \sigma^2 \cdot \mathbf{1}\right)$. We assume $\sigma$ is known at the the receiver, and $\mathrm{SNR} = \frac{1}{kwhc\sigma^2}\|\boldsymbol{x}\|_2^2$. $\hat{s}$ is denoised by the reverse process of diffusion with the diffusion parameter $\boldsymbol{\phi}$, i.e., $p_d\left(\boldsymbol{y}' \mid \boldsymbol{y}, \sigma^2, \boldsymbol{\phi}\right)$. The decoder transforms the reshaped vector $\boldsymbol{x}'$ from diffusion into the image, which can be represented as

$$\hat{\boldsymbol{s}} = D(\boldsymbol{y}'; \boldsymbol{\delta}), \tag{6.7}$$

where $D(\cdot; \boldsymbol{\delta})$ denotes the semantic decoder with the parameter $\boldsymbol{\delta}$. In our framework, the parameters of encoder $\boldsymbol{\theta}$, diffusion $\boldsymbol{\phi}$ and decoder $\boldsymbol{\delta}$ are utilized in the same as those of the Stable Diffusion [4].

### Adaptive Forward Process

In this section, we introduce the semantic encoder and decoder based on the adaptive forward process of diffusion, as shown in Fig. 6.3. In this design, the signal transmission process through the channel is mapped to the forward process of diffusion and utilize its reverse process to eliminate the noise. Additionally, the transformation of received signals into the features for the forward process of diffusion is extended to more complex wireless channels, including fading and MIMO channels.

For diffusion, the forward process involves gradually adding increasing levels of Gaussian noise to the data with a decreasing variance schedule $\alpha_1, \alpha_2, \ldots, \alpha_T$, which can be expressed as

$$p\left(\boldsymbol{x}_T \mid \boldsymbol{x}_{T-1}\right) \sim \mathcal{N}\left(\sqrt{\alpha_T} \cdot \boldsymbol{x}_{T-1}, \sqrt{1-\alpha_T} \cdot \mathbf{1}\right), \tag{6.8}$$

$$p\left(\boldsymbol{x}_T \mid \boldsymbol{x}_0\right) \sim \mathcal{N}\left(\sqrt{\bar{\alpha}_T} \cdot \boldsymbol{x}_0, \sqrt{1-\bar{\alpha}_T} \cdot \mathbf{1}\right), \tag{6.9}$$

where $\bar{\alpha}_T = \prod_{s=1}^{T} \alpha_s$, $\boldsymbol{x}_0 = \boldsymbol{x}$. $\boldsymbol{x}_T$ can be sampled directly from the conditional probability, that is

$$\boldsymbol{x}_T = \sqrt{\bar{\alpha}_T} \cdot \boldsymbol{x}_0 + \sqrt{1-\bar{\alpha}_T} \cdot \boldsymbol{\epsilon}_T, \tag{6.10}$$

where $\boldsymbol{\epsilon}_T \sim \mathcal{N}(\mathbf{0}, \mathbf{1})$. Note that $\bar{\alpha}_T$ is a hyperparameter that has been given in the diffusion-driven generation model.

The reverse process of diffusion involves reconstructing the distribution of $\boldsymbol{x}_{T-1}$ from $\boldsymbol{x}_T$. The deep learning network with parameters $\boldsymbol{\phi}$, such as U-Net, is employed to predict the posterior distribution, which can be represented as

$$p_{\boldsymbol{\phi}}\left(\boldsymbol{x}_{T-1} \mid \boldsymbol{x}_T\right) \sim \mathcal{N}\left(\boldsymbol{\mu}_{\boldsymbol{\phi}}\left(\boldsymbol{x}_T, T\right), \boldsymbol{\Sigma}_{\boldsymbol{\phi}}\left(\boldsymbol{x}_T, T\right)\right), \tag{6.11}$$

$$\boldsymbol{\mu}_{\boldsymbol{\phi}}\left(\boldsymbol{x}_T, T\right) = \frac{1}{\sqrt{\alpha_T}}\left(\boldsymbol{x}_T - \frac{1-\alpha_T}{\sqrt{1-\bar{\alpha}_T}} \cdot z_{\boldsymbol{\phi}}\left(\boldsymbol{x}_T, T\right)\right), \tag{6.12}$$

$$\boldsymbol{\Sigma}_{\boldsymbol{\phi}}\left(\boldsymbol{x}_T, T\right) = \frac{\left(1-\bar{\alpha}_{T-1}\right)\left(1-\alpha_T\right)}{1-\bar{\alpha}_T} \cdot \mathbf{1}, \tag{6.13}$$

where $z_{\boldsymbol{\phi}}\left(\boldsymbol{x}_T, T\right)$ is the predicted distribution of noise by the network given $\boldsymbol{x}_T$ and $T$.

As shown in Fig. 6.3, when the signal $\boldsymbol{x}$ is transmitted over the AWGN channel with the Gaussian noise of variance $\sigma^2$, it also can be represented as $\boldsymbol{y} = \boldsymbol{x} + \sigma \cdot \boldsymbol{\epsilon}$, $\boldsymbol{\epsilon} \sim \mathcal{N}(\mathbf{0}, \mathbf{1})$. The received signal follows a Gaussian distribution with mean

$\boldsymbol{x}$ and variance $\sigma^2 \cdot \mathbf{1}$. The transmission process through the wireless channel can be transformed into a forward process of diffusion. At the receiver, the reverse process of diffusion can be used to eliminate the noise and recover the signal. To transform the communication process into the diffusion process, the noise from the wireless channel is treated as the noise of the forward process, i.e., $p_c(\boldsymbol{y} \mid \boldsymbol{x}) \sim p(\boldsymbol{x}_u \mid \boldsymbol{x}_0, \bar{\alpha}_u)$. This can be expressed as:

$$\begin{aligned} \boldsymbol{y} &= \boldsymbol{x} + \sigma \cdot \boldsymbol{\epsilon} \\ &= \frac{1}{\sqrt{\bar{\alpha}_u}} \cdot \left( \sqrt{\bar{\alpha}_u} \cdot \boldsymbol{x}_0 + \frac{\sqrt{\bar{\alpha}_u}\sigma}{\sqrt{1-\bar{\alpha}_u}} \sqrt{1-\bar{\alpha}_u} \cdot \boldsymbol{\epsilon} \right). \end{aligned} \tag{6.14}$$

According to Eq. (6.14), when the variance of noise in wireless channel is $\sigma^2$, the transmission process over the channel is equivalent to the $u$-th step forward process. This can be expressed as

$$p_c(\boldsymbol{y} \mid \boldsymbol{x}) \sim p(\boldsymbol{x}_u \mid \boldsymbol{x}_0)|_{\bar{\alpha}_u = 1/(1+\sigma^2)} \cdot \tag{6.15}$$

The received signal $\boldsymbol{y}$ is denoised by the reverse process of diffusion, expressed as

$$p_d\left(\boldsymbol{y}_{u-1} \mid \boldsymbol{y}_u, \boldsymbol{\phi}\right) \sim p_{\boldsymbol{\phi}}(\boldsymbol{x}_{u-1} \mid \boldsymbol{x}_u)\big|_{\bar{\alpha}_u = 1/(1+\sigma^2)}, \tag{6.16}$$

where $\boldsymbol{y}_u$ denotes the $u$-th step feature of $\boldsymbol{y}$ in the reverse process of diffusion. Here, $\boldsymbol{y}_u = \boldsymbol{y}$, $\boldsymbol{y}_0 = \boldsymbol{y}'$. By transforming the noise of the wireless channel into the noise of the forward process, the impact of channel noise can be minimized.

For the Rayleigh fading channel, the received signals deviate from a Gaussian distribution, making them incompatible with the diffusion model's distribution requirements. To address this issue, MSE (Minimum Mean Square Error) equalization is employed. This process transforms the signal distribution into the Gaussian distribution, ensuring the features align with the requirements of the diffusion process. The received signal before equalization is given as

$$\boldsymbol{y} = \boldsymbol{h}\boldsymbol{x} + \sigma \cdot \boldsymbol{\epsilon}, \tag{6.17}$$

where $\boldsymbol{h}$ is the channel gain matrix. Using MSE equalization, the equalized received signal $\boldsymbol{y}_{\text{MSE}}$ is expressed as

$$\boldsymbol{y}_{\text{MSE}} = \boldsymbol{w}_{\text{MSE}}\boldsymbol{y}. \tag{6.18}$$

The MSE equalization coefficient $\boldsymbol{w}_{\text{MSE}}$ is defined as

$$\boldsymbol{w}_{\text{MSE}} = \frac{\boldsymbol{h}^*}{|\boldsymbol{h}|^2 + \frac{\sigma^2}{|\boldsymbol{x}|^2}}, \tag{6.19}$$

where $\boldsymbol{h}^*$ is the conjugate of $\boldsymbol{h}$. Thus, the received signal after equalization is expressed as

$$\begin{aligned} \boldsymbol{y}_{\text{MSE}} &= \frac{|\boldsymbol{h}|^2}{|\boldsymbol{h}|^2 + \frac{\sigma^2}{|\boldsymbol{x}|^2}} \cdot \boldsymbol{x} + \sigma \cdot \frac{\boldsymbol{h}^*}{|\boldsymbol{h}|^2 + \frac{\sigma^2}{|\boldsymbol{x}|^2}} \boldsymbol{\epsilon} \\ &= \frac{|\boldsymbol{h}|^2}{|\boldsymbol{h}|^2 + \frac{\sigma^2}{|\boldsymbol{x}|^2}} \cdot (\boldsymbol{x} + \sigma \cdot \boldsymbol{h}\boldsymbol{\epsilon}) \\ &= \frac{|\boldsymbol{h}|^2}{\sqrt{\bar{\alpha}_u}\left(|\boldsymbol{h}|^2 + \frac{\sigma^2}{|\boldsymbol{x}|^2}\right)} \cdot \left( \sqrt{\bar{\alpha}_u} \cdot \boldsymbol{x}_0 \right. \\ &\quad \left. + \sqrt{1-\bar{\alpha}_u} \cdot \frac{\sqrt{\bar{\alpha}_u}\sigma \cdot \boldsymbol{h}}{\sqrt{1-\bar{\alpha}_u}} \boldsymbol{\epsilon} \right). \end{aligned} \tag{6.20}$$

Based on Eq. 6.20, when $|\frac{\sqrt{\bar{\alpha}_u}\sigma \cdot \boldsymbol{h}}{\sqrt{1-\bar{\alpha}_u}}| = 1$, the transmission process over the Rayleigh fading channel is equivalent to the $u$-th step forward process in the diffusion process. This equivalence can be expressed as

$$p_c\left(\boldsymbol{y}_{\text{MSE}} \mid \boldsymbol{x}\right) \sim p(\boldsymbol{x}_u \mid \boldsymbol{x}_0)\big|_{\bar{\alpha}_u = \frac{1}{1+|\sigma \cdot \boldsymbol{h}|^2}} \cdot \tag{6.21}$$

On the LSUN Bedrooms and LSUN Churches datasets, wireless channel noise was incorporated as part of the complete forward process to adapt to varying SNR conditions, and the generative performance was systematically evaluated. To validate the effectiveness of the method, experiments were extended to Rayleigh channels. In these cases, the received signals were transformed into features compatible with the forward process of the diffusion model. MSE equalization was applied to Rayleigh channels.

The experimental results, summarized in Tables 6.1 and 6.2, demonstrate that MSE equalization allows the system to achieve performance comparable to AWGN channels under high SNR conditions for Rayleigh channels. These results underscore the adaptability and robustness of the method across diverse wireless channel environments.

To evaluate the semantic content of the generated images, the CLIP metric was employed. The experimental results indicate that diffusion-based semantic communication effectively transmits semantic information across varying SNR conditions. At higher SNR levels, semantic communication not only accurately conveys semantic information but also preserves finer image details, as reflected in pixel-level evaluation metrics such as PSNR and SSIM.

Figure 6.4 presents visualizations of the generated images under different SNR conditions. Even at lower SNR levels, semantic information is effectively preserved. For instance, when comparing SNR = 0 dB and SNR = 12 dB, notable differences in details such as church roofs, windows, and architectural styles are observed. Although some details are lost under low SNR conditions, the overall semantic content remains intact. As SNR increases, these details are progressively restored, leading to a significant improvement in the visual quality of the generated images.

**Table 6.1** The adaptive generation in LSUN-Bedrooms datasets [3]

| Metric | Channel | 0dB | 3dB | 6dB | 9dB | 12dB |
|---|---|---|---|---|---|---|
| PSNR ↑ | AWGN | 22.74 | 25.59 | 26.79 | 26.79 | 27.83 |
| | Rayleigh | 14.87 | 17.70 | 21.23 | 23.27 | 24.39 |
| SSIM ↑ | AWGN | 0.68 | 0.73 | 0.76 | 0.79 | 0.81 |
| | Rayleigh | 0.52 | 0.57 | 0.66 | 0.70 | 0.72 |
| CLIP ↑ | AWGN | 0.88 | 0.90 | 0.91 | 0.91 | 0.91 |
| | Rayleigh | 0.53 | 0.66 | 0.82 | 0.83 | 0.85 |
| FID ↓ | AWGN | 28.43 | 23.15 | 19.45 | 16.56 | 14.45 |
| | Rayleigh | 234.08 | 115.28 | 43.26 | 30.67 | 22.47 |
| LPIPS ↓ | AWGN | 0.22 | 0.18 | 0.15 | 0.12 | 0.11 |
| | Rayleigh | 0.78 | 0.58 | 0.34 | 0.25 | 0.21 |

**Table 6.2** The adaptive generation in LSUN-Churches datasets [3]

| Metric | Channel | 0dB | 3dB | 6dB | 9dB | 12dB |
|---|---|---|---|---|---|---|
| PSNR ↑ | AWGN | 20.16 | 21.39 | 22.52 | 23.52 | 24.38 |
| | Rayleigh | 13.84 | 16.15 | 19.39 | 21.14 | 22.01 |
| SSIM ↑ | AWGN | 0.58 | 0.62 | 0.67 | 0.70 | 0.73 |
| | Rayleigh | 0.41 | 0.45 | 0.55 | 0.60 | 0.63 |
| CLIP ↑ | AWGN | 0.86 | 0.88 | 0.89 | 0.89 | 0.89 |
| | Rayleigh | 0.45 | 0.56 | 0.77 | 0.82 | 0.83 |
| FID ↓ | AWGN | 27.17 | 22.43 | 19.12 | 16.52 | 14.72 |
| | Rayleigh | 272.50 | 151.58 | 43.80 | 37.96 | 29.50 |
| LPIPS ↓ | AWGN | 0.24 | 0.20 | 0.17 | 0.14 | 0.13 |
| | Rayleigh | 0.86 | 0.70 | 0.42 | 0.28 | 0.23 |

**Fig. 6.4** The visual results of the adaptation generation across various SNR levels. When the SNR is low, the semantic information of the image can be correctly transmitted, and as the SNR increases, more details can be recovered. Noticeably different regions are used for clear comparison to demonstrate how improved SNR contributes to more detailed reconstruction [3]

## 6.3 Channel Scenario Adaptive Semantic Communication

Building upon the attention-based channel adaptation framework introduced in Sect. 6.1, which integrates SNR optimization under AWGN channel conditions, this section extends the methodology to address more realistic and heterogeneous channel environments. While AWGN serves as a simplified baseline for theoretical analysis, practical wireless channels exhibit complex spatial and temporal variations. To bridge this gap, it is desired to generalize the framework to accommodate different channel models, e.g., clustered delay line (CDL), tapped delay line (TDL), urban macro-cell (UMa) and urban micro-cell (UMi) which are standard channel models defined by 3GPP.

Existing attention-based DJSCC methods [5], while effective for SNR adaptation, remain restricted to single channel-type specialization. Current implementations train different models for isolated environments, AWGN or Rayleigh fading channels, resulting in poor cross-scenario generalization. Although one could deploy multiple networks tailored to specific channel types (e.g., CDL, TDL, UMa, UMi), it incurs prohibitive computational and storage overhead for resource-limited transceivers. Thus, it calls for a unified framework capable of dynamically adapting to multiple channel scenarios.

It is desired to develop a single model that achieves superior performance across multiple channel scenarios compared to scenario-specific DJSCC implementations. Crucially, channel scenario information is incorporated not through explicit channel coefficients (a domain explored in channel-adaptive methods [6]), but via discrete numerical labels, e.g., 0 for CDL-A, 1 for CDL-B, 2 for CDL-C, 3 for CDL-D, and 4 for CDL-E.

Given the attention-based DJSCC framework for point-to-point image transmission under AWGN, we extend it to other channels and integrate scenario adaptation into both the encoder and the decoder, as shown in Fig. 6.5. It is assumed that the channel scenario label is known both at the transmitter and the receiver. An input image with the size of $C \times H \times W$ is represented by a vector $\boldsymbol{s} \in \mathbb{R}^n$, where $n = C \times H \times W$ and $\mathbb{R}$ denotes the set of real numbers. The joint encoder processes $\boldsymbol{s}$, channel scenario label $\zeta$, and the SNR $\mu$ by the encoding function $f_\theta : \mathbb{R}^n \times \mathbb{R} \times \mathbb{R} \rightarrow \mathbb{C}^k$, which leads to a vector of complex-valued channel input symbols $\boldsymbol{x} \in \mathbb{C}^k$. The encoding process can be expressed as

$$\boldsymbol{x} = f_\theta(\boldsymbol{s}, \mu, \zeta) \in \mathbb{C}^k, \tag{6.22}$$

where $k$ is the size of channel input symbols, $\theta$ is the parameter set of the joint semantic-channel encoder, and $\mathbb{C}$ denotes the set of complex numbers. The encoder maps the $n$-dimensional vector of real-valued image $\boldsymbol{s}$ to a $k$-dimensional vector of complex-valued channel input symbols $\boldsymbol{x}$. To satisfy the average power constraint at the encoder, $\frac{1}{k}\mathbb{E}(\boldsymbol{x}\boldsymbol{x}^*) \leq 1$ is enforced, where $\boldsymbol{x}^*$ is the complex conjugate of $\boldsymbol{x}$.

The encoded symbols in $\boldsymbol{x}$ are transmitted over a wireless fading channel. We consider five different types of CDL channels. The channel output symbols $\boldsymbol{y}$ received by the joint semantic-channel decoder can be expressed as

$$\boldsymbol{y} = h\boldsymbol{x} + \boldsymbol{\omega}, \tag{6.23}$$

where $\boldsymbol{\omega} \in \mathbb{C}^k$ is the common AWGN noise, $h \in \mathbb{C}$ is the channel gain. Remember that the power of the transmitting signal $\boldsymbol{x}$ has been normalized to one. An equalizer is used at the receiver to deal with channel fading.

The decoder applies a function $g_\phi : \mathbb{C}^k \times \mathbb{R} \times \mathbb{R} \rightarrow \mathbb{R}^n$ to map $\boldsymbol{y}$, $\mu$ and $\zeta$ as follows:

$$\hat{\boldsymbol{s}} = g_\phi(\boldsymbol{y}, \mu, \zeta), \tag{6.24}$$

where $\hat{\boldsymbol{s}} \in \mathbb{R}^n$ is an estimation of the original image $\boldsymbol{s}$, $\phi$ is the parameter set of the decoder. The distortion between the original image and the recovered image is defined as

$$d(\boldsymbol{s}, \hat{\boldsymbol{s}}) = \frac{1}{n}\sum_{i=1}^{n}\left(s_i - \hat{s}_i\right)^2, \tag{6.25}$$

where $s_i$ and $\hat{s}_i$ represent the intensity of each pixel in a color image corresponding to $\boldsymbol{s}$ and $\hat{\boldsymbol{s}}$ respectively.

In attention-based DJSCC framework, attention feature (AF) module is used to modify the feature importance by leveraging SNR to generate a multiplying factor. Motivated by this strategy, one could use a channel scenario label to generate a multiplying factor to adjust the encoded features. It could be achieved by employing the attention mechanism, which is a technique of DL widely used in natural language processing and computer vision [8, 9]. Attention mechanisms employ neural networks to dynamically identify and emphasize salient features within input data through learned weight distributions. These mechanisms operate through two principal approaches: feature selection via attention masks and feature weighting through scaling factors. Attention masks implement selective feature gating by suppressing irrelevant information while preserving critical features. This masking process effectively constrains the model's receptive field to task-relevant data regions, analogous to a cognitive focus mechanism in biological systems. The weighting process involves generating context-dependent scaling parameters through dedicated neural sub-networks. These scaling factors can be applied at varying granularities, for example single global scaling factor, channel-wise weighting and element-wise adjustments. The choice of weighting strategy depends on specific architectural requirements and the nature of the target task.

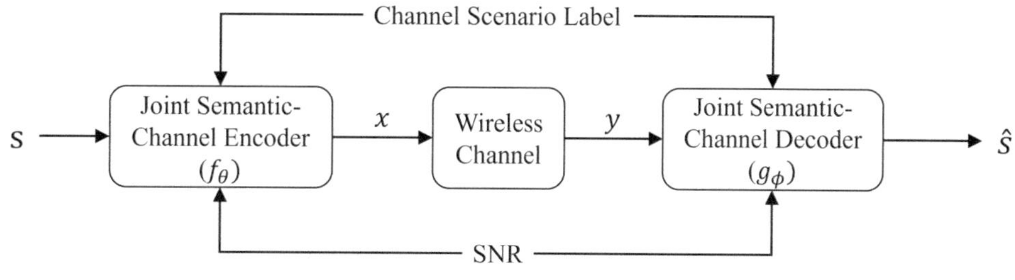


**Fig. 6.5** The model of the point-to-point image transmission system [7]

This scenario adaptive deep joint semantic-channel coding (SA-DJSCC) [7] relies on the designing of scenario learning (SL) module. SA-DJSCC includes an encoder, a wireless channel and a decoder, as shown in Fig. 6.6. Both the encoder and decoder are the combinations of FL module and SL module. FL module extracts features from the input, and SL module takes the output of the FL module as the input and then uses channel scenario label and SNR respectively to adjust the input of SL module. SL module consists of two parts shown in Fig. 6.7, one is another additional neural network, and the other is AF module. The design of FL module and AF module follows the same with attention-based DJSCC. The additional neural network tacking channel scenario label in SL module can be seen as a bridge connecting FL module and AF module. It follows the structure of multi-layer perceptron (MLP) to generate scaling factors with the same amount of the output of FL module feature channels, which is composed of five fully connected layers, and a LeakyReLU activation function follows each layer, except the last layer. The channel-wise multiplication of scaling factors and the output of FL module realizes the aim of scenario adaptive through channel label information. It should be noted that we directly use the channel scenario label to get scaling factors, but AF module concatenates SNR with the output of FL module and then predicts the scaling factors.

The bandwidth ratio is defined as $R = k/n$, which is the ratio of channel bandwidth and source bandwidth. We could change the last convolution layer's output channel size $c$ in joint semantic-channel encoder to obtain different bandwidth ratios. More advanced neural network structures may replace the fully convolutional network such as transformer [10].

The loss function chosen remains the Mean Squared Error (MSE), and the evaluating metric continues to be the Peak Signal-to-Noise Ratio (PSNR). Besides, the model is trained over a uniform distribution of SNR from 0 dB to 35 dB but evaluates its performance at a specific SNR value. The dataset used is CIFAR-10 [11].

The performance is evaluated at specific SNR values as indicated in Table 6.3, which compares the SA-DJSCC method with the attention-based DJSCC method at band-

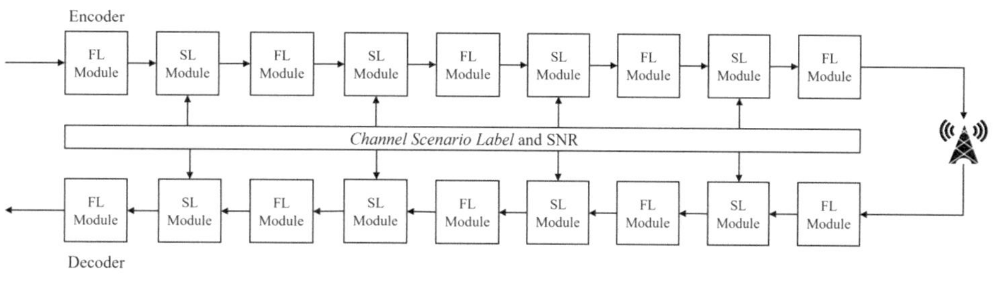


**Fig. 6.6** The architecture of SA-DJSCC. Each FL module is followed by an SL module except the last FL module in the encoder and the decoder [7]

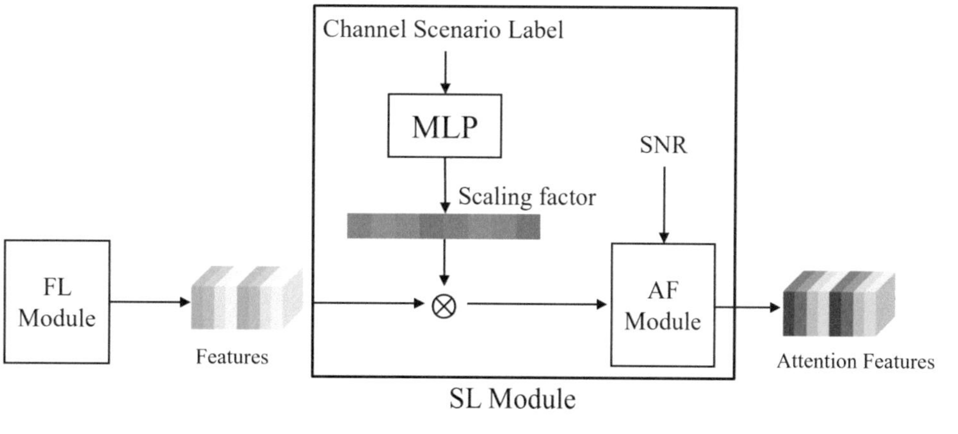


**Fig. 6.7** The relationship between FL modules and SL modules in SA-DJSCC. The structures and parameters of FL Module and AF Module stay the same with attention-based DJSCC [5]

**Table 6.3** Performance comparison of different models tested at different channel scenarios and SNRs in the metric of PSNR (dB) [7]

| Scenario | SNR (dB) | SA-DJSCC | Attention-based DJSCC trained in mixed models | Attention-based DJSCC trained in CDL-A | Attention-based DJSCC trained in CDL-B | Attention-based DJSCC trained in CDL-C | Attention-based DJSCC trained in CDL-D | Attention-based DJSCC trained in CDL-E |
|---|---|---|---|---|---|---|---|---|
| Mixture | 1 | **27.25** | 26.20 | 26.31 | 26.01 | 26.15 | 24.55 | 24.25 |
| | 8 | **28.85** | 27.96 | 28.06 | 27.69 | 27.98 | 27.39 | 27.21 |
| | 15 | **30.35** | 29.60 | 29.62 | 29.35 | 29.59 | 29.95 | 29.85 |
| | 23 | **31.67** | 31.15 | 31.41 | 30.99 | 31.26 | 31.54 | 31.44 |
| | 31 | **32.06** | 31.58 | 31.87 | 31.43 | 31.72 | 31.97 | 31.88 |
| CDL-A | 1 | **26.84** | 25.99 | 26.38 | 25.97 | 26.13 | 22.60 | 22.34 |
| | 8 | **28.48** | 27.85 | 28.16 | 27.71 | 28.03 | 25.91 | 25.65 |
| | 15 | **29.98** | 29.56 | 29.79 | 29.43 | 29.67 | 29.39 | 29.29 |
| | 23 | **31.64** | 31.21 | 31.50 | 31.06 | 31.35 | 31.46 | 31.37 |
| | 31 | **32.08** | 31.60 | 31.90 | 31.46 | 31.75 | 31.97 | 31.89 |
| CDL-B | 1 | **26.53** | 25.61 | 25.76 | 25.65 | 25.79 | 22.09 | 21.90 |
| | 8 | **28.20** | 27.53 | 27.69 | 27.45 | 27.72 | 25.39 | 24.97 |
| | 15 | **29.67** | 29.22 | 29.29 | 29.14 | 29.37 | 28.82 | 28.70 |
| | 23 | **31.37** | 30.97 | 31.25 | 30.85 | 31.11 | 31.10 | 31.03 |
| | 31 | **31.99** | 31.52 | 31.81 | 31.38 | 31.67 | 31.85 | 31.76 |
| CDL-C | 1 | **26.65** | 25.74 | 25.99 | 25.76 | 25.89 | 22.33 | 22.40 |
| | 8 | **28.29** | 27.67 | 27.89 | 27.52 | 27.83 | 25.50 | 25.33 |
| | 15 | **29.78** | 29.36 | 29.53 | 29.23 | 29.48 | 28.95 | 28.93 |
| | 23 | **31.49** | 31.05 | 31.35 | 30.94 | 31.20 | 31.24 | 31.18 |
| | 31 | **32.03** | 31.56 | 31.85 | 31.41 | 31.70 | 31.89 | 31.79 |
| CDL-D | 1 | **28.01** | 26.83 | 26.63 | 26.30 | 26.44 | 27.83 | 27.77 |
| | 8 | **29.58** | 28.39 | 28.29 | 27.87 | 28.17 | 29.68 | 29.64 |
| | 15 | **31.05** | 29.92 | 29.74 | 29.47 | 29.71 | 31.30 | 31.23 |
| | 23 | **31.90** | 31.27 | 31.49 | 31.05 | 31.33 | 31.94 | 31.86 |
| | 31 | **32.10** | 31.61 | 31.90 | 31.45 | 31.75 | 32.06 | 31.98 |
| CDL-E | 1 | **28.13** | 26.86 | 26.63 | 26.31 | 26.47 | 27.96 | 27.94 |
| | 8 | **29.70** | 28.41 | 28.27 | 27.87 | 28.18 | 29.84 | 29.81 |
| | 15 | **31.16** | 29.93 | 29.73 | 29.46 | 29.71 | 31.37 | 31.30 |
| | 23 | **31.91** | 31.27 | 31.48 | 31.05 | 31.32 | 31.95 | 31.87 |
| | 31 | **32.10** | 31.61 | 31.90 | 31.45 | 31.74 | 32.06 | 31.98 |

width ratios $R = 1/6$ [7]. The columns are SNR values and rows are different trained models, SA-DJSCC and attention-based DJSCC trained in mixed models are trained at mixed CDL channel scenarios, and the others are trained at a single channel scenario CDL-X respectively. Generally speaking, the best performance appears with matched test datasets, so the performance in descending order when tested under CDL-X model should be attention-based DJSCC trained in CDL-X, SA-DJSCC, and attention-based DJSCC trained in mixed models. The experiment results show that SA-DJSCC method which combines the channel scenario information, performs better than the attention-based DJSCC method either in mixed channel models or in a single channel model, when tested under different SNRs. In addition, the performance of SA-DJSCC is close to or exceeds that of attention-based DJSCC trained in CDL-X. With the increment of SNR, the performance keeps going up. All in all, the performance of our model indicates that our method can adapt to different channel scenarios simply with a label number.

## 6.4 Summary

This chapter addresses the performance degradation of semantic communication systems in dynamic wireless channels, where traditional DJSCC-based methods often fail due to their reliance on stable SNR conditions. To improve robustness against channel fluctuations—such as multipath fading, interference, and environmental changes—two adaptive mechanisms are introduced: (1) an attention-based strategy that dynamically adjusts feature importance based on real-time SNR variations, and (2) a diffusion-based scheme that models channel noise using a forward diffusion process and applies reverse denoising at the receiver. These approaches enhance adaptability and reliability in real-world wireless environments. In addition, a semantic communication framework that adapts to diverse channel scenarios is presented by incorporating discrete channel labels as model inputs, offering superior cross-scenario generalization with comparable or better performance than models trained on individual channel types.

## Problems

6.1 In the SA-DJSCC framework, "scenario label" in the SA-DJSCC framework is introduced to address what shortcoming of the traditional DISCC model?

A. Inability to dynamically adapt to multi-channel scenarios (e.g., CDL/TDL).
B. Over-reliance on channel state information (CSI) feedback.
C. High computational complexity on the decoding side.
D. Insufficient robustness to noise outside the AWGN channel.

6.2 Compared to traditional methods, the advantage of the Attention mechanism in channel adaptation is:

A. Adapting to different channel qualities to avoid mismatches between training and testing
B. Reducing computational complexity
C. Requiring no channel feedback

6.3 The core idea of Attention-based channel adaptation is:

A. Fixing coding priorities for all channel features
B. Dynamically modulating transmission features under different SNRs through attention weights
C. Relying solely on pilot signals for channel state estimation

6.4 The core idea of Diffusion-based channel adaptation is:

A. Gradually removing noise to recover transmission features
B. Increasing signal transmission power
C. Simplifying modem design

6.5 The distinctive feature of Diffusion models compared to Attention is:

A. Strong robustness to estimation errors
B. Higher real-time performance
C. Completely avoiding iterative processes

6.6 The main bottleneck in deploying Diffusion models is:

A. Increased RF front-end power consumption
B. Disruption to channel coding schemes
C. Processing latency caused by multiple iterations

## References


1. E. Bourtsoulatze, D.B. Kurka, D. Gündüz, Deep joint source-channel coding for wireless image transmission, in *ICASSP 2019—2019 IEEE International Conference on Acoustics, Speech and Signal Processing (ICASSP)* (2019), pp. 4774–4778
2. J. Xu, B. Ai, W. Chen, A. Yang, P. Sun, M. Rodrigues, Wireless image transmission using deep source channel coding with attention modules. IEEE Trans. Circuits Syst. Video Technol. **32**(4), 2315–2328 (2022)
3. L. Guo, W. Chen, Y. Sun, B. Ai, N. Pappas, T.Q.S. Quek, Diffusion-driven semantic communication for generative models with bandwidth constraints. IEEE Trans. Wirel. Commun. 1–14 (2025)
4. R. Rombach, A. Blattmann, D. Lorenz, P. Esser, B. Ommer, High-resolution image synthesis with latent diffusion models, in *2022 IEEE/CVF Conference on Computer Vision and Pattern Recognition (CVPR)* (2022), pp. 10674–10685
5. J. Xu, B. Ai, W. Chen, A. Yang, P. Sun, M. Rodrigues, Wireless image transmission using deep source channel coding with attention modules. IEEE Trans. Circuits Syst. Video Technol. **32**(4), 2315–2328 (2021)
6. H. Wu, Y. Shao, K. Mikolajczyk, D. Gündüz, Channel-adaptive wireless image transmission with ofdm. IEEE Wirel. Commun. Lett. **11**(11), 2400–2404 (2022)
7. S. Gao, W. Chen, Z. Song, B. Ai, G. Jiangyuan, SA-DJSCC: Scenario adaptive deep joint source-channel coding for wireless image transmission, in *2025 IEEE 101st Vehicular Technology Conference (VTC2025-Spring)* (2025)
8. J. Hu, L. Shen, G. Sun, Squeeze-and-excitation networks, in *Proceedings of the IEEE Conference on Computer Vision and Pattern Recognition* (2018), pp. 7132–7141
9. S. Woo, J. Park, J.-Y. Lee, I.S. Kweon, Cbam: convolutional block attention module, in *Proceedings of the European Conference on Computer Vision (ECCV)* (2018), pp. 3–19
10. J. Dai, S. Wang, K. Tan, Z. Si, X. Qin, K. Niu, P. Zhang, Nonlinear transform source-channel coding for semantic communications. IEEE J. Sel. Areas Commun. **40**(8), 2300–2316 (2022)
11. A. Krizhevsky, G. Hinton, et al., *Learning Multiple Layers of Features from Tiny Images* (2009)

# Bandwidth Adaptive Semantic Communication

# 7


## Abstract

This chapter explores solutions for designing bandwidth-adaptive semantic communication systems that can operate efficiently under varying bandwidth constraints. It addresses the limitations of using multiple fixed-bandwidth models, which introduce significant storage overhead and latency. The chapter presents two key adaptive strategies. Entropy-aware methods, inspired by traditional coding, allocate more symbols to features with higher information density, improving performance by focusing on important data. Semantic-aware methods, a learning-based approach, dynamically assign bandwidth based on the semantic importance of content for a specific task. These strategies enable a single system to achieve performance comparable to specialized models while drastically reducing overhead, paving the way for more practical, real-world applications.



## Keywords

Semantic communication • Bandwidth adaptation • Entropy-aware coding • Semantic-aware coding • Variable-length coding


## 7.1 Introduction

Deep learning-based joint semantic-channel coding has provided valuable guidance for designing end-to-end semantic communication systems, significantly improving their performance. However, the application of semantic communication in real-world scenarios, such as autonomous driving, immersive reality, and holographic communication, still faces many challenges. Different applications come with varying communication requirements and resource constraints, such as end-to-end latency, resulting in various available bandwidth across these use cases. Consequently, semantic communication systems must be capable of operating efficiently under various bandwidth constraints, ensuring reliable communication in diverse scenarios [1].

As shown in Fig. 7.1a, a classic solution is to train multiple semantic communication systems under various bandwidth constraints, with all systems stored locally for practical use. During testing, the most suitable semantic communication system is selected to perform the communication task based on the given available bandwidth. However, this solution obviously increases the storage burden and introduces additional latency due to model switching, which is not conducive to deployment on the device side. Therefore, it is necessary to design semantic communication systems capable of efficient and flexible transmission under varying bandwidth constraints, in other words, to achieve bandwidth adaptation for semantic communication systems. As shown in Fig. 7.1b, the bandwidth-adaptive solution is trained under a range of bandwidth constraints. During testing, the trained codec, i.e., a semantic encoder and semantic decoder, can operate under variable bandwidth conditions efficiently, avoiding the use of multiple fixed-bandwidth systems.

To ensure the effectiveness of bandwidth adaptation, the performance of the bandwidth-adaptive system under each specific bandwidth constraint should approach the performance of fixed-bandwidth system trained under the same bandwidth constraint. In the context of semantic coding, two critical problems are essential to achieve effective bandwidth adaptation:

- **Problem 1**: How to *allocate the available bandwidth* for different semantics efficiently?
- **Problem 2**: How to achieve *variable-length coding* for different semantics, i.e., allowing the source data to be encoded into different numbers of transmission symbols?

Traditional communication systems based on separate source-channel coding can provide insights for addressing these problems. On one hand, source coders, such as Joint Photographic Experts Group (JPEG), Better Portable



W. Chen and Z. Qin, *Semantic Communications*, Textbooks in Telecommunication Engineering,
https://doi.org/10.1007/978-3-032-11005-3_7

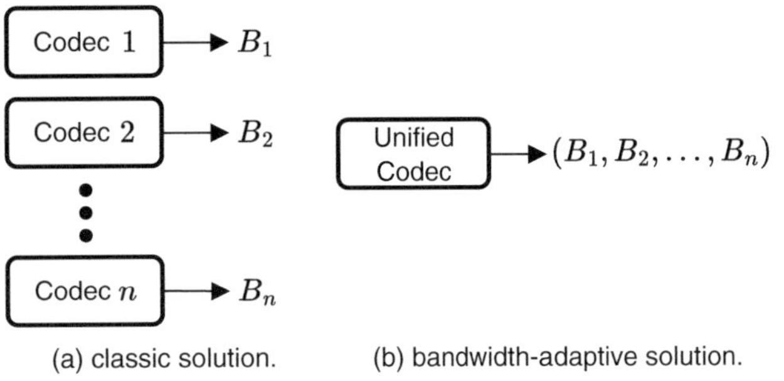


**Fig. 7.1** Feasible solution for meeting variable bandwidth requirements

Graphics (BPG) for images and H.264/H.265 for videos, offer options to adjust output bit rates, e.g., by modifying quantization levels. Moreover, to improve the coding efficiency, source coders typically divide the source data into multiple coding units and exploit the correlations, or *contexts*, that exist between these coding units to encode them into different number of bits. On the other hand, channel coders, e.g., low-density parity check (LDPC) and Turbo codes, support bandwidth adjustment through changes of channel coding rates. By employing different modulation schemes, the bandwidth usage can also be adjusted.

For **Problem 1**, traditional source encoders suggest leveraging entropy coding to achieve efficient bit allocation. Firstly, the contexts are exploited to eliminate internal redundancies within the source data, minimizing the entropy of all source coding units. Then, entropy coding is utilized to allocate bits for each coding unit based on its entropy, thereby eliminating statistical redundancies and achieving effective coding. For images or videos, low-frequency regions like backgrounds contain relatively simpler information with lower entropy, which need fewer bits for coding. In contrast, high-frequency regions such as textures contain richer and more complex information with higher entropy, requiring more bits for coding. Overall, *entropy* can serve as a potential criterion to achieve efficient bandwidth allocation.

For **Problem 2**, traditional source encoders suggest modifying quantization levels to change the total bits coding for the source data. Since entropy coding typically used by source encoders is suitable for discrete representations, quantization is introduced before entropy coding to map continuous values of the source data to a finite set of discrete values. Each quantization level corresponds to a quantization step size, indicating the degree of quantization. Taking H.265 as an example, the adopted quantization is close to scalar uniform quantization as follows:

$$b_i = floor(\frac{a_i}{Q_{\text{step}}} + f), \tag{7.1}$$

where $a_i$ and $b_i$ are the $i$-th element before and after quantization, respectively. $floor(\cdot)$ is the rounding down operation. $Q_{\text{step}}$ is the step size of quantization. $f$ is the offset to achieve the desired rounding. As the quantization step size increases, quantization goes from fine to coarse, and the total number of bits required for coding increases accordingly. By adjusting the quantization step size, the number of bits required for coding can be controlled, which can further change the required communication bandwidth. Additionally, by adjusting the channel code rate, e.g., $\frac{1}{2}$, $\frac{2}{3}$, $\frac{3}{4}$ for LDPC codes, and using different modulation schemes, e.g., BPSK, 4QAM, 16QAM, the communication bandwidth can also be controlled. Overall, both these available options of source coders and channel coders can provide a useful view for variable-length coding.

Although a semantic communication system could adopt either separate coding or joint coding architectures, these traditional methods have provided valuable insights for many studies and have facilitated the development of bandwidth adaptation. In summary, from the perspective of allocation principle, existing bandwidth adaptation methods can be roughly categorized into *entropy-aware* and *semantic-aware* bandwidth adaptation methods.

- **Entropy-Aware Bandwidth Adaptation**: Inspired by bit allocation in entropy coding, entropy-aware bandwidth adaptation typically utilizes an entropy model to estimate entropy in the feature domain. Features with higher entropy are considered to contain richer information of the source data, and thus, are encoded with more DJSCC symbols. There is an explicit mapping between the entropy and the number of DJSCC symbols for all features.
- **Semantic-Aware Bandwidth Adaptation**: Different from entropy-aware bandwidth adaptation, semantic-aware bandwidth adaptation is learning-based, which implicitly establishes a mapping between semantic importance and the number of DJSCC symbols. As an abstract concept, the semantic importance is highly task-dependent and is learned through a neural network. By optimizing under a given measurement, the network learns the importance of different semantics for specific tasks. More important semantics are allocated more DJSCC symbols for transmission, as they carry critical information that contributes more significantly to tasks, which aligns more closely with the objective of semantic communication.

From the perspective of implementation, existing bandwidth adaptation methods can be roughly categorized into *mask-based*, *mapping-based* and *refinement-based* bandwidth adaptation.

- **Mask-Based Bandwidth Adaptation**: A learnable mask is utilized to mask certain channels or features of the input, thereby discarding these information. Extra information needs to be transmitted to indicate the positions of the discarded information at the receiver.

- **Mapping-Based Bandwidth Adaptation**: A dynamic network as encoder is utilized to encode fixed-length features into different given dimensions. After dynamic encoding, the encoded features are of different lengths. Then, another dynamic network as a decoder is utilized to decode fixed-length features from the received features of different lengths. Extra information needs to be transmitted to indicate the dimensions of encoded features for decoding at the receiver.
- **Refinement-Based Bandwidth Adaptation**: An encoding network encodes the source data into multiple successive data streams, which are transmitted sequentially to the receiver. The receiver can select any number of data streams in order to reconstruct the source data. Each data stream is used to refine the reconstruction based on the previous data streams. The quality of reconstruction improves as more data streams are used. When the receiver is able to reconstruct the desired quality, it can choose to stop receiving the remaining data streams, thereby adjusting the communication bandwidth.

Next, we will introduce some classic bandwidth adaptation methods based on entropy-awareness and semantics-awareness, respectively.

## 7.2 Entropy-Aware Bandwidth Adaptation

Entropy-aware bandwidth adaptation methods achieve effective bandwidth allocation by incorporating entropy models. To illustrate the rationale behind entropy-based bandwidth allocation, we first review the communication process of semantic systems. Take images as a representative source, which are modeled as vectors of pixel intensities $\boldsymbol{s} \in \mathbb{R}^m$. The encoder $E(\cdot; \boldsymbol{\theta})$ parameterized by $\boldsymbol{\theta}$ encodes $\boldsymbol{s}$ into complex-valued channel input symbols $\boldsymbol{x} \in \mathbb{C}^n$:

$$\boldsymbol{x} = E(\boldsymbol{s}; \boldsymbol{\theta}). \tag{7.2}$$

In the encoding process, $R = \frac{n}{m}$ is defined as the channel bandwidth ratio (CBR), denoting the average number of channel symbols used for representing one value of the source. Typically, $R < 1$ is satisfied. The smaller the value of $R$, the greater the level of information compression. After DJSCC encoding, the channel input symbols $\boldsymbol{x}$ are transmitted over the wireless noisy channel $W(\cdot; \boldsymbol{\eta})$:

$$\boldsymbol{y} = W(\boldsymbol{x}; \boldsymbol{\eta}), \tag{7.3}$$

where $\boldsymbol{\eta}$ contains the parameters of the wireless channel. Given the distorted received signal $\boldsymbol{y}$, the receiver utilizes $D(\cdot; \boldsymbol{\phi})$ parameterized by $\boldsymbol{\phi}$ to reconstruct the original source.

$$\hat{\boldsymbol{s}} = D(\boldsymbol{y}; \boldsymbol{\phi}). \tag{7.4}$$

By establishing a distortion measurement $d(\cdot, \cdot)$ between $\boldsymbol{s}$ and $\hat{\boldsymbol{s}}$, the learnable parameters $\boldsymbol{\theta}$ and $\boldsymbol{\phi}$ can be jointly optimized to minimize the distortion:

$$(\boldsymbol{\theta}^{\star}, \boldsymbol{\phi}^{\star}) = \arg\min_{\boldsymbol{\theta}, \boldsymbol{\phi}} \mathbb{E}_{\boldsymbol{s} \sim p_{\boldsymbol{s}}} \mathbb{E}_{\hat{\boldsymbol{s}} \sim p_{\hat{\boldsymbol{s}}|\boldsymbol{s}}} (d(\boldsymbol{s}, \hat{\boldsymbol{s}})). \tag{7.5}$$

However, an obvious observation is that as the source dimension increases, e.g., for high-resolution images, the performance of semantic communication systems is gradually saturated. When the coding rate $R$ increases, the slope of performance curve slows down, and it no longer provides substantial performance gains. This issue arises from traditional semantic communication systems neglect of the source distribution, where channel symbols are allocated equally across all elements of the source. For example, as $R$ increases, more channel symbols can be used to encode the elements of the source. However, the coding performance of simpler elements is quickly saturated, and a significant portion of these channel symbols is wasted on encoding these elements, leading to severe bandwidth waste. In contrast, the encoding performance of more complex elements still has considerable potential for improvement.

The solution to this problem is to dynamically adjust the number of channel symbols allocated to each source element based on the source distribution, thus enabling variable-length encoding. An intuitive approach is to encode simpler elements with fewer channel symbols, while more complex elements are encoded with more channel symbols. In entropy coding, values with lower probabilities have higher entropy and thus require more bits for encoding. Conversely, values with higher probabilities have lower entropy and require fewer bits for encoding. Drawing from this idea, a feasible approach is to use entropy to quantify abstract concepts such as simplicity or complexity. Simple elements correspond to low entropy, while complex elements correspond to high entropy. Therefore, entropy can serve as the measurement to allocate channel symbols to each element of the source, ensuring efficient utilization of bandwidth.

This entropy-aware bandwidth adaptation method is based on the nonlinear transform semantic-channel coding (NTSCC) architecture [2], which integrates nonlinear transform coding (NTC) [3] and DJSCC techniques. Nonlinear transform coding (NTC) has been widely applied in image and video compression in recent years. Compared to linear transform coding, NTC can better approximate the source distribution, thereby achieving superior rate-distortion performance across various existing perceptual quality metrics, e.g., PSNR, MS-SSIM, LPIPS. NTC establishes a learnable

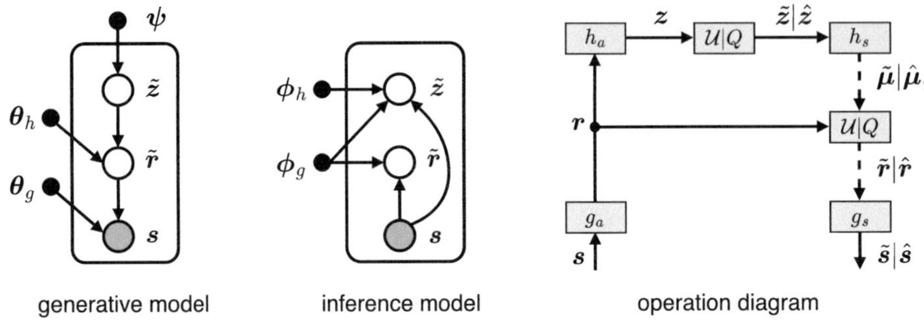


**Fig. 7.2** Left: The encoder and decoder of NTC, which are represented as a variational inference model and a generative Bayesian model, respectively. Right: Operation diagram of NTC. Arrows denote the flow of data, and boxes denote transformations of the data. Boxes labeled $\mathcal{U} \mid Q$ represent either addition of uniform noise applied during training, or quantization during testing

entropy model in the latent space to accurately estimate the probability distribution of the source. As a result, benefiting from this entropy model, semantic communication systems can perform efficient bandwidth allocation for entropy-aware bandwidth adaptation.

Specifically, NTSCC first maps $\boldsymbol{s}$ to a latent representation $\boldsymbol{r}$ in the latent space before a learnable entropy model with a hyperprior is used to estimate the distribution of $\boldsymbol{r}$. This distribution serves as a prior, assisting the selection of an appropriate coding scheme to encode each latent representation $\boldsymbol{r}_i$ with a variable length. In this way, the rate-distortion performance of each $\boldsymbol{r}_i$ can be optimized in a customized manner.

### 7.2.1 Neural Network Based Entropy Modeling

As a widely used lossy image compression framework, NTC includes an encoder to process the original image vector $\boldsymbol{s}$ into bit streams, and a decoder to generate a reconstruction $\hat{\boldsymbol{s}}$ from the bit streams. The architecture of NTC is shown in Fig. 7.2. The encoder is typically represented as a variational inference model, while the decoder is represented as a generative Bayesian model. Now we describe the process of compression and decompression in NTC.

The encoder, or analysis transform of NTC $g_a(\cdot;\boldsymbol{\phi}_g)$ parameterized by $\boldsymbol{\phi}_g$ firstly transforms $\boldsymbol{s}$ into a continuous-valued representation $\boldsymbol{r}$ in the latent space:

$$\boldsymbol{r} = g_a(\boldsymbol{s};\boldsymbol{\phi}_g). \tag{7.6}$$

However, the latent representation $\boldsymbol{r}$ extracted by $g_a(\cdot;\boldsymbol{\phi}_g)$ still contains significant spatial redundancy, leading to low coding efficiency. Therefore, a hyper encoder $h_a(\cdot;\boldsymbol{\phi}_h)$ parameterized by $\boldsymbol{\phi}_h$ are introduced to extract hyperprior $z$ from $\boldsymbol{r}$:

$$z = h_a(\boldsymbol{r};\boldsymbol{\phi}_h). \tag{7.7}$$

Then, $z$ is quantized into discrete-valued vector $\hat{z}$ and further encoded into bit streams:

$$\hat{z} = Q(z), \boldsymbol{b}_z = AE(\hat{z} \mid P_{\hat{z}}). \tag{7.8}$$

where $Q(\cdot)$ denotes the quantization by rounding in integers, and $AE(\cdot)$ denotes arithmetic encoding. $\boldsymbol{b}_z$ is the encoded bit streams and $P_{\hat{z}}$ is the estimated distribution of $\hat{z}$.

With the help of hyperprior $z$, $\boldsymbol{r}$ is quantized and encoded into bit streams $\boldsymbol{b}_r$ more efficiently:

$$\hat{\boldsymbol{r}} = Q(\boldsymbol{r}), \boldsymbol{b}_r = AE(\hat{\boldsymbol{r}} \mid P_{\hat{\boldsymbol{r}}|z}). \tag{7.9}$$

The coding rate $R_r$ denotes the expected length of $\boldsymbol{b}_r$, which is given by the estimated entropy of $\hat{\boldsymbol{r}}$. Similarly, $R_z$ is given by the estimated entropy of $\hat{z}$.

In the process of decompression, $\hat{z}$ is first decoded from $\boldsymbol{b}_z$, which are further transformed into parameters $\boldsymbol{\alpha}$. by a hyper decoder $h_s(\cdot;\boldsymbol{\theta}_h)$ parameterized by $\boldsymbol{\theta}_h$:

$$\hat{z} = AD(\boldsymbol{b}_z \mid P_{\hat{z}}), \boldsymbol{\alpha} = h_s(\hat{z};\boldsymbol{\theta}_h). \tag{7.10}$$

Actually, $\boldsymbol{\alpha}$ contains the mean and variance of $P_{\hat{\boldsymbol{r}}|z}$ to give distribution prior for decoding latent representation $\hat{\boldsymbol{r}}$. Then, $\hat{\boldsymbol{r}}$ is decoded from $\boldsymbol{b}_r$ with $\boldsymbol{\alpha}$. Finally, The decoder, or synthesis transform of NTC $g_s(\cdot;\boldsymbol{\theta}_g)$ parameterized by $\boldsymbol{\theta}_g$ generate the reconstructed image $\hat{s}$ from $\hat{\boldsymbol{r}}$:

$$\hat{s} = g_s(\hat{\boldsymbol{r}};\boldsymbol{\theta}_g). \tag{7.11}$$

Quantization results in a zero-gradient problem, making it impossible to optimize learnable parameters through gradient descent. Hence, a relaxed method is widely utilized to achieve end-to-end training. Specifically, $\tilde{\boldsymbol{r}}$ is used as a proxy of $\hat{\boldsymbol{r}}$ in model training, which is obtained by adding noise following a uniform distribution $\mathcal{U}(-\frac{1}{2},\frac{1}{2})$ to $\boldsymbol{r}$.

The compression structure of NTC can be viewed as a variational autoencoder (VAE), where a learnable distribution $q_{\tilde{\boldsymbol{r}},\tilde{z}|s}(\tilde{\boldsymbol{r}},\tilde{z} \mid s)$ is introduced to approximate the true posterior distribution $p_{\tilde{\boldsymbol{r}},\tilde{z}|s}(\tilde{\boldsymbol{r}},\tilde{z} \mid s)$. Then, the optimization problem is transformed into minimizing the expectation of Kullback-Leibler (KL) divergence over the data distribution $p_s$:

$$\min_{\boldsymbol{\phi}_g,\boldsymbol{\phi}_h,\boldsymbol{\theta}_g,\boldsymbol{\theta}_h} \mathbb{E}_{s\sim p_s} D_{\mathrm{KL}}\left[q_{\tilde{\boldsymbol{r}},\tilde{z}|s} \| p_{\tilde{\boldsymbol{r}},\tilde{z}|s}\right] = \min_{\phi_g,\phi_h,\theta_g,\theta_h} \mathbb{E}_{s\sim p_s} \mathbb{E}_{\tilde{\boldsymbol{r}},\tilde{z}\sim q_{\tilde{\boldsymbol{r}},\tilde{z}|s}} \left[\left[\log q_{\tilde{s},\tilde{z}|s}(\tilde{\boldsymbol{r}},\tilde{z}|s)\right] \underbrace{-\log p_{\tilde{\boldsymbol{r}}|\tilde{z}}(\tilde{\boldsymbol{r}}|\tilde{z})}_{\text{rate of latent}} \underbrace{-\log p_{\tilde{z}}(\tilde{z})}_{\text{rate of hyperprior}} \underbrace{-\log p_{s|\tilde{\boldsymbol{r}}}(s|\tilde{\boldsymbol{r}})}_{\text{weighted distortion}}\right] + \text{const.} \tag{7.12}$$

Optimizing this KL divergence in Eq. (7.12) can be viewed as optimizing the RD performance of NTC actually. The first term in Eq. (7.12) can be expressed as:

$$q_{\tilde{r},\tilde{z}|s}(\tilde{r},\tilde{z} \mid s) = \prod_i \mathcal{U}(\tilde{r}_i \mid r_i - \frac{1}{2}, r_i + \frac{1}{2}) \cdot \prod_j \mathcal{U}(\tilde{z}_j \mid z_j - \frac{1}{2}, z_j + \frac{1}{2}), \tag{7.13}$$

where $\boldsymbol{r}_i$ is the $i$-th element of $\boldsymbol{r}$ and $z_j$ is the $j$-th element of $z$. As each variable is modeled using a uniform distribution $\mathcal{U}$ over a unit-width interval, the first term evaluates to a constant and can be removed from the loss function. Similarly, the last term can also be removed from the loss function because this term is always a constant. The second term is to compute the cross-entropy between the approximation distribution $q_{\tilde{r}|\tilde{z}}(\tilde{\boldsymbol{r}}|\tilde{z})$ and the true prior distribution $p_{\tilde{r}|\tilde{z}}(\tilde{\boldsymbol{r}}|\tilde{z})$, which is related to the rate loss of encoding $\tilde{\boldsymbol{r}}$ following the true distribution $p_{\tilde{r}|\tilde{z}}(\tilde{\boldsymbol{r}}|\tilde{z})$ by the estimated distribution $q_{\tilde{r}|\tilde{z}}(\tilde{\boldsymbol{r}}|\tilde{z})$. Similarly, the third term is to compute the rate loss of encoding hyperprior $\tilde{z}$. The fourth term is to compute the reconstruction loss between the original image $\boldsymbol{s}$ and the reconstructed image $\hat{\boldsymbol{s}}$. If the squared error is used as the measurement, the fourth term can be expressed as:

$$p_{s|\tilde{r}}(s \mid \tilde{\boldsymbol{r}}) = \mathcal{N}(s \mid \hat{s}, (2\lambda)^{-1}\mathbf{I}_m), \tag{7.14}$$

where $\lambda$ denotes the weight of squared error and $\mathcal{N}$ denotes the Gaussian distribution. Minimizing this term is equivalent to minimizing the distortion of the reconstructed image. This is the variational modeling of NTC framework. To training the RD performance of NTC, the loss function is given by:

$$\mathcal{L} = \mathbb{E}_{s\sim p_s(s)}\left[\lambda(-\log p_{\tilde{r}|\tilde{z}}(\tilde{\boldsymbol{r}} \mid \tilde{z}) - \log p_{\tilde{z}}(\tilde{z})) + d(s,\hat{s})\right], \tag{7.15}$$

where $\lambda$ is the hyperparameter to trade-off rate cost and the expected distortion.

By introducing the hyperprior $z$, the spatial redundancy present in $\boldsymbol{r}$ can be effectively removed. Specifically, given the hyperprior, each element $\boldsymbol{r}_i$ of $\tilde{\boldsymbol{r}}$ is modeled as following a Gaussian distribution with its own mean $\tilde{\boldsymbol{\mu}}_i$ and standard deviation $\tilde{\boldsymbol{\sigma}}_i$. Then, the distribution of $\tilde{\boldsymbol{r}}$ given $\tilde{z}$ is formulated as:

$$p_{\tilde{r}|\tilde{z}}(\tilde{\boldsymbol{r}}|\tilde{z}) = \prod_i \left(\mathcal{N}(\tilde{\boldsymbol{\mu}}_i, \tilde{\boldsymbol{\sigma}}_i^2) * \mathcal{U}(-\frac{1}{2}, \frac{1}{2})\right)(\tilde{\boldsymbol{r}}_i). \tag{7.16}$$

The mean vector $\tilde{\boldsymbol{\mu}}$ and standard deviation vector $\tilde{\boldsymbol{\sigma}}$ of $\tilde{\boldsymbol{r}}$ are obtained from the decoding of $\tilde{z}$ by the hyper decoder $h_s(\cdot;\boldsymbol{\theta}_h)$. As there is no prior beliefs about $\tilde{z}$, the distribution of $\tilde{z}$ is modeled as fully factorized density model without learnable parameters:

$$p_{\tilde{z}|\boldsymbol{\psi}}(\tilde{z} \mid \boldsymbol{\psi}) = \prod_j \left(p_{z_j|\boldsymbol{\psi}^{(j)}}(z_j|\boldsymbol{\psi}^{(j)}) * \mathcal{U}(-\frac{1}{2}, \frac{1}{2})\right)(\tilde{z}_j), \tag{7.17}$$

where $\boldsymbol{\psi}^{(i)}$ include all parameters of $p_{z_j|\boldsymbol{\psi}^{(j)}}$.

### 7.2.2 Entropy-Based Bandwidth Allocation

Building upon the discussion in Sect. 7.2.1, combining NTC with DJSCC can fully leverage the advantages of both, which is the basic idea of NTSCC. NTSCC encodes the original image $\boldsymbol{s}$ into variable-length symbols $\boldsymbol{x}$ with the help of NTC at the transmitter. After wireless transmission, NTSCC generates the reconstruction $\hat{\boldsymbol{s}}$ from the received noisy symbols $\boldsymbol{y}$. The architecture of NTSCC with its variational modeling is shown in Fig. 7.3. Extended from NTC, NTSCC represents the combination of the analysis transform of NTC, the DJSCC encoder, and the wireless channel as a variational inference model, while the combination of the DJSCC decoder and the synthesis transform of NTC is a generative Bayesian model.

As illustrated at the top of Fig. 7.3, NTSCC extends analysis transform of NTC, JSCC encoding, and the communication channel as an inference model, while JSCC decoding and

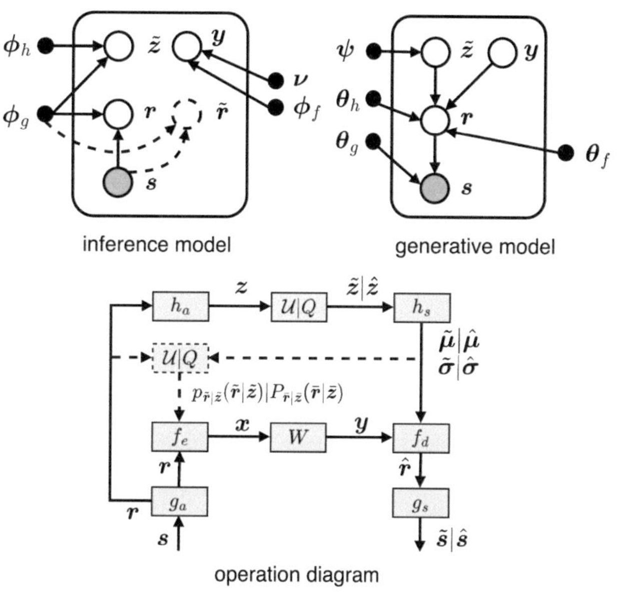


**Fig. 7.3** Top: Representation of analysis transform of NTC, JSCC encoding and the communication channel as an inference model, and JSCC decoding and synthesis transform of NTC as a generative model. Bottom: Operation diagram of NTSCC. Arrows denote the flow of data, and boxes denote transformations or transmission of the data. Boxes labeled $\mathcal{U} \mid Q$ represent either the addition of uniform noise applied during training, or quantization during testing

synthesis transform of NTC are represented as a generative model. The bottom of Fig. 7.3 shows the operation diagram of NTSCC. Now we describe the process of communication in NTSCC.

Similar to NTC, the transmitter first use analysis transform $g_a(\cdot; \boldsymbol{\phi}_g)$ parameterized by $\boldsymbol{\phi}_g$ to map the source vector $\boldsymbol{s} \in \mathbb{R}^m$ to a latent representation $\boldsymbol{r}$:

$$\boldsymbol{r} = g_a(\boldsymbol{s}; \boldsymbol{\phi}_g). \tag{7.18}$$

The latent representation $\boldsymbol{r}$ is still with spatial redundancies. Therefore, a hyperprior $z$ is extracted from $\boldsymbol{r}$ by a hyper encoder $h_a(\cdot; \boldsymbol{\phi}_h)$ parameterized by $\boldsymbol{\phi}_h$ to help establish a more effective entropy model:

$$z = h_a(\boldsymbol{r}; \boldsymbol{\phi}_h). \tag{7.19}$$

$z$ is served as side information, which should be quantized and encoded into bit streams $\boldsymbol{b}_z$ by entropy coding:

$$\hat{z} = Q(z), \boldsymbol{b}_z = AE(\hat{z}|P_{\hat{z}}). \tag{7.20}$$

The proxy $\tilde{z}$ is also employed in model testing, and the distribution of $\hat{z}$ and $\tilde{z}$ is modeled as a fully factorized density model without learnable parameters as Eq. (7.17):

$$P_{\hat{z}|\boldsymbol{\psi}}(\hat{z}|\boldsymbol{\psi}) = p_{\tilde{z}|\boldsymbol{\psi}}(\tilde{z}|\boldsymbol{\psi}) = \prod_j \left( p_{z_j|\boldsymbol{\psi}^{(j)}}(z_j|\boldsymbol{\psi}^{(j)}) * \mathcal{U}(-\frac{1}{2}, \frac{1}{2}) \right)(\tilde{z}_j), \tag{7.21}$$

where $\boldsymbol{\psi}^{(i)}$ include all parameters of $p_{z_j|\boldsymbol{\psi}^{(j)}}$. Note $\boldsymbol{b}_z$ are transmitted through the reliable wireless link to the receiver to share the same side information between the transmitter and the receiver.

The distribution of $\hat{\boldsymbol{r}}$ is independent Gaussian with the mean $\hat{\boldsymbol{\mu}}$ and the standard deviation $\hat{\boldsymbol{\sigma}}$ conditioned on $\hat{z}$ as Eq. (7.16), where $\hat{\boldsymbol{\mu}}$ and $\hat{\boldsymbol{\sigma}}$ are predicted by a hyper decoder $h_s(\cdot; \boldsymbol{\theta}_h)$ parameterized by $\boldsymbol{\theta}_h$ using $\hat{z}$:

$$\hat{z} = AD(\boldsymbol{b}_z|P_{\hat{z}}), [\hat{\boldsymbol{\mu}}, \hat{\boldsymbol{\sigma}}] = h_s(\hat{z}; \boldsymbol{\theta}_h). \tag{7.22}$$

Given $\hat{\boldsymbol{\mu}}$ and $\hat{\boldsymbol{\sigma}}$, the entropy model of $\hat{\boldsymbol{r}}$ is established. Instead of encoding $\boldsymbol{r}$ into bit streams, the transmitter uses a JSCC encoder $f_e(\cdot; \boldsymbol{\phi}_f)$ parameterized by $\boldsymbol{\phi}_f$ to map $\boldsymbol{r}$ into channel input symbols $\boldsymbol{x} \in \mathbb{C}^n$ of variable lengths with the help of the entropy model:

$$\boldsymbol{x} = f_e(\boldsymbol{r}, \hat{\boldsymbol{\mu}}, \hat{\boldsymbol{\sigma}}; \boldsymbol{\phi}_f) \tag{7.23}$$

The channel input symbols $\boldsymbol{x}$ are transmitted over the wireless noisy channel $W(\cdot, \boldsymbol{\eta})$:

$$\boldsymbol{y} = W(\boldsymbol{x}; \boldsymbol{\eta}), \tag{7.24}$$

where $\boldsymbol{y}$ denotes channel output symbols and $\boldsymbol{\eta}$ contains channel parameters.

Due to $\boldsymbol{b}_z$ are losslessly transmitted to the receiver, the receiver can obtain the same $\hat{\boldsymbol{\mu}}$ and $\hat{\boldsymbol{\sigma}}$ and further exploits them to recover latent representation $\hat{\boldsymbol{r}}$ from received symbols $\boldsymbol{y}$ by a JSCC decoder $f_d(\cdot; \boldsymbol{\theta}_f)$ parameterized by $\boldsymbol{\theta}_f$:

$$\hat{\boldsymbol{r}} = f_d(\boldsymbol{y}, \hat{\boldsymbol{\mu}}, \hat{\boldsymbol{\sigma}}; \boldsymbol{\theta}_f). \tag{7.25}$$

Finally, the receiver use a synthesis transform $g_s(\cdot; \boldsymbol{\theta}_g)$ parameterized by $\boldsymbol{\theta}_g$ to generate a reconstruction of source vector $\hat{\boldsymbol{s}}$:

$$\hat{\boldsymbol{s}} = g_s(\hat{\boldsymbol{r}}; \boldsymbol{\theta}_g). \tag{7.26}$$

The structure of NTSCC can also be viewed from a variational perspective. The inference model (the analysis transform of NTC, the DJSCC encoder, and the wireless channel) produces a learnable variational distribution $q_{\boldsymbol{y},\tilde{z}|\boldsymbol{s}}(\boldsymbol{y}, \tilde{z}|\boldsymbol{s})$ to approximate the true posterior distribution $p_{\boldsymbol{y},\tilde{z}|\boldsymbol{s}}(\boldsymbol{y}, \tilde{z}|\boldsymbol{s})$. The optimization of NTSCC can also be achieved by minimizing the KL divergence between these two distributions over the data distribution $p_s$, which is actually optimizing the transmission RD performance of NTSCC:

$$\begin{aligned}
&\min_{\phi_{\text{all}}, \theta_{\text{all}}} \mathbb{E}\left[D_{\text{KL}}\left[q_{\boldsymbol{y},\tilde{z}|\boldsymbol{s}} \| p_{\boldsymbol{y},\tilde{z}|\boldsymbol{s}}\right]\right] \\
&= \min_{\boldsymbol{\phi}_{\text{all}}, \boldsymbol{\theta}_{\text{all}}} \mathbb{E}\left[\log q_{\boldsymbol{y},\tilde{z}|\boldsymbol{s}}(\boldsymbol{y}, \tilde{z}|\boldsymbol{s}) - \log p_{\boldsymbol{y},\tilde{z}}(\boldsymbol{y}, \tilde{z}) - \log p_{\boldsymbol{s}|\boldsymbol{y},\tilde{z}}(\boldsymbol{s}|\boldsymbol{y}, \tilde{z})\right] \\
&= \min_{\phi_{\text{all}}, \theta_{\text{all}}} \mathbb{E}\left[\log q_{\boldsymbol{y},\tilde{z}|\boldsymbol{s}}(\boldsymbol{y}, \tilde{z}|\boldsymbol{s}) - \log p_{\tilde{z}}(\tilde{z}) - \log p_{\boldsymbol{y}|\tilde{z}}(\boldsymbol{y}|\tilde{z}) - \log \mathbb{E}_{\boldsymbol{r}\sim p_{\boldsymbol{r}|\boldsymbol{y},\tilde{z}}}\left[p_{\boldsymbol{s}|\boldsymbol{r}}(\boldsymbol{s}|\boldsymbol{r})\right]\right] \\
&\leq \min_{\phi_{\text{all}}, \theta_{\text{all}}} \mathbb{E}[\log q_{\boldsymbol{y},\tilde{z}|\boldsymbol{s}}(\boldsymbol{y}, \tilde{z}|\boldsymbol{s}) \underbrace{- \log p_{\tilde{z}}(\tilde{z})}_{\text{rate of } z} \underbrace{- \log p_{\boldsymbol{y}|\tilde{z}}(\boldsymbol{y}|\tilde{z})}_{\text{rate of } \boldsymbol{x}} \underbrace{- \mathbb{E}_{\boldsymbol{r}\sim p_{\boldsymbol{r}|\boldsymbol{y},\tilde{z}}}\left[\log p_{\boldsymbol{s}|\boldsymbol{r}}(\boldsymbol{s}|\boldsymbol{r})\right]}_{\text{weighted distortion}}],
\end{aligned} \tag{7.27}$$

where $\boldsymbol{\phi}_{\text{all}}$ contain $\boldsymbol{\phi}_g$, $\boldsymbol{\phi}_h$ and $\boldsymbol{\phi}_f$ while $\boldsymbol{\theta}_{\text{all}}$ contain $\boldsymbol{\theta}_g$, $\boldsymbol{\theta}_h$ and $\boldsymbol{\theta}_f$.

The first term in Eq. (7.27) computes the analysis transform and the DJSCC encoding of the source vector $\boldsymbol{s}$, and then adds the channel noise $\boldsymbol{n}$ if AWGN channel is considered, as follows:

$$q_{\boldsymbol{y},\tilde{z}|\boldsymbol{s}}(\boldsymbol{y}, \tilde{z}|\boldsymbol{s}) = \prod_i \mathcal{N}(\hat{y}_i|y_i, \sigma_{\boldsymbol{n}}^2) \cdot \prod_j \mathcal{U}(\tilde{z}_j|z_j - \frac{1}{2}, z_j + \frac{1}{2}). \tag{7.28}$$

This term is calculated as a constant and can be removed from the equation. The second term computes the cross-entropy between the marginal distribution $q_{\tilde{z}}(\tilde{z}) = \mathbb{E}_{s\sim p_s}\mathbb{E}_{\boldsymbol{y}\sim q_{\boldsymbol{y}|s}}[q_{\boldsymbol{y},\tilde{z}|\boldsymbol{s}}(\boldsymbol{y}, \tilde{z}|\boldsymbol{s})]$ and the true prior $p_{\tilde{z}}(\tilde{z})$, which represents the rate cost for encoding the side information $\hat{z}$. Similarly, the third term computes the cross-entropy between $q_{\boldsymbol{y}|\tilde{z}}(\boldsymbol{y}|\tilde{z})$ and true conditional distribution $p_{\boldsymbol{y}|\tilde{z}}(\boldsymbol{y}|\tilde{z})$, which represents the rate cost for encoding $\boldsymbol{y}$ given side information $\hat{z}$. Specifically, $p_{\boldsymbol{y}|\tilde{z}}(\boldsymbol{y}|\tilde{z})$ can represented as:

$$p_{\boldsymbol{y}|\tilde{z}}(\boldsymbol{y}|\tilde{z}) = \prod_i \left( p_{\boldsymbol{x}_i|\tilde{z}}(\boldsymbol{x}_i|\tilde{z}) * \mathcal{N}(0, \sigma_{\boldsymbol{n}}^2) \right)(\boldsymbol{y}_i), \tag{7.29}$$

where $*$ denotes the convolution operation. The sum of the second and third terms constitutes the transmission cost of NTSCC. Finally, the last term computes the reconstruction loss between the original image $\boldsymbol{s}$ and the reconstructed image $\hat{\boldsymbol{s}}$. This term is determined by $p_{s|\hat{r}}(\boldsymbol{s}|\hat{\boldsymbol{r}})$ and $p_{r|y,\tilde{z}}(\boldsymbol{r}|\boldsymbol{y},\tilde{z})$ if the squared error is used as the measurement, $p_{s|\hat{r}}(\boldsymbol{s}|\hat{\boldsymbol{r}})$ and $p_{r|y,\tilde{z}}(\boldsymbol{r}|\boldsymbol{y},\tilde{z})$ can be represented as:

$$p_{s|\hat{r}}(\boldsymbol{s}|\hat{\boldsymbol{r}}) = \mathcal{N}(\boldsymbol{s}|\hat{\boldsymbol{s}},(2\lambda_1)^{-1}\mathbf{I}_m) \tag{7.30}$$

and

$$\begin{aligned} p_{r|y,\tilde{z}}(\boldsymbol{r}|\boldsymbol{y},\tilde{z}) &\geq p_{r|y}(\boldsymbol{r}|\boldsymbol{y}) \cdot_{r|\tilde{z}} (\boldsymbol{r}|\tilde{z}) \\ &= \prod_i \mathcal{N}(\boldsymbol{r}_i|\hat{\boldsymbol{r}}_i,(2\lambda_2)^{-1}) \cdot \prod_i \mathcal{N}(\boldsymbol{r}_i|\tilde{\boldsymbol{\mu}}_i,\tilde{\sigma}_i^2), \end{aligned} \tag{7.31}$$

respectively. Therefore, this term indicates the weighted distortion between origin and reconstruction. Up to this point, we have presented the variational modeling and operational diagram of NTSCC.

By utilizing the entropy model $P_{\hat{r}|\hat{z}}(\hat{\boldsymbol{r}}|\hat{z})$, NTSCC can easily encode the latent representation $\boldsymbol{r}$ into the channel input symbols $\boldsymbol{x}$ of variable lengths. The JSCC coding of NTSCC adopts Vision Transformer architecture as the backbone, which means the JSCC encoder $f_e(\cdot;\boldsymbol{\phi}_f)$ and JSCC decoder $f_d(\cdot;\boldsymbol{\theta}_f)$ is composed of stacked Vision Transformer blocks.

For JSCC encoding, the JSCC encoder $f_e(\cdot;\boldsymbol{\phi}_f)$ firstly partitions the latent representation $\boldsymbol{r}$ into a sequence with $l$ patch embeddings $(\boldsymbol{r}_1,\boldsymbol{r}_2,...\boldsymbol{r}_l)$, where each patch embedding $\boldsymbol{r}_i$, $i\in(1,l)$ is a $c$-dimensional vector. With the learned entropy model $P_{\hat{r}|\hat{z}}(\hat{\boldsymbol{r}}|\hat{z})$, $P_{\hat{r}_i|\hat{z}}(\hat{\boldsymbol{r}}_i|\hat{z})$ can be computed by summing the entropy of each $\boldsymbol{r}_i$ along the channel dimension. Specifically, the channel bandwidth cost $\bar{k}_{r_i}$ required for transmitting $y_i$ can be expressed as:

$$\begin{aligned} \bar{k}_{r_i} &= Q'(k_{r_i}) \\ &= Q'(-\nu\log P_{\hat{r}_i|\hat{z}}(\hat{\boldsymbol{r}}_i|\hat{z})) = Q'(-\nu\log p_{\tilde{r}_i|\tilde{z}}(\tilde{\boldsymbol{r}}_i|\tilde{z})), \end{aligned} \tag{7.32}$$

where $Q'$ denotes a scalar quantization whose range consists $2^q$ integers ($q=1,2,...$) and $\nu$ is a scaling factor. $\nu$ contains the physical meaning: the performance of DJSCC encoder/decoder, which can be represented as:

$$\nu = \frac{\nu'}{C}, \tag{7.33}$$

where $C$ denotes the channel capacity and $\nu'$ denotes the capability of DJSCC encoder/decoder. When $\nu'$ is 1, the JSCC encoder/decoder is ideal which adopts entropy-achievable source coding combined with channel capacity-achievable channel coding.

Based on Eq. (7.32), the candidate channel bandwidth cost $\bar{k}_{r_i}$ is determined by $q$, $\nu$ and Lagrange multiplier $\lambda$. In order to map a fixed $c$-dimensional embedding $y_i$ into given $\bar{k}_{r_i}$-dimensional channel input symbols $\boldsymbol{x}_i$, one solution is to train $2^q$ encoder-decoder pairs of JSCC encoder $f_e(\cdot;\boldsymbol{\phi}_f)$, where each encoder-decoder is only responsible for encoding and decoding of one channel bandwidth rate. However, this approach significantly increases the training and storage costs of the network. As quantization becomes more refined (with $q$ gradually increasing), these number of encoder-decoder pairs required for training grow exponentially.

NTSCC introduces the dynamic neural network structure to avoid multiple fixed-rate JSCC encoder/decoder training and encode every $\boldsymbol{r}_i$ into given rate adaptively. Compared to static models with fixed computation graphs and parameters, dynamic networks can adapt their structure or parameters to different inputs, offering significant advantages in terms of performance and computational efficiency. A light fully-connected (FC) layer serves as the dynamic network for variable-length DJSCC coding. There are additional several Transformer blocks are utilized to extract feature of each $\boldsymbol{r}_i$ before inputting in the light FC layer, where the parameters of Transformer blocks are shared between all embedding $(\boldsymbol{r}_1,\boldsymbol{r}_2,...,\boldsymbol{r}_l)$.

Basically, the JSCC encoder $f_e(\cdot;\boldsymbol{\phi}_f)$ includes following two steps:

1. Use several Transformer blocks to extract features from each embedding $\boldsymbol{r}_i$ given the prior information, i.e., the channel bandwidth cost $\bar{k}_{r_i}$ through the positional embedding process.
2. Use the light FC layer to encode each fixed $c$-dimensional $\boldsymbol{r}_i$ into given $\bar{k}_{r_i}$-dimensional channel input symbols $\boldsymbol{x}_i$ with variable rate.

For Step 1, the method of incorporating pre-computed channel bandwidth cost as prior information during the feature extraction process is derived from the position encoding in the original Transformer. Given the candidate channel bandwidth set $\mathcal{V}=\{v_1,v_2,...,v_{2^q}\}$, a rate token set $\mathcal{R}=\{\boldsymbol{a}_1,\boldsymbol{a}_2,...,\boldsymbol{a}_{2^q}\}$ is developed to indicate the transmission rate during the feature extraction. A "rate embedding" operation is employed before feature extraction by adding a corresponding rate token $\boldsymbol{a}_{\bar{k}_{r_i}}$ to each embedding $\boldsymbol{r}_i$. Therefore, these Transformer blocks can adapt the feature extraction to different transmission rates, ultimately helping the following FC layer (Step 2) to encode $\boldsymbol{r}_i$ into the given dimension effectively. Note that additional $q$ bits should be transmitted losslessly to indicate the transmission rate of $\boldsymbol{r}_i$ at the receiver.

At the receiver, the received noisy symbols $\boldsymbol{y}_i$ related to $\boldsymbol{r}_i$ of variable lengths are firstly mapped to fixed $c$-dimensional vectors through a similar FC layer. Then, similar to DJSCC encoding, these fixed $c$-dimensional vectors are added with corresponding rate tokens and further fed into several Transformer blocks for latent representation reconstruction. Actually, there are two versions for latent representation reconstruction, i.e., decoding with or without a hyperprior.

For decoding without a hyperprior, the output of Transformer blocks is the reconstructed latent representation $\check{\boldsymbol{r}}$, which is further processed by the synthesis transform $g_s(\cdot; \boldsymbol{\theta}_g)$:

$$\check{\boldsymbol{r}} = f_d(\boldsymbol{y}; \boldsymbol{\theta}_f). \tag{7.34}$$

This reconstruction is at a coarse-grained level, as only the transmission rate of each $\boldsymbol{r}_i$ is known at the receiver.

For decoding with hyperprior, as stated in Eq. (7.34), the JSCC decoder can utilize $\hat{\boldsymbol{\mu}}$ and $\hat{\boldsymbol{\sigma}}$ to further refine $\check{\boldsymbol{r}}$ to $\hat{\boldsymbol{r}}$, which can be achieved by a light MLP:

$$\hat{\boldsymbol{r}} = \text{MLP}(\check{\boldsymbol{r}}, \hat{\boldsymbol{\mu}}, \hat{\boldsymbol{\sigma}}). \tag{7.35}$$

These refinements are at a fine-grained level, as the distribution of $P_{\hat{\boldsymbol{r}}|\hat{\boldsymbol{z}}}(\hat{\boldsymbol{r}}|\hat{\boldsymbol{z}})$ is known at the receiver. However, to ensure that the receiver can obtain accurate $\hat{\boldsymbol{\mu}}$ and $\hat{\boldsymbol{\sigma}}$, $\hat{\boldsymbol{z}}$ needs to be transmitted through a lossless link, which consequently increases the channel bandwidth cost.

Following optimizing RD performance, the training loss of NTSCC can be expressed as:

$$\mathcal{L}_{\text{NTSCC}} = d(\boldsymbol{s}, \hat{\boldsymbol{s}}) + \lambda R, \tag{7.36}$$

where $d(\cdot;\cdot)$ can be reconstruction distortion or perceptual distortion (LPIPS). $R$ denotes the bandwidth cost. For refinement without a hyperprior, $R$ only includes bandwidth cost for transmitting latent representation $\boldsymbol{r}$. For refinement with hyperprior, $R$ includes bandwidth cost for transmitting both latent representation $\boldsymbol{r}$ and hyperprior $\hat{\boldsymbol{z}}$. $\lambda$ controls the trade-off between distortion and bandwidth cost. Models trained under a lower $\lambda$ tend to learn latent representation $\boldsymbol{r}$ with higher entropy and allocate more bandwidth for transmitting $\boldsymbol{r}$. In contrast, models trained under higher $\lambda$ tend to learn $\mathbf{y}$ with lower entropy and allocate less bandwidth for transmitting $\boldsymbol{r}$.

Although we use images as an example source to explain the operation of NTSCC, this transmission framework can be easily extended to the video source [4]. A video can be seen as a series of consecutive frames with temporal correlations. These frames are sequentially encoded, transmitted, and decoded under the conditional coding structure. Therefore, NTSCC can be applied repeatedly for the transmission of each frame, with bandwidth allocated based on the entropy of each frame to achieve DJSCC variable-length coding. Given the additional temporal correlations present in videos, these correlations should be captured to build an effective entropy model. In fact, this has already been widely studied in neural video compression.

Figure 7.4 demonstrates quantified RD performance of different transmission schemes under different average CBRs. The AWGN channel is adopted and SNR is 10 dB in testing. PSNR is used as a reconstruction metric to measure the performance of different models. Both the CIFAR-10 dataset with low-resolution and the Kodak dataset with high-resolution are served as testing datasets. For different RD trade-offs, NTSCC are trained under different Lagrange multiplier $\lambda$s ($\lambda \in \{1024, 256, 64, 16, 4\}$). For "BPG + LDPC" scheme, 2/3 rate (4096, 6044) LDPC code with 16-ary quadrature amplitude modulation (16QAM) is employed. For "BPG + Capacity" scheme, channel capacity transmission rate is achievable, which can be derived from the Shannon capacity theorem. For the DJSCC scheme, each latent representation is encoded into a fixed dimension. CBR denotes the average channel input symbols used for transmitting one value of the image source.

A basic observation is that NTSCC outperforms the DJSCC method under all CBRs. The gap between the NTSCC and DJSCC methods stems from entropy-aware bandwidth allocation, which increases with increasing image resolution (from CIFAR-10 to Kodak) and increasing CBR. Besides, NTSCC also outperforms the "BPG + LDPC"

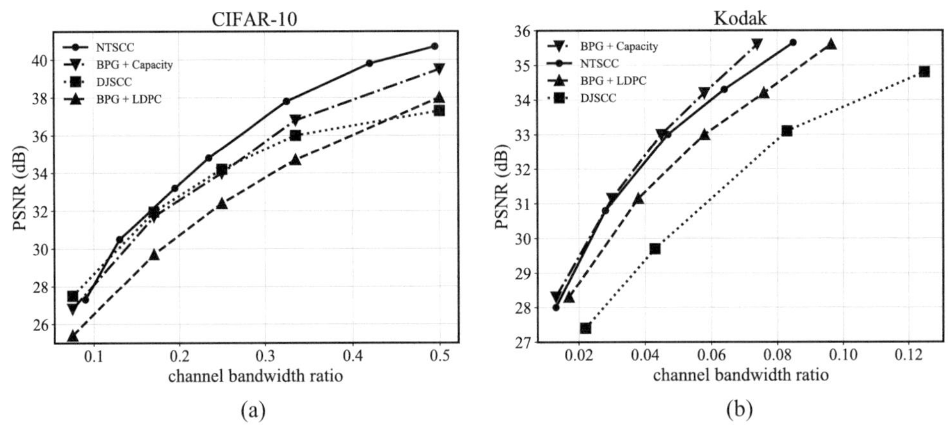


**Fig. 7.4** PSNR performance of different transmission schemes versus the average CBR over the AWGN channel when SNR is 10 dB

scheme adopted separate source and channel coding, and achieves performance comparable to the ideal "BPG + Capacity" scheme.

Benefiting from entropy-aware bandwidth adaptation, NTSCC can overcome the performance saturation issue that arises with DJSCC as the CBR increases, providing significant coding gains. However, when the available bandwidth is abundant, NTSCC still exhibits some degree of performance saturation. This is due to the network architecture of NTSCC, where the channel dimension of the analysis transform $g_a(\cdot; \boldsymbol{\phi}_g)$ is set to 256. This means that the DJSCC encoder $f_e(\cdot; \boldsymbol{\phi}_f)$ can support encoding each embedding $y_i$ into a maximum of 256 dimensions. When the CBR is very high, most of the embeddings are encoded into 256 dimensions, rendering the bandwidth allocation ineffective, which leads to the performance saturation phenomenon.

Additionally, when the CBR is very low, NTSCC performs even worse than the DJSCC method on the low-resolution CIFAR-10 dataset. This is because when the image resolution is low, and the bandwidth is severely constrained, the cost of transmitting the hyperprior occupies a significant portion of the total transmission cost, thereby diminishing the coding gains from entropy-aware bandwidth adaptation. However, for high-resolution images, the cost of transmitting the hyperprior becomes negligible, allowing NTSCC to benefit from entropy-aware bandwidth adaptation.

## 7.3 Semantic-Aware Bandwidth Adaptation

Unlike entropy-aware bandwidth adaptation that relies on the entropy model, semantic-aware bandwidth adaptation is typically achieved by learning-based bandwidth allocation, which learns the semantic importance distribution through optimizing the specific objective. Semantics that contribute more to the task can be allocated more transmission bandwidth, ensuring promising task performance. This approach is highly task-dependent and more aligned with the goal of semantic communication. Besides, other methods achieve bandwidth adaptation through successive transmission, where the transmission of the source data decomposes into the transmission of multiple semantic streams. The receiver can use any subset of these received semantic streams to reconstruct the image, thereby enabling bandwidth adaptation. The more semantic streams used for reconstruction, the better the reconstruction quality. Unless otherwise stated, we still use image source as an example to introduce these methods.

### 7.3.1 Progressive Semantic Transmission

Instead of traditional transmission methods, a classic progressive transmission scheme, DJSCC-*l* [5] encodes an image into a sequence of semantic streams, which are transmitted sequentially or in parallel to the receiver, enabling flexible transmission. For example, these semantic streams can be transmitted over different frequency bands, and the receiver can tune into any subset of these bands to receive the corresponding semantic streams. The receiver can then use the received semantic streams to reconstruct the image. Alternatively, these semantic streams can be transmitted sequentially over time, allowing the receiver to stop receiving the remaining semantic streams as soon as the image quality meets the desired level, thus saving time and energy.

The above application scenarios correspond to *multiple descriptions problems*, where the compression and transmission of images is decomposed into multiple layers. The goal of DJSCC-*l* is to get the best reconstruction quality, whatever the subset of these semantic streams are used. The *successive refinement problem* is a special case of *multiple descriptions problems*, where the semantic streams are transmitted sequentially, and the receiver can only sequentially use these semantic streams to reconstruct the image. In this case, the first received semantic stream is used to obtain a basic quality reconstruction of the image, and the remaining received semantic streams are sequentially used to refine the previous reconstruction, gradually improving its quality. DJSCC-*l* has made corresponding designs for both cases to achieve bandwidth adaptation.

Firstly, DJSCC-*l* encodes an image $\boldsymbol{s} \in \mathbb{R}^m$ into a sequence of semantic streams, or channel input vectors $\{\boldsymbol{x}_1, \boldsymbol{x}_2, ..., \boldsymbol{x}_k\}$, where $\boldsymbol{x}_i \in \mathbb{C}^{n_i}$ denotes the $i$-th channel input vector. These semantic streams are transmitted over $k$ parallel channels, with each used for transmitting one semantic stream. The $i$-th received latent representation denotes as $\boldsymbol{y}_i \in \mathbb{C}^{n_i}$ at the receiver. For *successive refinement problems*, the receiver uses any subset of these semantic streams corresponding to sequential and consecutive channels, i.e., $\mathcal{Y} = \{\boldsymbol{y}_i\}$ for some $1 \leq i \leq k$ to make a reconstruction $\hat{\boldsymbol{s}}_{\mathcal{Y}} \in \mathbb{R}^m$. For *multiple descriptions problems*, any subset can be used to make a reconstruction without constraints. The channel bandwidth ratio (CBR) for $i$-th channel is represented as $n_i/m$.

DJSCC-*l* has one encoder and multiple decoders to perform progressive transmission. The encoder $f(\cdot; \boldsymbol{\theta})$ parameterized by $\boldsymbol{\theta}$ is responsible for encoding $\mathbf{x}$ into channel input vectors, which can be expressed as:

$$\boldsymbol{x} = f(\boldsymbol{s}; \boldsymbol{\theta}) \in \mathbb{C}^n. \tag{7.37}$$

$\boldsymbol{x}$ denotes the combination of all $\boldsymbol{x}_i$ and $n = \sum_{i=1}^{k} n_i$ is the total bandwidth usage. $\boldsymbol{x}_i$ is transmitted over the $i$-th channel. For each subset $\mathcal{Y}$, a different decoder $g_{\mathcal{Y}}(\cdot; \boldsymbol{\phi})$ is utilized to generate a reconstruction $\hat{\boldsymbol{s}}_{\mathcal{Y}}$:

$$\hat{\boldsymbol{s}}_{\mathcal{Y}} = g_{\mathcal{Y}}(\boldsymbol{y}_{\mathcal{Y}}; \boldsymbol{\phi}), \tag{7.38}$$

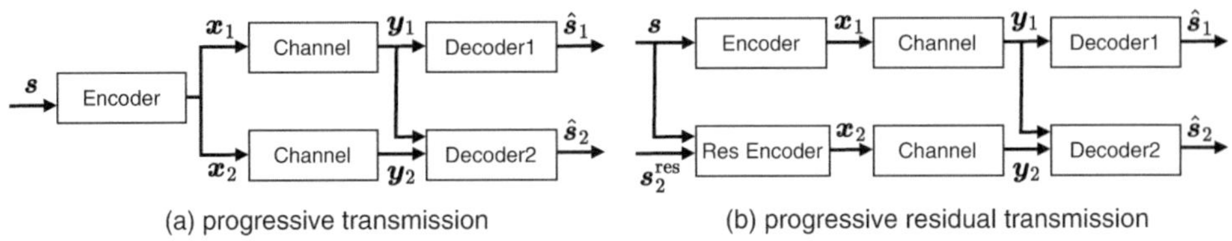


**Fig. 7.5** Two progressive transmission schemes for successive refinement problems. For convenience, $k = 2$ is adopted for example

where $\boldsymbol{y}_{\mathcal{Y}}$ denotes the combination of chosen received channel output vectors corresponding to subset $\mathcal{Y}$.

$\boldsymbol{\theta}$ and $\boldsymbol{\phi}$ are jointly optimized to minimize the average distortion between original image $\boldsymbol{s}$ and a partial reconstructions $\hat{\boldsymbol{s}}_{\mathcal{Y}}$:

$$(\boldsymbol{\theta}^*, \boldsymbol{\phi}^*) = \sum_{\mathcal{Y}} \arg\min_{\boldsymbol{\theta}, \boldsymbol{\phi}, \mathcal{X}} \mathbb{E}_{p(\boldsymbol{s}, \hat{\boldsymbol{s}}_{\mathcal{Y}})}[d(\boldsymbol{s}, \hat{\boldsymbol{s}}_{\mathcal{Y}})], \tag{7.39}$$

where $\boldsymbol{\phi}$ is the combination of all parameters of decoders and $d(\cdot, \cdot)$ is the distortion measurement.

For *successive refinement problems*, there are $k$ decoders for decoding. The $i$-th decoder is used for generate reconstruction $\hat{\boldsymbol{s}}_{\mathcal{Y}}$ based on subset $\mathcal{Y} = \{\boldsymbol{y}_1, ..., \boldsymbol{y}_i\}$. The loss function is defined as the average loss over all possible $\mathcal{Y}$:

$$\mathcal{L} = \frac{1}{k} \sum_{i=1}^{k} d(\boldsymbol{s}, \hat{\boldsymbol{s}}_i), \tag{7.40}$$

where $\hat{\boldsymbol{s}}_i$ denotes the output of $i$-th decoder. For convenience, equal weight is applied for distortions of all decoders. Actually, applying different weights on distortions of different decoders does not affect the overall performance of DJSCC-$l$.

Two feasible progressive transmission schemes when $k = 2$ are shown in Fig. 7.5. For the first transmission scheme illustrated in Fig. 7.5a, the encoder simultaneously generates $\boldsymbol{x}_1$ and $\boldsymbol{x}_2$. The receiver can use $\boldsymbol{x}_1$ to reconstruct image $\hat{\boldsymbol{s}}_1$ with base quality by decoder 1. Furthermore, $\boldsymbol{x}_2$ can be utilized for refinement by decoder 2, further improving the quality of reconstruction.

For the second transmission scheme illustrated in Fig. 7.5b, residual coding and transmission are employed. There is a base encoder to firstly generate $\boldsymbol{x}_1$ from $\boldsymbol{s}$. Then, other semantic streams $\boldsymbol{x}_i (i \neq 1)$ are generated by residual coding. Specifically, $i$-th ($i \neq 1$) encoder use the original image $\boldsymbol{s}$ and an estimation of residual $\boldsymbol{s}_i^{\text{res}}$ to generate $\boldsymbol{x}_i$. $\boldsymbol{x}_i$ can be viewed as the residual information to compensate for the error between the original image $\boldsymbol{s}$ and the reconstructed image $\hat{\boldsymbol{s}}_{i-1}$ based on all previous semantic streams:

$$\boldsymbol{s}_i^{\text{res}} = \boldsymbol{s} - \check{\boldsymbol{s}}_{i-1}, \tag{7.41}$$

where $\check{\boldsymbol{s}}_{i-1}$ is an estimation of $\hat{\boldsymbol{s}}_{i-1}$ based on the statistics of the dataset and the channel at the transmitter, because $\hat{\boldsymbol{s}}_{i-1}$ cannot be obtained exactly without feedback. Actually, $\check{\boldsymbol{s}}_{i-1}$ can be obtained by the transmitter simulating the same operation (wireless transmission and image reconstruction) that should be performed subsequently $T$ times:

$$\check{\boldsymbol{s}}_{i-1} = \frac{1}{T} \sum_{j=1}^{T} \tilde{\boldsymbol{s}}_{i-1}^{j}, \tag{7.42}$$

where $\tilde{\boldsymbol{s}}_{i-1}^{j}$ denotes the $j$-th simulation at the transmitter. Through the progressive residual transmission framework, each semantic stream contains missing information that was not captured by previous semantic streams, and thus can be used at the receiver to refine the reconstructed image progressively.

For *multiple descriptions problems*, DJSCC-$l$ with a single encoder and multiple decoders architecture can also be adopted. Take $k = 2$ for example, as illustrated in Fig. 7.6, there are single encoder and three decoders. Both decoder $01_2$ and $10_2$ use only one received semantic stream to reconstruct image, generating $\hat{\boldsymbol{s}}_{01_2}$ and $\hat{\boldsymbol{s}}_{10_2}$, respectively. Decoder $11_2$ uses both $\boldsymbol{y}_1$ and $\boldsymbol{y}_2$ to reconstruct image, generating $\hat{\boldsymbol{s}}_{11_2}$. The single encoder is used to generate $\boldsymbol{x}$ including multiple semantic streams. Due to the architecture that includes $k$ semantic streams, there are $2^k - 1$ decoders used to generate reconstruction $\hat{\boldsymbol{s}}_{\mathcal{Y}}$ based on any subset $\mathcal{Y}$. In the $k = 2$ case, decoder $10_2$ can be disabled to transform multiple descriptions problems into successive refinement problems, which suggests that successive refinement problems are the special cases of multiple descriptions problems and multiple descriptions problems are a general extension of successive refinement problems. It is evident that the multiple descriptions problems are more challenging, as they allow any combination of unordered semantic streams to reconstruct the image.

To train DJSCC-$l$ for multiple descriptions problems, the loss function in Eq. (7.40) can be modified as:

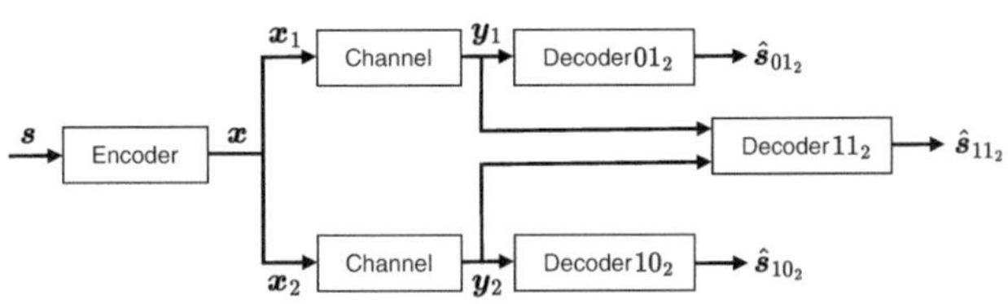


**Fig. 7.6** DeepJSCC-$l$ for multiple descriptions problems. For convenience, $k = 2$ is adopted for example

$$\mathcal{L} = \frac{1}{2^k - 1} \sum_{i=1}^{2^k-1} d(s, \hat{s}_i). \tag{7.43}$$

It should be noted that the performance for multiple descriptions problems is inferior to that for successive refinement problems, which can be understood from the operational mechanism. Due to the unordered nature, each semantic stream generated by the encoder for multiple descriptions problems must be capable of reconstructing the image independently, which leads to considerable information overlap when using multiple semantic streams for reconstruction, resulting in a reduction in total information. In contrast, the first semantic stream for successive refinement problems is used to reconstruct an image with basic quality, while the remaining semantic streams provide additional reconstruction information to refine the image. When multiple semantic streams are used for image reconstruction, the total information is equivalent to the sum of the information carried by each semantic stream, leading to better reconstruction performance.

### 7.3.2 Semantic-Based Bandwidth Allocation

Semantic-based bandwidth allocation is typically a built-in mechanism that dynamically determines the active features for transmission through a learnable policy network, enabling multi-rate communication with a single network. The policy network can both consider the source contents and channel conditions to give a reasonable rate control, which is trained with semantic communication systems in an end-to-end manner.

The architecture of a classic semantic communication system [6] with rate control is shown in Fig. 7.7, where a policy network is introduced to adjust the number of transmission features dynamically. The semantic encoder firstly extract features $f$ from an image $s$. Then, the joint semantic-channel (JSC) encoder adjusts $f$ with the help of channel SNR to generate $(G_s + G_n)$ tokens, each with the channel dimension $d$, which are adaptive to channel conditions. The first $G_s$ tokens, denoted as $\mathbf{F}_s \in \mathbb{R}^{G_s \times d}$, are *selective* tokens, which can be either active or inactive, controlled by the policy network. Another $G_n$ tokens are *non-selective* tokens, denoted as $\mathbf{F}_n \in \mathbb{R}^{G_n \times d}$, which are always active. The policy network learns to generate a mask $\mathbf{M} \in \{0, 1\}^{G_s \times 1}$ conditioned on the channel SNR and image contents, which are used for selecting active features. Specifically, given selective tokens $\mathbf{F}_s$, active tokens are selected by:

$$\mathbf{F}_s \odot \mathbf{M} \rightarrow \mathbf{F}_s. \tag{7.44}$$

Then, the transmitter combines selected $\mathbf{F}_s$ and non-selective $\mathbf{F}_{ns}$ as $\mathbf{F}_{\text{all}} \in \mathbb{R}^{G_{\text{all}} \times d}$ for transmission, where total number of transmitted tokens is given by:

$$G_{\text{all}} = G_n + \sum_{i=1}^{G_s} \mathbf{M}_i \tag{7.45}$$

Each two real values of $\mathbf{F}_{\text{all}}$ are grouped to form one complex symbol, generating channel input vectors $\mathbf{X} \in \mathbb{C}^{G_{\text{all}} \times d/2}$. The received $\mathbf{Y}$ are decomposed into $\hat{\mathbf{F}}$ and processed by JSC decoder and semantic decoder sequentially at the receiver.

However, $\mathbf{M}$ is typically generated by a discrete sampling operation, which is non-differentiable. It hinders the policy network from being involved in the end-to-end training of the system. To overcome the problem, the classical Gumbel-Softmax trick [7] is employed to implement differentiable sampling operations. The policy network first generate a prediction vector $\mathbf{P} \in \mathbb{R}^{G_s \times 2}$, where $\mathbf{P}_i$ represents the active probability of the $i$-th token ($\mathbf{P}_{i,1}$ is active probability and $\mathbf{P}_{i,2}$ is inactive probability).

Given the prediction vector $\mathbf{P}$, we want to draw a one-hot sample $\mathbf{P}_i^{\text{hard}} \in \mathbb{R}^2$ for each token to form the mask. Firstly, the Gumbel-Max trick formulates the sampling process as:

$$\mathbf{P}_i^{\text{hard}} = \text{onehot}(\arg\max_j(\log(\mathbf{P}_{i,j}) + \mathbf{G}_{i,j})), \ j \in [1, 2], \tag{7.46}$$

where all elements of $\mathbf{G} \in \mathbb{R}^{G_s \times 2}$ follow the Gumbel distribution Gumbel$(0, 1)$ and onehot$(\cdot)$ is the one-hot encoding function. $\mathbf{G}_{i,j}$ is given by:

$$\mathbf{G}_{i,j} = -\log(-\log(\mathbf{U}_{i,j})), \tag{7.47}$$

where $\mathbf{U} \in \mathbb{R}^{G_s \times 2}$ consists independent and identically distributed (i.i.d.) samples drawn from Uniform$(0, 1)$. Then, the softmax function with temperature coefficient $\tau$ is utilized

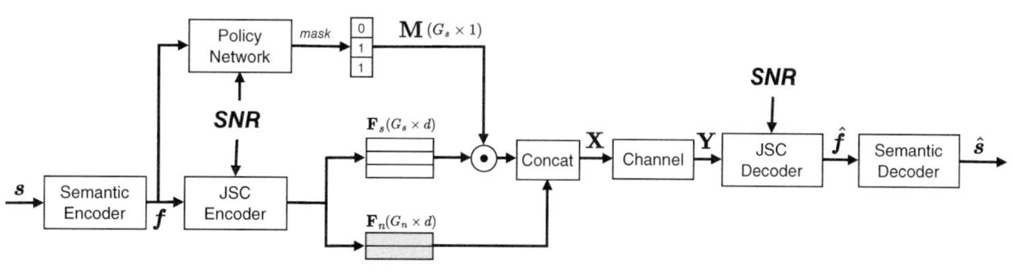


**Fig. 7.7** Semantic communication with rate control

as a continuous, differentiable approximation to arg max $(\cdot)$, obtaining the soft version $\mathbf{P}_i^{\mathrm{soft}} \in \mathbb{R}^2$ of $\mathbf{P}_i^{\mathrm{hard}}$:

$$\mathbf{P}_i^{\mathrm{soft}} = \frac{e^{(\log(\mathbf{D}_{i,j})+\mathbf{G}_{i,j})/\tau}}{\sum_{j=1}^{2} e^{(\log(\mathbf{D}_{i,j})+\mathbf{G}_{i,j})/\tau}}. \tag{7.48}$$

By using the Gumbel-Softmax trick, $\mathbf{P}_i^{\mathrm{hard}}$ is involved in the forward propagation of the network, while during backpropagation, the parameters of the network are updated by computing the gradient of $\mathbf{P}_i^{\mathrm{soft}}$. As the temperature coefficient $\tau$ decreases, the soft version $\mathbf{P}_i^{\mathrm{soft}}$ becomes closer to the hard version $\mathbf{P}_i^{\mathrm{hard}}$, which aligns the forward and backward propagation processes of the network progressively. However, a small $\tau$ can lead to instability of training. Therefore, a large temperature coefficient $\tau$ is typically used at the beginning and is decayed progressively during the training process. Furthermore, the training loss is expressed as:

$$\mathcal{L} = d(\boldsymbol{s}, \hat{\boldsymbol{s}}) + \lambda \sum_{i=1}^{G_s} M_i, \tag{7.49}$$

where $d(\cdot, \cdot)$ is mean squared error.

The quantified results are shown in Fig. 7.8. The comparison methods include:

- Rate-adaptive: DJSCC with rate control.
- Fixed Rate: DJSCC with fixed rate.
- BPG+Capacity: BPG for source coding with capacity-achievable channel.

channel usage per pixel (CPP) is defined as $\frac{G_{\mathrm{all}} \times d}{2H \times W}$, which is marked near corresponding point. For a small $\lambda = 5 \times 10^{-4}$, the rate-adaptive model tends to allocate a relatively high CPP and outperforms "BPG+Capacity" method under low SNRs. For a high $\lambda = 1.5 \times 10^{-3}$, the rate-adaptive model tends to allocate a relatively low CPP and outperforms "BPG+Capacity" method under all SNRs. This demonstrates the effectiveness of rate control, which achieves performance gain especially under low SNRs or low CPPs. Moreover, one rate-adaptive model achieves a similar performance for each comparison point compared with multiple fixed-rate models, which significantly reduces both training and storage costs for models.

A better rate control framework is predictive and adaptive deep coding (PADC) [8], which consists of a variable-length semantic codec, an Oracle Network (OraNet) for predicting the quality of reconstruction given channel SNR and compression ratio (CR), and a CR optimizer to minimize CR given the quality constraints.

Given the original image $\boldsymbol{s} \in \mathbb{R}^m$, semantic encoder $f(\cdot; \boldsymbol{\theta})$ parameterized by $\boldsymbol{\theta}$ encodes it into complex-valued channel input vector $\boldsymbol{x} \in \mathbb{C}^n$ with additional inputs channel SNR $\eta$ and CR $R = n/m$:

$$\boldsymbol{x} = f(\boldsymbol{s}, \eta, R; \boldsymbol{\theta}), \tag{7.50}$$

where $\boldsymbol{x}$ is adaptive to channel SNR and CR. This semantic encoding is divided into two steps. Firstly, latent representation $z \in \mathbb{C}^{n_{\max}}$ with maximal dimension $n_{\max}$ is generated:

$$z = f(\boldsymbol{s}, \eta; \boldsymbol{\theta}), \tag{7.51}$$

where $n_{\max}$ is typically set to $m/2$.

Then, a simper mask $\boldsymbol{m} \in \{0, 1\}^{n_{\max}}$ is employed to transform $\mathbf{z}$ into channel input vector $\boldsymbol{x}$ based on given CR, where only the first $n$ values are selected for transmitting while other values are discarded. The element of $\boldsymbol{m}$ is set as:

$$\boldsymbol{m}_i = \begin{cases} 1, & \text{if } i \le n \\ 0, & \text{otherwise.} \end{cases}, \tag{7.52}$$

where $\boldsymbol{m}_i$ is $i$-th element of $\boldsymbol{m}$. The mask operation is expressed by:

$$\boldsymbol{x} = \boldsymbol{m} \odot z. \tag{7.53}$$

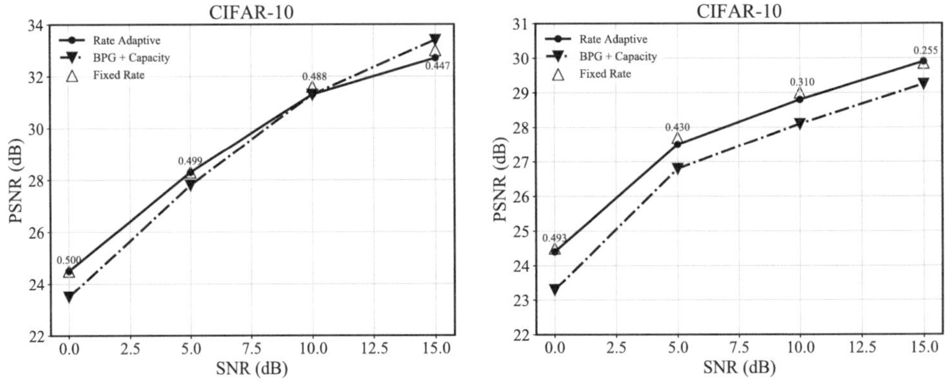


(a) $\lambda = 5 \times 10^{-4}$ (b) $\lambda = 1.5 \times 10^{-3}$

**Fig. 7.8** PSNR performance of different transmission schemes versus SNR over the AWGN channel with the same bandwidth usage

After transmitting over wireless channels, The semantic decoder $g(\cdot;\boldsymbol{\phi})$ parameterized by $\boldsymbol{\phi}$ reconstructs $\hat{s}$ from received channel output vector $\boldsymbol{y}$ with additional inputs channel SNR $\eta$ and CR $R = n/m$:

$$\hat{s} = g(\boldsymbol{y}, \eta, R; \boldsymbol{\phi}). \tag{7.54}$$

The semantic decoding is also divided into two steps. Firstly, $\boldsymbol{y}$ is padded with $n_{\max} - n$ zeros to form $\hat{z}$:

$$\hat{z} = [\boldsymbol{y}, \boldsymbol{0}]^T, \tag{7.55}$$

where $\boldsymbol{0} \in \mathbb{C}^{n_{\max}-n}$ is zero vector. Then, $g(\cdot;\boldsymbol{\phi})$ reconstructs $\hat{s}$ based on $\hat{z}$:

$$\hat{s} = g(\hat{z}, \eta; \boldsymbol{\phi}). \tag{7.56}$$

Given this context, the selection of CR significantly impacts the performance of PADC. Therefore, PADC develops an OraNet and a CR optimizer to select the optimal CR for each transmission. The OraNet is responsible for considering channel SNR, CR, and image contents to predict the quality, e.g., PSNR of the reconstructed image $\hat{\mathbf{x}}$. If the shape of $z$ is $h \times w \times d$ channel-wise mean $\boldsymbol{\mu}_z$ and channel-wise standard deviation $\boldsymbol{\sigma}_z$ of $z$ are used to represent image content features for prediction, whose shapes are $h \times w$. Let $o(\cdot;\boldsymbol{\omega})$ parameterized by $\boldsymbol{\omega}$ be the OraNet, the PSNR prediction of the reconstructed image is given by:

$$\text{PSNR} = o(\boldsymbol{\mu}_z, \boldsymbol{\sigma}_z, \eta, R; \boldsymbol{\omega}) \tag{7.57}$$

If $\boldsymbol{\mu}_z$, $\boldsymbol{\sigma}_z$, and $\eta$ are all fixed, $o(\cdot;\boldsymbol{\omega})$ is a monotonically increasing function to $R$. Then, the CR optimizer divide CR range into multiple levels ($R_1, R_2, ..., R_L$) and selects the minimal CR that satisfies the PSNR constraint, which is the optimal CR for the transmission.

The training of PADC follows as:

1. Train variable-length semantic codec $f(\cdot;\boldsymbol{\theta})$ and $g(\cdot;\boldsymbol{\phi})$. For each training sample, channel SNR $\eta$ and CR $R$ are randomly generated.
2. Train OraNet $o(\cdot;\boldsymbol{\omega})$. The objective is the mean squared error between the predicted PSNR and the actual PSNR.

## 7.4 Summary

This chapter addresses the operational inefficiencies of semantic communication systems when faced with fluctuating bandwidth availability, where reliance on multiple fixed-bandwidth models typically results in considerable storage demands and switching delays. To ensure efficient and reliable communication across diverse bandwidth scenarios, the development of Unified bandwidth-adaptive systems is crucial. The chapter details primary adaptive strategies, including entropy-aware bandwidth adaptation, which uses entropy as a criterion to allocate more transmission symbols to features with richer information content, and semantic-aware bandwidth adaptation, a learning-based method that implicitly maps semantic importance to the number of symbols for a specific task. These approaches aim to enhance transmission flexibility and performance, approaching that of specialized systems without their associated overhead.

## Problems

7.1 Which approach avoids training and storing multiple models for different bandwidths?

A. Fixed-bandwidth coding.
B. Separate source and channel coding.
C. A unified bandwidth-adaptive codec trained across bandwidths.

7.2 Which of the following is NOT a reason for adjusting bandwidth in a semantic communication system?

A. Dynamic channel conditions.
B. Reducing security risks.
C. Variations in semantic importance of data.

7.3 How does bandwidth adaptation improve communication efficiency?

A. By using a fixed-bandwidth allocation method.
B. By compressing all transmitted data equally.
C. By dynamically adjusting bandwidth based on entropy or semantic importance.

7.4 In entropy-aware bandwidth adaptation, what does higher entropy of a feature typically indicate?

A. The feature is less important and can be discarded.
B. The feature contains richer information and requires more transmission symbols.
C. The feature has already been encoded.

7.5 Which technique is used to estimate the entropy of latent representations in entropy-aware bandwidth adaptation?

A. Learnable entropy models with hyperpriors.
B. Static lookup tables.
C. Max pooling.

7.6 In entropy-aware bandwidth adaptation, what is the purpose of the hyperprior vector $z$?

A. To provide side information that improves entropy estimation.
B. To eliminate the need for quantization.
C. To replace the latent representation.

7.7 Which component adapts the latent embeddings to different transmission rates in NTSCC?

A. Quantizer only.
B. Lightweight fully connected (FC) layers informed by rate tokens.
C. Activation function layers.

7.8 What does the Lagrange multiplier $\lambda$ control during training of bandwidth-adaptive semantic systems?

A. Noise levels in the channel.
B. Batch size of training.
C. Trade-off between distortion and bandwidth usage.

7.9 How does the system determine which semantic features are important in semantic-aware bandwidth adaptation?

A. Using a Gaussian noise model in the decoder.
B. By training a neural network end-to-end to learn task-dependent feature relevance.
C. By measuring the length of input vectors.

7.10 Why is the Gumbel-Softmax trick used in semantic-aware bandwidth adaptation?

A. To enable differentiable selection of active transmission features during training.
B. To normalize the entropy model.
C. To generate uniform noise for quantization.

## References

1. J. Xu, T.-Y. Tung, B. Ai, W. Chen, Y. Sun, D. Gündüz, Deep joint source-channel coding for semantic communications. IEEE Commun. Mag. **61**(11), 42–48 (2023)
2. J. Dai, S. Wang, K. Tan, Z. Si, X. Qin, K. Niu, P. Zhang, Nonlinear transform source-channel coding for semantic communications. IEEE J. Sel. Areas Commun. **40**(8), 2300–2316 (2022)
3. J. Ballé, D. Minnen, S. Singh, S.J. Hwang, N. Johnston, Variational image compression with a scale hyperprior (2018)
4. S. Wang, J. Dai, Z. Liang, K. Niu, Z. Si, C. Dong, X. Qin, P. Zhang, Wireless deep video semantic transmission. IEEE J. Sel. Areas Commun. **41**(1), 214–229 (2023)
5. D.B. Kurka, D. Gündüz, Bandwidth-agile image transmission with deep joint source-channel coding. IEEE Trans. Wireless Commun. **20**(12), 8081–8095 (2021)
6. M. Yang, H.-S. Kim, Deep joint source-channel coding for wireless image transmission with adaptive rate control, in (2022), pp. 5193–5197
7. E. Jang, S. Gu, B. Poole, Categorical reparameterization with gumbel-softmax (2017)
8. W. Zhang, H. Zhang, H. Ma, H. Shao, N. Wang, V.C.M. Leung, Predictive and adaptive deep coding for wireless image transmission in semantic communication. IEEE Trans. Wireless Commun. **22**(8), 5486–5501 (2023)

# Semantic Communication with Privacy Preservation

# 8

**Abstract**

Semantic communication, while driving technological innovation, also faces significant privacy and security challenges, especially in the joint semantic channel coding framework. During data sharing, the leakage of sensitive information can expose user privacy, creating potential security risks. This chapter will explore the network attack patterns that could lead to privacy breaches in semantic communication systems, as well as common countermeasures. Privacy protection methods tailored to different tasks and encryption strategies for private information in various scenarios are introduced.

**Keywords**

Privacy protection • Encryption strategies • Semantic communication • Joint semantic-channel coding

## 8.1 Introduction

The rapid advancement of wireless communication technologies has deeply integrated smart devices and networks into daily life, enabling seamless interaction for personal work, and other activities. Devices like smart appliances, autonomous vehicles, and wearable technologies enhance convenience, but also collect vast amounts of sensitive data, such as user behavior, location, and health information. While these data facilitate service quality, improper protection can lead to privacy breaches. Unlike traditional syntax-based models, semantic communication focuses on transmitting task-relevant semantic information rather than raw data, improving transmission efficiency, reducing latency, and enhancing network reliability.

However, along with the technological innovations brought about by semantic communication, significant privacy and security challenges arise. The implementation of semantic communication relies heavily on large-scale data training, and these training data are often distributed between different users and institutions. To construct and optimize semantic communication neural networks, participants must share their respective knowledge bases and data. Although data sharing can significantly enhance the model's representational power and generalization ability, it also introduces the risk of privacy leakage. In particular during the data sharing process, the exposure of sensitive information can lead to the disclosure of user privacy, forming potential security risks.

Therefore, protecting privacy in semantic communication is of paramount importance. During model training and data sharing, adversaries or unauthorized third parties may employ model inversion techniques or other data recovery methods to recover the original content of transmitted data or extract sensitive privacy information. Specifically, the sender often possesses private information that they do not wish to be known by the receiver or other participants, and the leakage of this information may occur through excessive inferences made by the receiver based on the transmitted semantic features, resulting in the exposure of private data. In this context, privacy protection technologies must not only safeguard the sender's personal privacy but also ensure that the data shared in public knowledge bases does not leak the private information of its owners.

With the increasing awareness of data privacy protection and the growing stringency of data compliance requirements across various nations, privacy protection in semantic communication has become a focal point of research in both academia and industry. Designing efficient and secure privacy protection mechanisms has become a critical issue in promoting the widespread application of semantic communication technologies. In response to these challenges, innovative privacy protection technologies and strategies require in-depth investigation to ensure the security and reliability of semantic communication in practical applications.


W. Chen and Z. Qin, *Semantic Communications*, Textbooks in Telecommunication Engineering,
https://doi.org/10.1007/978-3-032-11005-3_8

The privacy and security risks in semantic communication are complex, stemming from the high correlation and sharing mechanisms of data during extraction, transmission, and decoding stages [1]. Attackers can exploit these risks to infer or recover sensitive information. During repeated transmission, sniffers may intercept data and use public background knowledge to deduce the sender's private information. Simultaneously, the receiver may over-recover the transmitted data using shared and private knowledge bases, exacerbating privacy leakage. As the sender's and receiver's private knowledge bases are independent and unknowable, the sender cannot accurately estimate the receiver's background knowledge, potentially leading to malicious inferences of unexposed information. Additionally, semantic communication faces privacy threats from model and data vulnerabilities, including attribute inference attacks, membership inference attacks, and data reconstruction attacks, all of which can reveal sensitive user data such as demographics, health status, and personal preferences.

The following will elaborate on the typical privacy leakage attack patterns and corresponding defense mechanisms that a semantic communication system may face, given in Fig. 8.1.

In the context of semantic communication, privacy leakage risks emerge due to the nature of information transmission and the interaction between the sender, receiver, and external environments. Several types of privacy leakage risks have been identified, including Property Inference Attacks, Membership Inference Attacks, Feature Reconstruction (FR) Attacks, and Imitation Learning Attacks. These risks highlight the vulnerabilities inherent in the process of transmitting semantically encoded information, especially when communication occurs over untrusted channels or involves multiple parties with potentially differing levels of access to background knowledge.

- **Property Inference Attack**: Property Inference Attack aims to infer whether a given instance in the training data or model possesses a certain specific attribute. In semantic communication, such attacks can be used to infer sensitive information about the source, such as demographic data, health status, financial condition, and personal preferences. For example, attackers can leverage background knowledge from knowledge graphs or social networks to infer privacy information from anonymized data. Experiments based on real social network data show that attackers can de-anonymize social networks using knowledge graphs, revealing individuals' private data. In semantic communication, attackers may use public background knowledge and metadata obtained during transmission to infer privacy attributes that the sender wishes to keep confidential.
- **Membership Inference Attack**: Membership Inference Attack aims to determine whether a specific piece of data is part of a dataset by analyzing how the model responds to known and unknown data. This can lead to the leakage of sensitive information, especially when the data involves personal privacy. In semantic communication, attackers may exploit the receiver's background knowledge and the transmitted semantic features, which could inadvertently expose the sender's privacy. For example, by accessing the semantic feature vector or the model's response, an attacker may infer if a particular data record was part of the sender's communication, revealing private information.
- **Feature Reconstruction Attack**: Feature Reconstruction Attack involves attackers accessing intermediate features of a deep neural network (such as gradients and activation values), and using optimization algorithms or inference techniques to reverse-engineer a close approximation of the original input data. In semantic communication, attackers may recover the original data by intercepting or observing intermediate features in the semantic encoding and decoding process. This type of attack is particularly effective when there is a strong correlation between feature representations and the original data, such as in images or text data. Even if the data has been transformed, the close relationship between intermediate features and original data in deep learning models may expose sensitive information. This poses a significant challenge to privacy protection mechanisms, as it undermines the security of encoded semantic features during transmission.

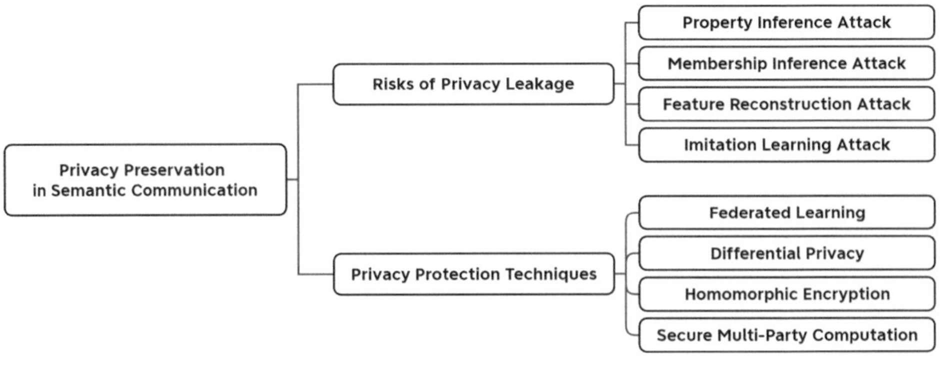


**Fig. 8.1** Privacy preservation in semantic communication schematic diagram

- **Imitation Learning Attack**: Imitation Learning Attack occurs when attackers observe the input and output data of an encrypted network or semantic encoding-decoding system and train a mimic model to replicate its behavior. By learning the input-output mapping, attackers can reproduce the decryption or decoding process without knowing the original encryption algorithm or key. In semantic communication, this allows attackers to reveal the sender's original information, especially in applications involving sensitive data, such as medical records or financial information.

In summary, the privacy and security issues in semantic communication primarily stem from the data sharing mechanisms, the asymmetry of knowledge bases, and the leakage of privacy features in deep learning models.

With the continuous development of semantic communication technology, ensuring the privacy and security of participants while maintaining efficient communication has become a critical issue that needs to be addressed. To this end, researchers have proposed various privacy protection techniques aimed at mitigating the risks of privacy leakage during data sharing and collaborative learning. These techniques not only enable effective data analysis and model training without revealing sensitive data but also effectively prevent privacy leakage [2]. Below are several privacy protection techniques widely applied in semantic communication.

- **Federated Learning**: Federated learning is a technique that allows collaborative model training in a distributed environment, where each participant trains the model on local data without sharing the raw data [3]. Through this approach, participants can share local model parameters while protecting data privacy and aggregate these parameters to construct a global model. Since no raw data is transmitted during the process, federated learning significantly reduces the risk of data leakage. While federated learning promotes collaborative learning and privacy protection, it still faces the risk of gradient leakage. To enhance its security, private models or other encryption methods must be introduced.
- **Differential Privacy**: Differential privacy is a method designed to protect individual privacy when analyzing data. It works by ensuring that the inclusion or exclusion of any single person's data does not significantly affect the overall result of an analysis. This is achieved by adding random noise to the data, which makes it difficult for attackers to determine whether a particular individual's information was used, even if they have access to the analysis results. The core idea is to guarantee that the privacy of each individual is preserved, regardless of what others know about the data. In semantic communication, differential privacy can be applied by introducing noise into the semantic features during data transmission. This makes it harder for attackers to extract sensitive information from the transmitted data.
- **Homomorphic Encryption**: Homomorphic encryption is a technique that allows computation to be performed directly on encrypted data. With homomorphic encryption, users can perform various computations on encrypted data without first decrypting it. This provides a high level of security for data privacy, as the encrypted data remains confidential throughout the computation process. In semantic communication, homomorphic encryption can be applied to protect sensitive semantic features. By encrypting information during semantic data transmission, participants can share encrypted semantic features without exposing the raw data. However, full homomorphic encryption is computationally complex and inefficient, so in low-latency semantic communication tasks, partial homomorphic encryption may be a more appropriate choice. Although partial homomorphic encryption supports only specific types of computation, it can protect semantic information privacy with higher efficiency.
- **Secure Multi-Party Computation**: Secure Multi-Party Computation (MPC) is a privacy-preserving collaborative computation method that allows multiple parties to jointly compute a result, but each participant only learns a portion of the result and cannot access the input data of other participants. In semantic communication, participants can use MPC for collaborative training and inference while ensuring that their knowledge base data remains confidential. This way, participants can share their model parameters or feature data, collectively enhancing the performance of the semantic communication model without exposing their private information. MPC provides a robust solution for privacy protection, particularly in complex scenarios that require multiple participants to collaborate.

With the ongoing advancement and refinement of privacy protection technologies, semantic communication offers more effective and intelligent solutions across various industries, all while maintaining data security. How to enhance communication efficiency while ensuring privacy protection remains an important direction for future research.

## 8.2 Privacy-Enhanced Joint Source and Channel Coding

In separate source/semantic coding and channel coding systems, there are two common strategies to protect source information:

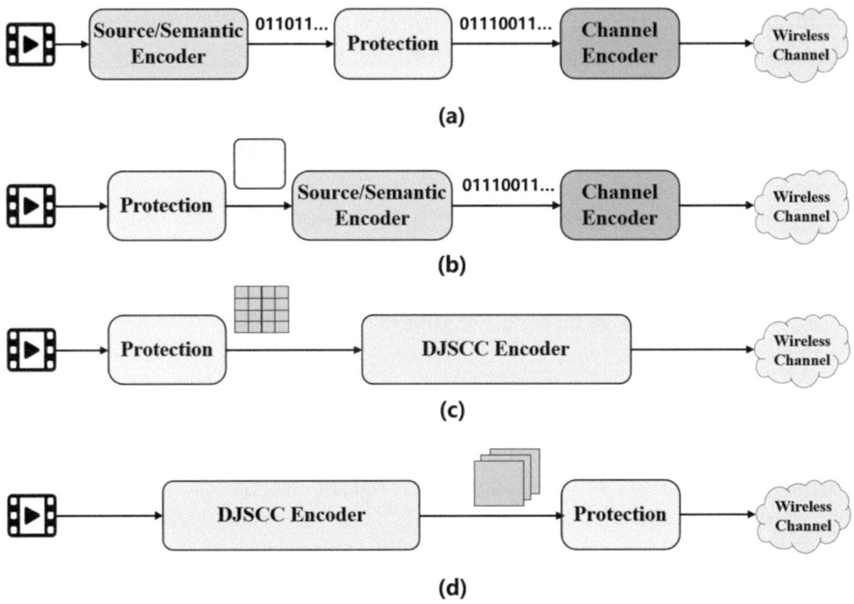


**Fig. 8.2** Encryption strategies in source/semantic coding and channel coding

- **Strategy 1: Source/Semantic Encoding Followed by Encryption** as shown in Fig. 8.2a, the source/semantic encoding followed by encryption strategy first compresses the source data into binary form using source/semantic encoding, and then applies a bit-level encryption algorithm to the compressed data. Common encryption algorithms include DES, AES, and RSA. The advantage of this method is that it effectively reduces the amount of data that need to be encrypted, thereby improving encryption efficiency.
- **Strategy 2: Encryption Followed by Source/Semantic Encoding** Fig. 8.2b shows the encryption followed by source/semantic encoding strategy. In this approach, the source data is first protected by an encryption algorithm, and then the encrypted data is transmitted and processed through the separate source/semantic-channel coding system. In this mode, the source owner mainly focuses on protecting the privacy of the source information, while the network side focuses on improving the channel utilization. Common encryption algorithms include the Arnold encryption algorithm, Hill encryption algorithm, and encryption methods based on three-dimensional chaotic mappings.

Unlike the separate source/semantic-channel coding architecture in traditional syntax communication systems, semantic communication systems use a source/semantic-channel joint coding architecture. In this architecture, the source is directly mapped into semantic feature symbols at the input of the wireless channel through a semantic encoder. This design not only simplifies the network protocol stack but also establishes a direct connection between the application layer and the physical layer, avoiding the privacy leakage risks associated with multi-layer transmission in traditional syntax communication systems.

However, during the transmission of the source through the source/semantic-channel encoder, the transmission path may involve untrusted wired networks (such as the Internet) and public physical channels between wireless transmission servers. This makes the source information vulnerable to leakage or theft during transmission over the wired network, and the semantic feature information transmitted over the physical channel may also be intercepted and analyzed by eavesdroppers.

To address this issue, the joint source/semantic-channel coding system has two encryption strategies to protect privacy information:

- **Strategy 1: Encrypt the Source First, then Perform Semantic Encoding** shown in Fig. 8.2c, this strategy applies encryption technology before DJSCC (Deep Learning Assisted Joint Source-Channel Coding) from the user's perspective. The user can freely manipulate the source signal at the application layer and encrypt the source information before providing it to the wireless service provider, thus preventing the source information from being leaked or stolen during transmission over the wired network.
- **Strategy 2: Extract Semantic Features First, then Encrypt** as shown in Fig. 8.2d, this strategy is implemented by the wireless service provider, applying encryption technology after DJSCC. In this process, DJSCC extracts task-relevant and channel-robust features, and these semantic features transmitted into the channel

are encrypted. This not only preserves the advantages of semantic communication but also prevents eavesdroppers from intercepting users' private information.

### 8.2.1 DJSCC with Encryption Mechanisms

In the DJSCC task, the key objective is to achieve an end-to-end reconstruction task. In this process, the sender and receiver deploy joint source-channel encoders and decoders, respectively, mapping the source signal to transmission feature symbols in the complex domain and decoding to recover the reconstructed source information.

Considering the privacy leakage risks faced by semantic communication systems based on joint source-channel coding, especially before the source information is transmitted to the joint source-channel encoder, a "encrypt-then-compress" strategy can be adopted for reconstruction tasks using DJSCC. This method involves encrypting the source information for protection, followed by encoding through a DJSCC-based semantic encoder. This approach ensures the semantic communication system architecture while providing encryption protection for the entire source information.

Taking image sources as an example, traditional syntax-based communication systems commonly use the "encrypt-then-compress" strategy. Common encryption algorithms include the Arnold transform, Hill cipher, and encryption algorithms based on three-dimensional chaotic mappings. However, these encryption algorithms alter the structural information of the source. If applied directly to semantic communication systems, they may affect the performance of joint source-channel encoding in the encrypted domain.

Therefore, for semantic communication systems targeting reconstruction tasks, it is essential to consider the characteristics of joint source-channel coding. An appropriate source protection and de-protection mechanism should be designed to hide the privacy information of the original source while enabling the extraction of beneficial features from the protected source for subsequent transmission via deep joint source-channel encoding. This process must ensure that the source is effectively protected from leakage or theft during transmission through an unreliable wired network, without compromising the performance of the semantic communication system.

To effectively implement privacy protection in joint source-channel coding systems, protection and de-protection modules can be appropriately deployed before and after the joint source-channel encoders and decoders at both the sender and receiver ends. In the DJSCC approach for reconstruction tasks, the design of the protection and de-protection modules must follow two principles: First, the *information scrambling principle*: at the sender, the source is processed by the protection module before being input to the DJSCC encoder. All information should be scrambled, making it unreadable to eavesdroppers. Similarly, at the receiver, the decoded information should also be scrambled before passing through the de-protection module. Second, the *information recovery principle*: even after the source undergoes protection and de-protection, the information reconstructed at the receiver should still align with the advantages of the semantic communication system, ensuring the system's effectiveness and performance.

As shown in Fig. 8.3, the model for DL-based joint protection and source-channel coding (DJPSCC) describes the transmission process of data between the sender and receiver. The sender, Alice, transmits the source signal $\boldsymbol{s}$, and the protection module transforms the source information into protected information $\boldsymbol{x}$, expressed as

$$\boldsymbol{x} = e_{\text{encoder}}(\boldsymbol{s}), \tag{8.1}$$

where $e_{\text{encoder}}(\cdot)$ represents the protection function, and $\boldsymbol{x}$ is the protected source information. The protected source $\boldsymbol{x}$ has the same size as the original source $\boldsymbol{s}$. After the source signal is protected, the joint source-channel encoder encodes the protected signal $\boldsymbol{x}$, with the encoding process expressed as

$$z = f_{\text{encoder}}(\boldsymbol{x}), \tag{8.2}$$

where $z$ is the input signal for the channel. After transmission through the wireless channel, the receiver receives the channel output signal $\hat{z}$. The joint source-channel decoder decodes the channel output signal $\hat{z}$ as

$$\boldsymbol{y} = f_{\text{decoder}}(\hat{z}), \tag{8.3}$$

where $f_{\text{decoder}}(\cdot)$ is the joint source-channel decoding function, and $\boldsymbol{y}$ is the recovered protected source, which has the same size as the protected signal $\boldsymbol{x}$. The de-protection module, similar to the protection module, transforms the decoded

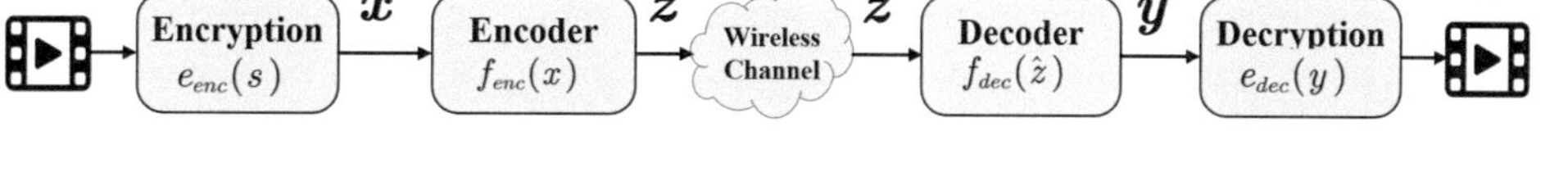


**Fig. 8.3** The architecture of DJPSCC strategy

protected signal $\hat{y}$ back into the deprotected signal $\hat{s}$, with the de-protection process expressed as

$$\hat{s} = e_{\text{decoder}}(\boldsymbol{y}), \tag{8.4}$$

where $e_{\text{decoder}}(\cdot)$ represents the de-protection function, and $\hat{s}$ is the reconstructed original source.

The optimization task of this scheme is to minimize the distortion between the transmitted source and the reconstructed source at the receiver, while ensuring that the correlation between the source information after passing through the protection module and the original source information, as well as the correlation between the encrypted information and the deprotected information, is as small as possible [4]. During the training phase of the protection module and the joint source-channel encoder, the feature extraction module $h_\psi$ extracts features from the source signals information $\boldsymbol{s}$, the ciphertext information $\boldsymbol{x}$, and the decoded information $\boldsymbol{y}$. The feature loss between the source signals $\boldsymbol{s}$ and the ciphertext $\boldsymbol{x}$ is denoted as

$$L_e = \|h_\psi(\boldsymbol{s}) - h_\psi(\boldsymbol{x})\|^2, \tag{8.5}$$

and the feature loss between the plaintext $\boldsymbol{s}$ and the decoded information $\boldsymbol{y}$ is denoted as

$$L_d = \|h_\psi(\boldsymbol{s}) - h_\psi(\boldsymbol{y})\|^2, \tag{8.6}$$

where $h_\psi(\boldsymbol{s})$, $h_\psi(\boldsymbol{x})$, and $h_\psi(\boldsymbol{y})$ represent the features of the source signals, ciphertext, and decoded information extracted by the feature extraction module, respectively.

Once the feature extraction module is selected, its parameters are fixed during the training phase of the DJSCC system and do not update during training. The reconstruction loss $L_r$ between the source signals $\boldsymbol{s}$ and the reconstructed signal $\hat{s}$ is expressed as

$$L_r = d(\boldsymbol{s}, \hat{\boldsymbol{s}}) = \|\boldsymbol{s} - \hat{\boldsymbol{s}}\|^2, \tag{8.7}$$

where $\boldsymbol{s}$ and $\hat{s}$ represent the source signals and reconstructed signals, respectively.

For the encryption and decryption modules, the goal is to maximize $L_e$ and $L_d$ to obscure the visual content of the encrypted image $\boldsymbol{x}$ and the decoded image $\boldsymbol{y}$. The total loss function used for training the DJSCC network is expressed as

$$L_{\text{total}} = L_r - \lambda_e L_e - \lambda_d L_d, \tag{8.8}$$

where $\lambda_e$ and $\lambda_d$ are the weights of $L_e$ and $L_d$, and $\lambda_e, \lambda_d \in \mathbb{R}^+$.

By minimizing the total loss function, the parameters of the encryption module and the deep joint source-channel encoder-decoder are optimized. Once the network training is complete, the protection module $e_{\text{encoder}}$ and de-protection module $e_{\text{decoder}}$ are distributed to the sender and receiver through secure protocols, such as the Secure Socket Layer (SSL), for secure transmission in the specific DJSCC network.

As an example of image source signals, the performance of the DJPSCC scheme with a protection module is verified through simulation experiments. In this experiment, the DJPSCC adopts a DJSCC architecture based on Convolutional Neural Networks (CNNs) for image transmission. The architecture includes normalization layers, five alternating convolutional layers with PReLU activation functions, reshaping layers, and power normalization layers.

The protection and de-protection modules are constructed using U-Net networks. The shallow U-Net network architecture includes a shortcut connection. The left path of UNet-S1 is the contracting path, responsible for compressing the image size. It consists of one normalization layer, two convolutional layers, one max-pooling layer, and two convolutional layers. The right path is the expanding path, which is used to expand the feature size. It includes one transposed convolutional layer, one concatenation layer, three convolutional layers, and one de-normalization layer. In addition to the serial connection between the contracting and expanding paths, the output of the second convolutional layer in the contracting path is concatenated with the output of the transposed convolutional layer in the expanding path, and the combined output is then fed into the first convolutional layer of the expanding path.

In this experiment, the U-Net-S1 network is used as the protection and de-protection module network, and the DJPSCC method is referred to as DJPSCC-UNet-S1. During the training phase, the feature extraction network of DJPSCC-UNet-S1 adopts the pre-trained VGG16 network on the ImageNet dataset. The number of training epochs is set to 500, with a batch size of 64, and the Adam optimizer with an initial learning rate of $10^{-3}$ is used. When learning stagnates for 10 epochs, the learning rate of the Adam optimizer is reduced to $10^{-4}$.

In the evaluation phase, the CIFAR-10 test set is used, and the performance of the DJPSCC-UNet-S1 network is evaluated at specific test SNR values $SNR_{\text{test}} \in [0, 20]$ dB. To mitigate the randomness introduced by the wireless channel, each image in the CIFAR-10 test set is transmitted 10 times. Peak Signal-to-Noise Ratio (PSNR) is used to assess the reconstruction performance between the original image and the de-protected image. For simplicity, the losses for visual protection images (including protected and decoded images) are assigned the same weight, i.e., $\lambda_e = \lambda_d = \lambda$.

Figure 8.4 compares the image reconstruction performance of the DJPSCC-UNet-S1 network under different loss weights $\lambda$ (specifically 0, $10^{-6}$, $10^{-5}$) when the bandwidth $R = 1/6$ (the ratio of the number of channels in the last convolutional layer to the first convolutional layer). For reference, the reconstruction performance of the DJSCC network, which does not use the protection and de-protection modules, is also pre-

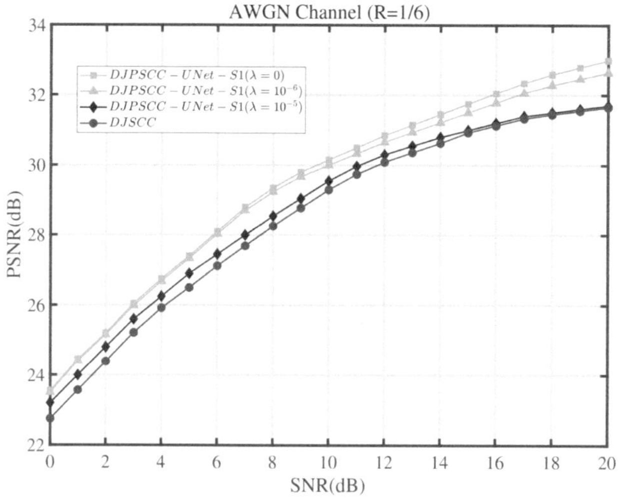


**Fig. 8.4** Performance of DJPSCC strategy

sented in Fig. 8.3. As the loss weight $\lambda$ increases, the DJPSCC-UNet-S1 network places more emphasis on the image protection task, which leads to a decrease in the image reconstruction performance. Since the DJPSCC-UNet-S1 network is deeper and has more parameters than the DJSCC network, within the range of $SNR_{\text{test}} \in [0, 20]$ dB, the DJPSCC-UNet-S1 (with $\lambda = 0$, no feature loss constraint) outperforms the DJSCC network by approximately 2 dB in terms of reconstruction performance. Even when the feature loss weight is increased during training to prioritize image protection, the DJPSCC-UNet-S1 (with $\lambda = 10^{-5}$) still demonstrates superior image reconstruction performance compared to the DJSCC network.

In order to prevent eavesdroppers from intercepting sensitive private information through eavesdropping signals and semantic decoders, an end-to-end semantic communication system aims to optimize the joint source-channel coding system and its associated privacy module. This system minimizes the distortion between the transmitted source and the reconstructed source at the target user while simultaneously minimizing the correlation between private sensitive information and the information inferred by the eavesdropper. This approach significantly enhances the confidentiality of user private data in semantic communication. However, such methods may result in excessive distortion in the information decoded by the eavesdropper, which could arouse suspicion and prompt them to initiate jamming attacks that could disrupt the communication process.

To address this, steganographic techniques can be utilized to embed sensitive information within publicly available data, misleading potential eavesdroppers. Specifically, steganography works by generating a container signal that can hide sensitive content while preserving visual features similar to the private information itself [5]. This effectively "parasitizes" the sensitive data within the container information. The sender then transmits the container image over a physical channel. Although the eavesdropper can intercept and decode the container signal through the eavesdropping channel, they cannot decipher the hidden private information within it, thus preventing the leakage of sensitive data while simultaneously misleading the eavesdropper.

The semantic communication model based on steganographic techniques is depicted in Fig. 8.5. In this model, both the sender Alice and the target user Bob deploy a pre-trained joint source-channel encoder and decoder at their respective ends. The system introduces a signal steganographic module at both ends—Alice and Bob—where it embeds the private information into the host signal. The private information, denoted as $\boldsymbol{s}_p$, is to be protected, while the host information used to conceal the private data is represented by $\boldsymbol{s}_h$. These two types of data are input into the joint source-channel encoder, resulting in corresponding channel input signals $\boldsymbol{x}_h$ and $\boldsymbol{x}_p$. The mathematical representation is given by

$$\boldsymbol{x}_h = E(\boldsymbol{s}_h), \quad \boldsymbol{x}_p = E(\boldsymbol{s}_p). \tag{8.9}$$

Here, $E(\cdot)$ denotes the encoding process. These semantic feature signals are then processed by the steganographic module to generate the channel input signal for the container image, denoted as $\boldsymbol{x}_c$. The mathematical expression for this transformation is

$$\boldsymbol{x}_c, l = S(\boldsymbol{x}_h, \boldsymbol{x}_p), \tag{8.10}$$

where $S(\cdot)$ represents the steganographic module based on a Deep Neural Network (DNN), and $l$ indicates the information lost during the steganographic operation. The signal $\boldsymbol{x}_c$ is transmitted over the communication channel. The signals received by Bob and the eavesdropper Eve are denoted as $\boldsymbol{y}_c$ and $\boldsymbol{y}_p$, respectively.

At Bob's end, a reverse operation of the steganographic module, also based on DNN, is applied to reconstruct the original channel input signals. The corresponding mathematical expression is

$$\boldsymbol{y}_h, \boldsymbol{y}_p = S^{-1}(\boldsymbol{y}_c, \overline{\boldsymbol{c}}), \tag{8.11}$$

where $S^{-1}(\cdot)$ denotes the inverse steganographic operation, and $\overline{\boldsymbol{c}}$ represents the missing information, typically predefined as a constant value or sampled from a given distribution. Subsequently, Bob can use the source-channel decoder to reconstruct both the host information $\hat{\boldsymbol{s}_h}$ and the private information $\hat{\boldsymbol{s}_p}$.

In contrast, the eavesdropper Eve can only attempt to decode the received signal $\boldsymbol{y}_p$. Even if Eve uses the same parameters for the source-channel decoder, she will only be able to decode the host information and will not be able to uncover the hidden private information.

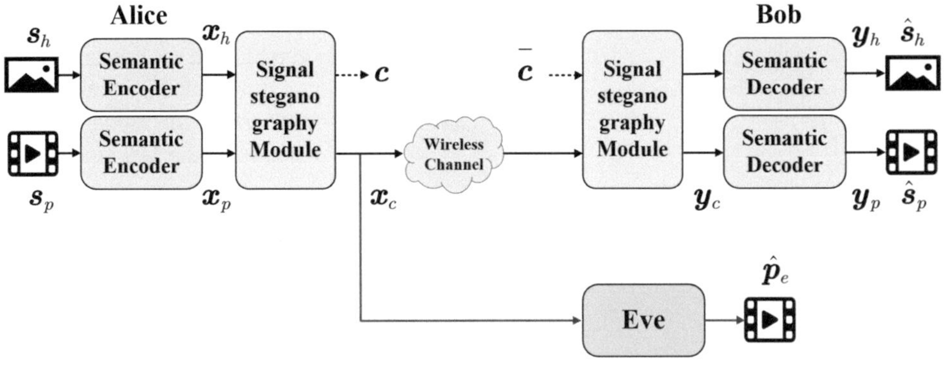


**Fig. 8.5** Strategy for semantic information with signal steganography

The optimization goal for the privacy protection mechanism is to ensure that Bob can successfully reconstruct the original channel input signals $\boldsymbol{x}_h$ and $\boldsymbol{x}_p$ through the reverse process of the steganographic module. The first objective is to minimize the semantic feature discrepancy between the public information $\boldsymbol{x}_c$ and $\boldsymbol{x}_h$ after passing through the steganographic module, while hiding the private information $\boldsymbol{x}_p$ as much as possible. Since the lost information $l$ is not transmitted to the target user Bob, it is also necessary to minimize the difference between $l$ and the estimated lost information $\hat{l}$. Consequently, the loss function for the forward process of the steganographic module can be expressed as

$$L_{\text{forward}} = \lambda_1 \|\boldsymbol{x}_c - \boldsymbol{x}_h\|^2 + \lambda_2 \|l - \hat{l}\|^2. \tag{8.12}$$

Here, $\lambda_1 > 0$ and $\lambda_2 > 0$ are hyper-parameters used to balance these two terms.

At the target user's end, the goal is to reconstruct the steganographic module's input signals $\boldsymbol{x}_p$ and $\boldsymbol{x}_h$. Therefore, the loss function for the reverse process is

$$L_{\text{backward}} = \lambda_3 \|\boldsymbol{x}_p - \boldsymbol{y}_p\|_2^2 + \lambda_4 \|\boldsymbol{x}_h - \boldsymbol{y}_h\|_2^2, \tag{8.13}$$

where $\lambda_3 > 0$ and $\lambda_4 > 0$ are hyper-parameters used to balance these two terms.

In addition, to ensure that the target user Bob successfully reconstructs the private image, a loss term is included to minimize the difference between $\boldsymbol{s}_p$ and $\hat{\boldsymbol{s}}_p$, representing the private information. This term is given by

$$L_{\text{privacy}} = \lambda_5 \|\boldsymbol{s}_p - \hat{\boldsymbol{s}}_p\|_2^2, \tag{8.14}$$

where $\lambda_5 > 0$ controls the weight of this privacy loss term.

Finally, the total loss function for the proposed method is

$$L_{\text{total}} = L_{\text{forward}} + L_{\text{backward}} + L_{\text{privacy}}, \tag{8.15}$$

The total loss is computed during both the forward and backward operations of the steganographic module. The parameters of the steganographic module are then trained based on the task-specific function. Once trained, the steganographic module is deployed within the joint source-channel encoding system. This allows for the embedding of a private image into a non-sensitive host image using steganography, thereby deceiving the eavesdropper. This approach not only protects private information but also reduces the likelihood of eavesdropping attacks, enhancing the security of image transmission.

For classification tasks, privacy protection mechanisms are increasingly emphasized, especially when dealing with sensitive data such as medical images, surveillance videos, and more. To effectively prevent personal privacy leakage, researchers have proposed innovative methods aimed at ensuring the privacy of data during both training and inference processes. These privacy protection strategies not only safeguard the sensitive information of the original images but also allow deep learning models to train and infer effectively on encrypted data without compromising classification performance. Through these mechanisms, classification tasks can continue to achieve efficient learning and accurate predictions while preserving data privacy. LE (Learnable Encryption) method protects privacy by encrypting images in a way that makes them unintelligible to humans, while allowing deep learning models to train on these encrypted images. The method uses a block-wise pixel shuffling algorithm to encrypt the images, making them appear random to humans but still providing the necessary feature information for the network to learn. In addition, LE ensures that no decryption is needed during the training process. The network structure is adapted so that the model can directly learn and make inferences from encrypted images. This allows data holders to share their data with network trainers while protecting privacy, as there is no risk of privacy violations from decrypting the images. PE (Pixel-based Encryption) method protects visual information by encrypting images at the pixel level, ensuring their privacy. The encryption process uses random binary transformations and optional color shuffling, making the encrypted images unrecognizable to unauthorized observers while still retaining enough feature information for deep neural networks (DNNs) to process. Furthermore, PE supports data augmentation in the encrypted domain, meaning that data augmentation can be performed on the encrypted images without decryption. This strategy allows privacy to be preserved while leveraging data augmentation techniques to improve model performance, overcoming the limitations of traditional methods that require augmentation before encryption.

### 8.2.2 DJSCC Against Eavesdrop Channels

In a DJSCC-based semantic communication system, the DJSCC encoder is located at the transmitter and is responsible for extracting, compressing, and protecting task-relevant information while resisting channel noise and adversarial attacks. Placing the encryption module before the DJSCC encoder may affect its ability to extract channel-relevant and task-relevant features. Therefore, it is more reasonable to deploy the protection mechanism after the source has passed through the joint source-channel encoder.

However, the inherent correlation between source samples and channel inputs makes the semantic features after DJSCC more susceptible to eavesdropping attacks. An eavesdropper could induce the transmitter to transmit a known dataset and then steal the semantic encoder-decoder network model from both the transmitter and receiver. Using the known dataset, the eavesdropper could infer the parameters of the source-channel decoder, allowing the stolen information from the physical layer channel to be reconstructed into the original source information via the inferred decoder.

Therefore, when considering the application of encryption techniques after DJSCC, the design of modules to protect task-relevant and channel-robust features extracted by DJSCC must account for the possibility that an eavesdropper [6], after stealing the semantic features, could infer the semantic decoder and launch an attack.

Considering the scenario where an eavesdropper may monitor the channel, the system based on source-channel joint coding can be illustrated as shown in Fig. 8.4. For example, in transmitting image data over a wireless channel, Alice (the sender) wants to send an image to Bob without the presence of an eavesdropper (Eve). Alice first encodes the source information through a semantic feature encoder $f_{\text{encoder}}(\boldsymbol{x})$, obtaining the channel input vector $\boldsymbol{x}$. After being influenced by the wireless channel, the receiver Bob obtains the semantic feature information $\boldsymbol{y}$, and then decodes it using the pre-trained semantic decoder $f_{\text{dec}}(\boldsymbol{y})$ to reconstruct the source information $\hat{s}$.

Meanwhile, the eavesdropper Eve can access the transmitted message $\boldsymbol{x}$ by monitoring the channel. After being affected by the channel conditions observed by Eve, she obtains the intercepted semantic feature information $z$. Eve can exploit a known-plaintext attack to infer the semantic decoder parameters by accessing known $(s, z)$ pairs and try to estimate the decoder $\overline{\boldsymbol{p}} = g_{\text{dec}}(z)$. In this process, the attacker can select a set of plaintexts, encrypt them, and obtain the corresponding ciphertext. The goal is to gain information about the encryption algorithm, making it easier to break the encryption of future communications using the same encryption algorithm. In the worst case, the attacker could directly obtain the decryption key (Fig. 8.6).

To address the potential known-plaintext attacks by the eavesdropper, a public-key encryption scheme can be deployed after the source-channel joint encoder to protect user private information. Public-key encryption is an asymmetric encryption method that uses a pair of keys: a public key for encryption and a private key for decryption. Encrypted data can only be decrypted by the recipient holding the corresponding private key, ensuring that even if the attacker intercepts the encrypted data, they cannot recover the original plaintext, as they do not have access to the private key.

In this encryption scheme, the signal after the source-channel joint encoding is first encrypted. The mathematical principles used in public-key encryption ensure the security of the encryption process. Even if the eavesdropper selects and sends specific plaintext data and receives the encrypted signal, they cannot deduce the plaintext from the encrypted result. This is because the design of public-key encryption ensures that even if the attacker analyzes the ciphertext or selects plaintext, they cannot reverse-engineer the original information. Moreover, the strength of public-key encryption is based on complex mathematical problems, ensuring that it is not reliant on the eavesdropper's channel conditions or their usage of the data. Therefore, even if the eavesdrop-

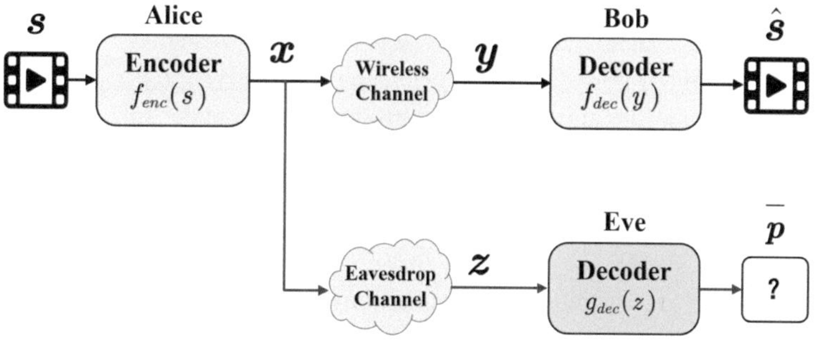


**Fig. 8.6** DJSCC against the Eavesdropper for reconstruction

per tries to interfere with the signal or reveal the transmitted content through a known-plaintext attack, they will not be able to recover the original information from the intercepted encrypted data. Through this encryption method, the confidentiality of communication is effectively protected, thus preventing the risk of information leakage during transmission.

By deploying a public-key encryption module after DJSCC, this method not only preserves the advantages of DJSCC but also provides security against the eavesdropper's known-plaintext attack, without requiring assumptions about the eavesdropper's channel conditions or the intended use of the intercepted signals.

In joint source-channel coding systems, privacy protection mechanisms can be designed by encrypting all information of the original source or by encrypting all the semantic information extracted from the source, which effectively protects the original source. However, the goal of privacy protection is not to encrypt all information but to focus on protecting specific private information [7]. For example, in image transmission scenarios for facial recognition, race, and gender are considered sensitive information. Encrypting the entire source will introduce additional computational and storage overhead. Therefore, the privacy protection mechanism in joint source-channel coding systems should focus on protecting sensitive information in the source, preventing eavesdroppers from inferring sensitive information by intercepting the signal.

As shown in Fig. 8.7, the source user Alice intends to transmit her source $\boldsymbol{s}^m \in \boldsymbol{s}$ to the target user Bob through a wireless channel, while the passive eavesdropper Eve attempts to infer the potential sensitive information $\boldsymbol{p}$ about the source $\boldsymbol{s}^m$. For example, $\boldsymbol{s}^m$ could be an image or video taken by Alice, and $\boldsymbol{p}$ might be the presence of specific objects or activities in the scene. The source data is encoded into a codeword $\boldsymbol{x}^n$ using a semantic encoding function $f_{\text{enc}}(\boldsymbol{s}^m) = \boldsymbol{x}^n$, where $\boldsymbol{x}^n \in x_1, \ldots, x_n$, and $\boldsymbol{x}^n$ and $z^n$ represent the signals received by Bob and Eve, respectively. Bob receives the noisy codeword and decodes it using the decoding function $f_{\text{dec}}(\boldsymbol{x}^n) = \boldsymbol{s}^m$, while Eve, using her intercepted signal $z^n$, tries to infer the sensitive information in the source using $g_{\text{dec}}(z^n) = \overline{\boldsymbol{p}}$.

In this scenario, the goal of the joint source-channel coding system is to balance the reconstruction distortion and the leakage of sensitive information. Reconstruction distortion refers to the difference between the signal received by Bob and the original source, while privacy leakage is measured by the mutual information (MI) between the sensitive information $\boldsymbol{p}$ and the signal observed by the eavesdropper. To achieve this balance, the objective function can be written as

$$\min \mathbb{E}[d(\boldsymbol{s}^m, \hat{\boldsymbol{s}^m})] - I(\boldsymbol{s}^m; \boldsymbol{x}^n) + \lambda I(\boldsymbol{p}; z^n), \tag{8.16}$$

where $d(\cdot, \cdot)$ represents the reconstruction distortion between the original source and the reconstructed signal, $I(\cdot; \cdot)$ is mutual information, and $\lambda$ is a parameter that adjusts the level of privacy protection. In addition to minimizing distortion, we also aim to maximize the mutual information between the source $\boldsymbol{s}^m$ and the signal received by Bob, $I(\boldsymbol{s}^m; \boldsymbol{x}^n)$, to enhance communication utility.

Since the exact calculation of mutual information (MI) is difficult due to the unknown data distribution, we approximate $I(\boldsymbol{s}^m; \boldsymbol{x}^n)$ and $I(\boldsymbol{p}; Z^n)$ using variational representations. Due to the intractability of the true posteriors $P(\boldsymbol{s}^m|\boldsymbol{x}^n)$ and $P(\boldsymbol{p}|z^n)$, we use their variational approximations: $f_{\text{dec}}(\boldsymbol{x}^n) = P(\boldsymbol{s}^m|\boldsymbol{x}^n)$ and $g_{\text{dec}}(z^n) = P(\boldsymbol{p}|z^n)$. Here, we assume that the eavesdropper attempts to predict the sensitive information $\overline{\boldsymbol{p}}$ as $g_{\text{dec}}(z^n) = \overline{\boldsymbol{p}}$. We can express $I(\boldsymbol{s}^m; \boldsymbol{x}^n)$ as

$$\begin{aligned} I(\boldsymbol{s}^m; \boldsymbol{x}^n) &= H(\boldsymbol{s}^m) - H(\boldsymbol{s}^m|\boldsymbol{x}^n) \\ &= H(\boldsymbol{s}^m) + D(P(\boldsymbol{s}^m|\boldsymbol{x}^n)||f_{\text{dec}}(\boldsymbol{x}^n)) + \mathbb{E}[\log f_{\text{dec}}(\boldsymbol{x}^n)] \end{aligned} \tag{8.17}$$

where $D(\cdot||\cdot)$ represents the Kullback-Leibler (KL) divergence, and $H(\boldsymbol{s}^m)$ is constant. For a given value of $\boldsymbol{x}^n$ in 8.17, any distribution $f_{\text{dec}}(\boldsymbol{x}^n)$ on $\boldsymbol{s}^m$ is valid.

Based on the above optimization objective, the parameters of the joint source-channel encoder can be obtained by training deep neural networks (DNNs). Privacy leakage is measured by the mutual information between the sensitive information and the noisy codeword observed by the eavesdropper. In order to enable the joint source-channel encoder output to approximate the distribution of both sensitive and eavesdropper information, we leverage generative networks like variational autoencoders to learn a mapping of joint source-channel encoding. Variational autoencoders use variational inference techniques to model the implicit distribution of the signal, allowing the system to adjust the encoding of the source signal during training to minimize the amount of useful information accessible to the eavesdropper, thereby protecting communication privacy. Finally, the trained model is deployed into the joint source-channel coding system, effectively preventing eavesdroppers from attacking user privacy.

This optimization method is based on the idea of privacy-utility equilibrium. It balances the information leakage to the eavesdropper and the distortion experienced by the legitimate receiver, ensuring communication over a compromised channel. The optimization goal is to minimize the privacy information acquired by the eavesdropper from the signal while ensuring the legitimate receiver can accurately recover the original message with minimal distortion. This method effectively achieves the balance between privacy protection and communication quality, demonstrating good adaptability and robustness, especially in complex eavesdropping environments.

In certain scenarios, the interception and analysis of semantic features output by the joint source-channel encoder are not independent among different eavesdroppers. Multiple eavesdroppers may collaborate in a coordinated attack

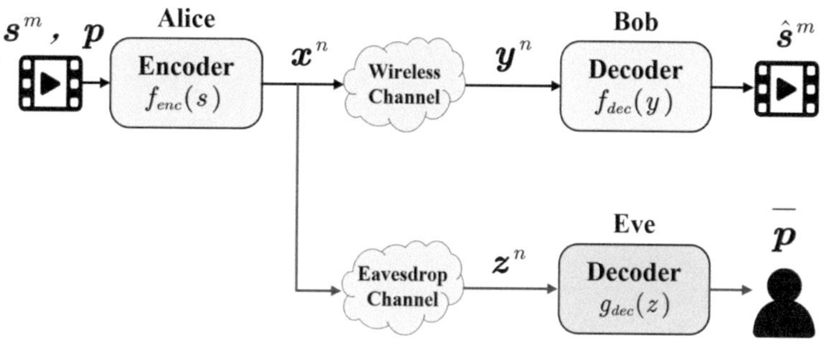


**Fig. 8.7** DJSCC against the Eavesdropper for classifier

to steal the user's private information [8]. In the case of collaborative attacks, eavesdroppers share their respective partial information and use ensemble learning methods to collectively infer the transmitted private information, thereby increasing the success rate of the attack.

In the scenario of multiple eavesdroppers collaborating in an attack on the joint source-channel coding system, which can be seen in Fig. 8.8 the sender user Alice transmits the source $s^m$ to the target user Bob, where $s$ represents the alphabet of the source information. Considering the presence of multiple eavesdroppers, they are represented by Eve1 ($E_1$), ..., EveM ($E_M$), who eavesdrop on the communication through their respective channels and attempt to infer the private information from the source.

Specifically, the sender Alice maps the source $s^m$ to the channel input codeword $x^n$ via the encoding function $f_A : s^m \rightarrow x^n$, where $x^n = f_{\text{enc}}(s^m)$. The target user Bob receives the output $x^n$, and the eavesdroppers $Eve_m$ (where $m \in \{1, \ldots, M\}$) receive the channel outputs $z_M^n$. After $n$ channel transmissions, the target user Bob decodes the signal using the decoding function $f_B$ to obtain the estimated source value $\hat{s^m} = f_B(y^n)$. Meanwhile, each eavesdropper Eve attempts to extract the sensitive information $\overline{p}$ from their observed signal $z_M^n$ and, through collaboration with other eavesdroppers, uses the shared stolen information to infer the user's private information.

In the context of multiple eavesdroppers coordinating an attack, the design goal of the joint source-channel coding system is to ensure that when transmitting images to the target user Bob, the highest possible fidelity is maintained while limiting the amount of sensitive information leaked to each eavesdropper. To achieve this, the core idea of privacy protection can be adopted by designing a data transformation mechanism, non-privacy information is retained as much as possible during transmission, while minimizing the leakage of private data. Specifically, the goal is to minimize the distortion $d(s^m, \hat{s^m})$ between the transmitted image and the received image between the sender Alice and the target user Bob, while minimizing the amount of private information each eavesdropper can infer from the received signals, i.e., minimizing the mutual information $I(s; z_m^k)$, where $s$ is the private information and $z_m^k$ is the signal received by eavesdropper $m$. This leads to the following optimization objective:

$$\min_{f_{\text{enc}}, f_{\text{dec}}} \mathbb{E}[d(s^m, \hat{s^m})] + \frac{1}{M} \sum_{m \in [M]} w_m I(s; z_M^n) \quad (8.18)$$

where $w_m$ shapes the privacy funnel and adjusts the trade-off between information leakage and distortion. Due to the presence of mutual information terms, solving the numerical solution of the optimization Equation 8.18 is challenging. To address this, we rewrite the objective function using a variational approximation of mutual information:

$$\min_{f_{\Omega_A}, f_{\Omega_B}} \mathbb{E}[d(s^m, \hat{s^m})] + \frac{1}{M} \sum_{m \in [M]} w_m \max_{q_{S|Z_M^n}} \mathbb{E}[\log q_{S|Z_M^n}(s|z_m)] \quad (8.19)$$

where $q_{S|Z_M^n}(s|z_m)$ represents the posterior distribution of $s$ given the observation $\boldsymbol{Z}_m^k = z_m$. In the data-driven approach, the encoder $f_{\text{enc}}$ and the decoder $f_{\text{dec}}$ are parameterized by deep neural networks (DNNs) with parameters $\Omega_A$ and $\Omega_B$, respectively. The approximation in the formula can be interpreted as the negative cross-entropy between the adversary's predicted distribution and the true distribution of the sensitive attribute.

To solve Eq. 8.19 in a data-driven manner, we assume that the eavesdroppers also use adversarially trained DNNs and attempt to infer the sensitive attributes of the transmitted image. The parameters $\Theta_E$ parameterize the adversarial network of eavesdropper $E_m$. Based on this, we define the following optimization goal:

$$L(\Omega_A, \Omega_B, \Theta_E) = \mathbb{E}[d(s^m, f_{\Omega_B}(x^n))] \quad (8.20)$$

$$+ \frac{1}{M} \sum_{m \in [M]} w_m \max_{\Theta_{E,m}} \mathbb{E}[\log q_{\Theta_{E,m}}(s|z_m)] \quad (8.21)$$

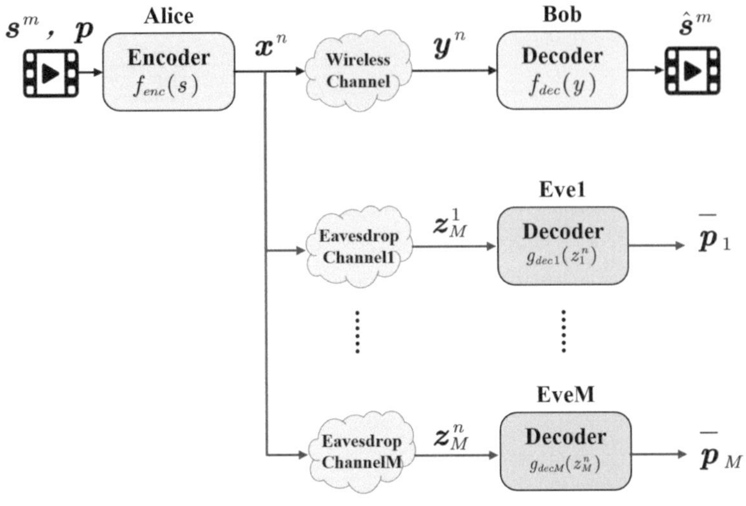


**Fig. 8.8** DJSCC against multiple Eavesdroppers

where $\Theta_E = \{\Theta_{E,1}, \ldots, \Theta_{E,M}\}$, and $q_{\Theta_{E,m}}(\boldsymbol{p}|z_m)$ represents the adversarial likelihood estimate of the correct value $\boldsymbol{p} = p$ made by the DNN of the $m$-th adversary.

To protect privacy in the joint source-channel coding system, an innovative source-channel encoder is designed to reduce the semantic correlation between low-sensitivity information of the user and the codewords observed by eavesdroppers, while still allowing the reconstruction of normal source information. This approach effectively prevents eavesdroppers from inferring sensitive information of the user by exploiting leaked information. However, the design of this optimization task leads to a trade-off between privacy protection and the reconstruction quality of the target user's source information. During the semantic information transmission, this design may result in the loss of private information, making it inaccessible to legitimate users and severely affecting the reconstruction quality.

Therefore, it is necessary to decouple private and public information. By separating private and public information, the encryption mechanism for the post-encoding privacy protection module focuses on protecting only sensitive information features, while ensuring the semantic characteristics of the public information remain unchanged. To ensure optimal performance for the semantic communication system, the task minimizes the reconstruction distortion between the transmitting user and the target user, while maximizing the mutual information between the private subcodeword and the private information to separate private and public information [9]. It also minimizes the mutual information between the private and public subcodewords to ensure that the encryption and decryption modules effectively protect sensitive semantic information without affecting the target user's recognition of private information (Fig. 8.9).

As shown in Fig. 8.8, in the joint source-channel coding system design for this optimization task, the source user Alice wishes to transmit her source $s$ to the target user Bob through a wireless channel. The source $\boldsymbol{s}$ contains the user's private information $\boldsymbol{p}$, while the eavesdropper Eve attempts to infer sensitive information $\boldsymbol{p}$ from the source $\boldsymbol{s}$. The source-channel encoder protects two sub-level semantic encoders, with the source data encoded by the source-channel encoder $f_{enc}(\boldsymbol{s}) = \boldsymbol{x}$ into a codeword $\boldsymbol{x}$, which is then decoupled into public subcodeword $\boldsymbol{x}_t$ and private subcodeword $\boldsymbol{x}_s$, ensuring that the two are independent. The public subcodeword $\boldsymbol{x}_t$ is transmitted directly through the wireless channel without protection, while the private subcodeword $\boldsymbol{x}_s$ undergoes encryption to generate the protected feature $\boldsymbol{x}_s^p$, which is transmitted through the wireless channel.

After transmission through the wireless channel, the target user Bob and the eavesdropper Eve receive the encrypted private information feature $\boldsymbol{x}_s^p$ and the public information feature $\boldsymbol{x}_t$, respectively. Bob uses the decryption module to decode the noisy codeword and retrieve the private semantic feature $\overline{\boldsymbol{y}_s^p}$, then decodes all semantic features using the decoding function $f_{dec}(\overline{\boldsymbol{y}}) = \hat{\boldsymbol{s}}$ to reconstruct the source. Eve, on the other hand, attempts to infer the sensitive information $\boldsymbol{p}$ by decoding the stolen signal $z$ through the decoding function $g_{dec}(z) = \hat{\boldsymbol{p}}$.

The optimization task aims to maximize the mutual information between the private subcodeword $\boldsymbol{x}_s$ and the sensitive information $\boldsymbol{p}$, while minimizing reconstruction distortion and the mutual information between the private and public subcodewords. This design ensures that the privacy-preserving joint source-channel coding system can decouple private and public information into independent subcodewords. The objective function of this optimization task is as follows:

$$\min \mathbb{E}_p(\boldsymbol{s}, \boldsymbol{p}) \left(d(\boldsymbol{s}, \hat{\boldsymbol{s}})\right) + \alpha I(\boldsymbol{x}_t; \boldsymbol{x}_s) - \beta I(\boldsymbol{x}_s; \boldsymbol{p}) \quad (8.22)$$

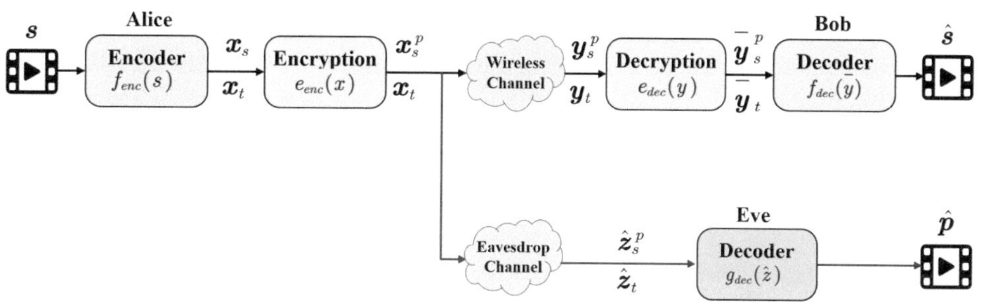


**Fig. 8.9** DJSCC against the Eavesdropper with encryption

where the first term minimizes the error between the original information $s$ sent by the source user and the reconstructed information $\hat{s}$ by the target user, improving the performance of the communication system. The second term minimizes the mutual information between the public information feature $\boldsymbol{x}_t$ and the private information feature $\boldsymbol{x}_s$, thus encouraging the independence of public and private information. The third term maximizes the mutual information between the private information feature $\boldsymbol{x}_s$ and the sensitive information $\boldsymbol{p}$, ensuring that the semantic features of sensitive information are retained in $\boldsymbol{x}_s$.

By jointly minimizing $I(\boldsymbol{x}_t; \boldsymbol{x}_s)$ and maximizing $I(\boldsymbol{x}_s; \boldsymbol{p})$, this design ensures that the private information is removed from the public information feature $\boldsymbol{x}_t$, allowing it to be transmitted directly through the channel. The privacy protection algorithm only needs to protect the private information feature $\boldsymbol{x}_s$, rather than the entire information, thus simplifying the encryption process.

Based on the above optimization goal, the parameters of the joint source-channel encoder can be learned by training deep neural networks (DNNs). The trained model can then be deployed in the joint source-channel coding system to effectively defend against joint attacks by multiple eavesdroppers, thereby enhancing the system's security.

This optimization method transforms the source data through the design of privacy mappings, ensuring that the data retains high utility for the target user Bob while minimizing leakage to the eavesdroppers. Since the coordinated attack by multiple eavesdroppers increases the complexity of information leakage, the optimization process needs to balance the trade-off between information utility and private data leakage. Ultimately, the core of the optimization task is to reasonably control the balance between privacy leakage and information utility, ensuring that the privacy of the data is adequately protected against multi-eavesdropper attacks.

## 8.3 Summary

In this chapter, various privacy protection techniques for DJSCC are discussed. It introduces four different types of privacy leakage risks associated with DJSCC and reviews six major privacy protection techniques, highlighting their advantages and disadvantages. These techniques include Signal Steganography and protection against multiple eavesdroppers, among others.

## Problems

8.1 Which of the following is a type of attack in semantic communication that aims to infer sensitive information, such as demographic data, health status, or financial condition, from anonymized data?

A. Property Inference Attack
B. Membership Inference Attack
C. Feature Reconstruction Attack
D. Imitation Learning Attack

8.2 What is the primary goal of a Membership Inference Attack in semantic communication?

A. To infer whether a specific data record exists in the user's dataset.
B. To recover the original data by observing intermediate features.
C. To replicate the behavior of an encrypted network or encoding system.
D. To learn the private attributes of the sender using background knowledge.

8.3 Which of the following privacy protection techniques in semantic communication allows model training on local data without sharing raw data?

A. Differential Privacy
B. Secure Multi-party Computation
C. Federated Learning
D. Homomorphic Encryption

8.4 What is the main purpose of adding random noise to data in the context of differential privacy?

A. To speed up data transmission
B. To ensure that attackers cannot recover individual sensitive information
C. To reduce the computational complexity of encrypted data
D. To perform computations directly on encrypted data

8.5 Which of the following strategies in semantic communication encrypts the source information before semantic encoding to prevent leakage during transmission over the wired network?

A. Strategy 1: Encrypt the Source First, then Perform Semantic Encoding
B. Strategy 2: Extract Semantic Features First, then Encrypt
C. Use of Homomorphic Encryption
D. Use of Differential Privacy

8.6 Which encryption strategy in semantic communication involves the wireless service provider encrypting the semantic features after the DJSCC process?

A. Strategy 1: Encrypt the Source First, then Perform Semantic Encoding
B. Strategy 2: Extract Semantic Features First, then Encrypt
C. Secure Multi-party Computation
D. Federated Learning

8.7 Which of the following privacy protection methods in classification tasks allow deep learning models to train on encrypted images without requiring decryption during the training process?

A. LE (Learnable Encryption)
B. PE (Pixel-based Encryption)
C. Homomorphic Encryption
D. Differential Privacy

8.8 In the DJSCC network, the total loss function used for training the network is given by

$$L_{\text{total}} = L_r - \lambda_e L_e - \lambda_d L_d$$

where $L_r$ represents the reconstruction loss, $L_e$ represents the feature loss between the source signals and ciphertext, and $L_d$ represents the feature loss between the source signals and decoded information. Which of the following is true about the parameters $\lambda_e$ and $\lambda_d$ in the total loss function?

A. $\lambda_e$ and $\lambda_d$ are negative weights that reduce the influence of feature losses.
B. $\lambda_e$ and $\lambda_d$ are positive weights that control the influence of feature losses on the total loss function.
C. $\lambda_e$ and $\lambda_d$ are zero, meaning feature losses do not influence the training.
D. $\lambda_e$ and $\lambda_d$ are arbitrary values chosen randomly.

8.9 In the joint source-channel coding system, the optimization objective for balancing the reconstruction distortion and the leakage of sensitive information is expressed as

$$\min \mathbb{E}[d(s^m, \hat{s^m})] - I(s^m; \boldsymbol{x}^n) + \lambda I(\boldsymbol{p}; z^n)$$

Where $d(\cdot, \cdot)$ represents reconstruction distortion, and $I(\cdot; \cdot)$ represents mutual information. What is the role of the parameter $\lambda$ in the objective function?

A. $\lambda$ controls the trade-off between privacy leakage and communication utility.
B. $\lambda$ adjusts the reconstruction distortion during signal transmission.
C. $\lambda$ directly influences the size of the sensitive information $\boldsymbol{p}$.
D. $\lambda$ determines the signal encoding method.

8.10 In the joint source-channel coding system design with privacy protection, the optimization task aims to balance between reconstruction distortion and privacy leakage. Which of the following terms in the optimization objective function ensure that private and public information are decoupled and independent?

A. $d(\boldsymbol{s}, \hat{\boldsymbol{s}})$, which minimizes reconstruction distortion.
B. $I(\boldsymbol{x}_t; \boldsymbol{x}_s)$, which minimizes the mutual information between the public and private subcodewords.
C. $I(\boldsymbol{x}_s; \boldsymbol{p})$, which maximizes the mutual information between the private subcodeword and the sensitive information.
D. All of the above.

## References

1. M. Shen, J. Wang, H. Du, D. Niyato, X. Tang, J. Kang, Y. Ding, L. Zhu, Secure semantic communications: Challenges, approaches, and opportunities. IEEE Netw. (2023)
2. S. Guo, A. Zhang, Y. Wang, C. Feng, T.Q. Quek, Trustworthy semantic-enabled 6g communication: A task-oriented and privacy-preserving perspective. arXiv preprint arXiv:2408.04188 (2024)

3. H. Du, J. Wang, D. Niyato, J. Kang, Z. Xiong, M. Guizani, D.I. Kim, Rethinking wireless communication security in semantic internet of things. IEEE Wirel. Commun. **30**(3), 36–43 (2023)
4. J. Xu, B. Ai, W. Chen, N. Wang, M. Rodrigues, *Deep joint source-channel coding for image transmission with visual protection* (IEEE Trans. Cogn. Commun, Netw, 2023)
5. S. Tang, C. Liu, Q. Yang, S. He, D. Niyato, Secure semantic communication for image transmission in the presence of eavesdroppers (2024)
6. T.-Y. Tung, D. Gündüz, Deep joint source-channel and encryption coding: Secure semantic communications, in (2023), pp. 5620–5625
7. E. Erdemir, P.L. Dragotti, D. Gündüz, in *Privacy-aware communication over a wiretap channel with generative networks.* (IEEE, 2022), pp.2989–2993
8. S.A. Ameli Kalkhoran, M. Letafati, E. Erdemir, B.H. Khalaj, H. Behroozi, D. Gündüz, Secure deep-jscc against multiple eavesdroppers, in (2023), pp.3433–3438
9. L. Sun, Y. Yang, M. Chen, C. Guo, Disentangled information bottleneck guided privacy-protective joint source and channel coding for image transmission. IEEE Trans. Commun. **72**(11), 6992–7005 (2024)

# 9 Task-Oriented Semantic Communication

**Abstract**

Task-oriented semantic communication is an important paradigm for machine-type communication in the 6G era, which integrates task-specific goals into system design and optimization. Task-oriented communication bypasses the need to reconstruct the original source, transmitting only task-relevant features to perform the downstream intelligent tasks, which reduces unnecessary data exchange and promotes communication efficiency. This chapter gives an overview of task-oriented semantic communications, with a practical case for comprehension. Moreover, advanced generative models can serve task-oriented communication to achieve higher efficiency.

**Keywords**

Semantic communication • Task-oriented communication • Generative model

## 9.1 Introduction

In classical communication systems, data is transmitted without regard to the semantic content or relevance to the task. Shannon's information theory, while revolutionary, approached communication from the perspective of maximizing data fidelity, independent of the ultimate use of the data. In contrast, semantic communication focuses on the meaning behind data, rather than raw bits. Semantic communication introduces the concept of context awareness, aiming to prioritize data that carried intrinsic meaning to the receiver.

Task-oriented semantic communication takes this concept further by embedding task-specific intelligence into the communication process and integrating task-centric goals into the data transmission process [1]. This focus aligns communication mechanisms with the specific requirements of the task at hand, reducing unnecessary data exchange and promoting efficiency. Task-oriented semantic communication emphasizes goal-driven communication, where data is selectively shared based on its contribution to the accomplishment of predefined tasks. For example, in a distributed sensor network, task-oriented semantic communication ensures that only critical alerts (e.g., temperature spikes or system failures) are transmitted, filtering out routine status updates that do not affect ongoing tasks. In the broader context of 6G networks, task-oriented semantic communication moves beyond traditional models focused merely on data transfer, prioritizing the communication of information essential for task completion [2]. This paradigm not only optimizes data transmission but also adapts to the specific needs of diverse applications, making it a key element of highly specialized, application-driven networks.

The essence of task-oriented semantic communication lies in transforming the communication process from "information transfer" to "task-driven optimization". Unlike traditional systems that prioritize source and channel coding to ensure high-fidelity data transmission, task-oriented semantic communication focuses on transmitting only the minimum information required to complete a task. This shift is grounded in the following principles:

- **Task-First Design**: Communication systems are designed around the task's requirements rather than data reconstruction.
- **Task-Relevant Feature Transmission**: Only transmit task-relevant features, avoiding redundant raw data.
- **End-to-End Optimization**: Use deep learning models to optimize the entire communication pipeline, from feature extraction to task execution.

By reducing data redundancy, task-oriented semantic communication achieves lower latency and bandwidth requirements, making it suitable for resource-constrained environments.

The typical workflow of task-oriented semantic communication involves:


W. Chen and Z. Qin, *Semantic Communications*, Textbooks in Telecommunication Engineering,
https://doi.org/10.1007/978-3-032-11005-3_9

- **Task Awareness and Analysis**: At the transmitter, the system analyzes the task, such as image classification, object detection, or speech recognition. The task-specific requirements determine the type and quality of features to extract.
- **Feature Extraction and Selection**: Deep learning models extract critical features from raw data, such as edge information in images or semantic components in speech. Features are filtered and compressed, retaining only those essential for task execution.
- **Efficient Transmission**: Features are encoded and transmitted using advanced techniques, e.g., DJSCC, for efficient utilization of bandwidth. Furthermore, channel adaptation dynamically adjusts transmission strategies based on real-time channel conditions.
- **Task Execution**: At the receiver, the task is directly executed on the transmitted features, such as classification, decision-making, or generation tasks.

A key technology enabling task-oriented semantic communication is edge computing [3]. By placing computational resources closer to the source of data generation, edge computing reduces latency and helps to process information more efficiently in real-time. In 6G networks, where applications demand ultra-low latency and high bandwidth, edge computing becomes critical. It allows task-oriented semantic communication systems to handle complex data processing tasks such as image recognition, AI-based decision-making, or real-time sensor data analysis, all of which require localized, efficient processing to meet strict performance requirements. Additionally, the integration of artificial intelligence (AI) and machine learning (ML) is crucial for the dynamic nature of task-oriented semantic communication. These technologies help the network predict task demands and optimize resource allocation, ensuring that communication resources are used in the most efficient manner possible. AI can also facilitate intelligent task scheduling, ensuring that the data required for specific tasks is delivered promptly without unnecessary delay.

One of the significant challenges in task-oriented semantic communication lies in ensuring seamless communication across heterogeneous networks. 6G is expected to integrate various types of networks, such as terrestrial, satellite, and aerial networks, which may have different communication protocols and capabilities. This heterogeneity can create issues related to compatibility and data transmission efficiency. A robust framework is necessary to handle the dynamic nature of tasks across different networks while ensuring the reliability and security of communication. Additionally, task-oriented semantic communication systems must be designed to handle the complexity of dynamically adjusting to the changing nature of tasks. For instance, in a smart city application, the communication system may need to prioritize traffic management data during peak hours and switch to environmental monitoring tasks during off-peak periods. The system's ability to adapt to such changes, while maintaining efficiency and low latency, is a challenging but crucial aspect of task-oriented semantic communication.

Security and privacy are also significant concerns for task-oriented semantic communication [4]. Given that this communication model involves transferring sensitive and mission-critical information, such as medical data or financial transactions, it is imperative to secure the communication channels and ensure data integrity. Privacy-preserving techniques, such as encryption and secure multi-party computation, need to be integrated into the network architecture to protect the information being exchanged. Furthermore, as more devices and applications become interconnected, the task-oriented semantic communication system must also ensure the protection of user data, preventing unauthorized access and ensuring compliance with data privacy regulations.

The potential applications of task-oriented semantic communication in 6G are vast. In smart manufacturing, for example, task-oriented semantic communication can enable real-time coordination of robots and machines, ensuring that tasks are executed in a highly efficient and synchronized manner. In autonomous driving, vehicles can communicate with each other and with infrastructure to optimize route planning, avoid accidents, and share critical information in real-time. Similarly, in healthcare, task-oriented semantic communication can facilitate remote surgeries or real-time patient monitoring, where timely and accurate transmission of medical data is essential for patient safety and treatment outcomes. Additionally, the integration of task-oriented semantic communication in smart cities can help optimize the management of resources like traffic, energy, and waste, providing real-time insights that improve urban living.

However, implementing task-oriented semantic communication on a large scale presents challenges. One such challenge is ensuring the scalability of the system as the number of connected devices increases exponentially in 6G environments. The system must be capable of handling millions of devices, each with varying communication requirements, without compromising on efficiency. Furthermore, energy consumption is a significant concern, especially when considering the power requirements of 6G networks that support high data rates and real-time processing. task-oriented semantic communication will need to balance high performance with energy efficiency to ensure sustainability in the long term.

In conclusion, task-oriented semantic communication is a crucial element in the evolution of 6G networks, offering a more intelligent, efficient, and adaptive approach to communication. By focusing on the specific needs of tasks, this communication model has the potential to transform a wide range of industries, from autonomous vehicles to smart cities. However, challenges such as network heterogeneity, security, and scalability must be addressed to fully realize its potential. As the development of 6G continues, task-oriented semantic communication will play a central role in enabling the next generation of intelligent, application-specific networks.

## 9.2 Semantic Communication for Video Question Answering

In this section, we consider semantic communication for a specific intelligent task: Video Question Answering. Typical video applications in 6G include autonomous driving, immersive Extended Reality, etc. These various emerging applications bring different communication requirements and constraints (e.g., AI service accuracy$\geq$90%, air-interface latency$\leq$10 ms.). For wireless video transmission, there exists two main problems: first problem is compared to other modal information, videos have much larger volume of data with the extra temporal dimension. The second problem is there are limited bandwidth and other constraints for different video-based applications. Therefore, it is necessary to design the wireless transmission system for videos with specific tasks.

Video question answering (VideoQA), where machines automatically answer natural language questions with video contents, is an intelligent task in the popular visual-language understanding domain. Compared to image-based visual question answering, VideoQA encompasses a wider variety of question types, which involves not only recognition of visual objects, actions, and events but also reasoning about spatiotemporal and causal relationships, making it more challenging. VideoQA can support many important applications in human-machine communications such as virtual reality, smart cities, and the metaverse. The rise of multimedia applications and the extensive deployment of cameras have led to a significant presence of videos in machines, which have significant impact on both human-machine and machine-machine communications.

The key of VideoQA is to understand video contents given the questions and answers, which needs efficient video-text information extraction and fusion for inference. However, on the one hand, traditional video transmission methods most focus on video reconstruction, with few investigations and developments for other intelligent tasks. These methods are not effective for extracting high-level video semantics, which leads to significant waste of transmission resources in communications for VideoQA. On the other hand, the video features extracted by conventional VideoQA methods are typically high-dimensional, which may not be suitable for bandwidth-limited wireless networks. Additionally, channel fading and noise can interfere with the accurate transmission of these features, leading to a deterioration in VideoQA performance.

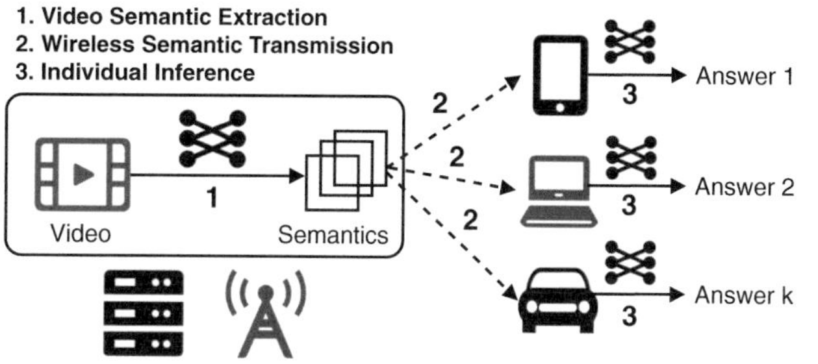


**Fig. 9.1** A potential application scenario of VideoQA-SC

Typically, the development of SC systems for VideoQA tasks faces two main challenges:

1. How to model the spatiotemporal relations present in videos to achieve efficient semantic extraction?
2. How to reduce the impact of wireless channel degradation and satisfy bandwidth limitations while preserving the performance of VideoQA tasks?

Given this context, we introduce VideoQA-SC [5], an end-to-end semantic communication system for VideoQA. Basically, VideoQA-SC tackles the aforementioned challenges through two key customized modules: a *spatiotemporal* video semantic encoder and a *learning-based bandwidth-adaptive* joint source-channel (JSC) encoder/decoder, which enable noise-robustness and bandwidth efficiency of semantic communication for VideoQA.

A potential application scenario of VideoQA-SC is shown in Fig. 9.1. Multiple terminal devices simultaneously request access to the same surveillance videos with different questions. The transmitter, such as an edge server, extracts the video semantics that encapsulate the entire content and sends them to all terminal devices. Each terminal device then independently performs VideoQA to generate its respective answer.

Given the video $\boldsymbol{S}_v \in \mathbb{R}^{l_v \times 3 \times x \times y}$ with $l_v$ frames and $x \times y$ resolution, and the question $\boldsymbol{S}_q \in \mathbb{R}^{l_q \times d_q}$ with $l_q$ tokens, The process of VideoQA is to predict an answer $a$ by utilizing both video and text information:

$$a^\star = \arg\max_{a \in \mathcal{A}} o_{\boldsymbol{\omega}}(a|\boldsymbol{S}_q, \boldsymbol{S}_v, \mathcal{A}), \quad (9.1)$$

where $a^\star$ is the predicted answer chosen from the candidate answers, i.e., multiple choices or a predefined answer set, denoted as $\mathcal{A}$, and $o_{\boldsymbol{\omega}}(\cdot)$ is the VideoQA model with parameters $\boldsymbol{\omega}$.

As shown in Fig. 9.2, the process of VideoQA is divided into three parts, i.e., transmitter, wireless channel, receiver. The transmitter extracts video semantics $\boldsymbol{G}_v \in \mathbb{R}^{l_v \times d}$ from the original video $\boldsymbol{S}_v$ by the spatiotemporal semantic encoder

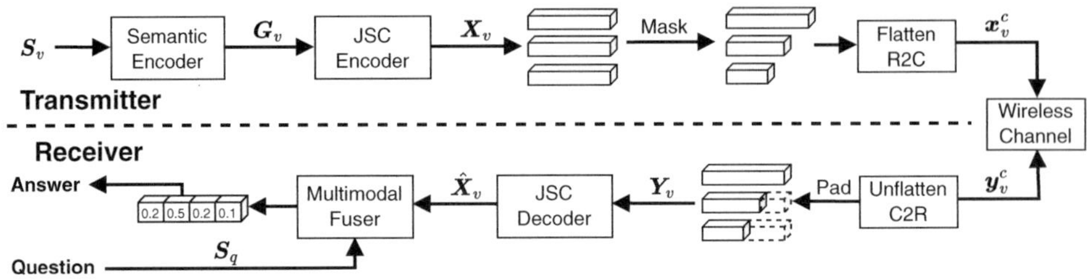


**Fig. 9.2** Overview architecture of VideoQA-SC

$g(\cdot;\boldsymbol{\zeta})$ parameterized by $\boldsymbol{\zeta}$. Then, the JSC encoder $f_e(\cdot;\boldsymbol{\theta},\boldsymbol{\epsilon})$ parameterized by $\boldsymbol{\theta}$ and $\boldsymbol{\epsilon}$ encodes $\boldsymbol{G}_v \in \mathbb{R}^{l_v \times d}$ into $\boldsymbol{X}_v \in \mathbb{R}^{l_v \times d}$ and generate a learnable mask $\boldsymbol{M} \in \mathbb{R}^{l_v \times 1}$. Channel masking is applied to mask a portion of channels of $\boldsymbol{X}_v$ as zero based on $\boldsymbol{M}$ for variable-length coding. The transmitter needs to transmit some additional bits losslessly to indicate the coding length of each token for receiver. Finally, non-zero channels of $\boldsymbol{X}_v$ are flattened and further generate complex channel input vector $\boldsymbol{x}_v^c \in \mathbb{C}^n$. $R = \frac{n}{l_v \times 3 \times x \times y} \leq 1$ is typically defined as channel bandwidth ratio (CBR), which represents the average length of channel usage for transmitting one video value. The process of $\boldsymbol{S}_v$ at the transmitter is formulated as:

$$\boldsymbol{x}_v^c = \mathrm{R2C}(\mathrm{Flatten}(f_e(g(\boldsymbol{S}_v;\boldsymbol{\zeta});\boldsymbol{\theta},\boldsymbol{\epsilon}))), \tag{9.2}$$

where R2C$(\cdot)$ and Flatten$(\cdot)$ are the real-to-complex and flattening operations, respectively.

For AWGN channels, the received channel output vector is represented as:

$$\boldsymbol{y}_v^c = \boldsymbol{x}_v^c + \boldsymbol{n}, \tag{9.3}$$

where $\boldsymbol{n} \in \mathbb{C}^n$ consists of independent and identically distributed (i.i.d.) samples following complex Gaussian distribution $\mathcal{CN}(0, \sigma_n^2 \boldsymbol{I})$. $\sigma_n^2$ denotes the average noise power. The channel SNR is given by $\frac{1}{\sigma_n^2}$. For Rayleigh block fading channels, the received channel output vector is represented as:

$$\boldsymbol{y}_v^c = h\boldsymbol{x}_v^c + \boldsymbol{n}, \tag{9.4}$$

where $h \in \mathbb{C}^1$ follows the Rayleigh distribution Rayleigh$(\sigma_h)$ with fading parameter $\sigma_h$. There are two SNRs for each $\boldsymbol{x}_v^c$, i.e., the statistical SNR and the current SNR, which are given by $\frac{h^2}{\sigma_n^2}$ and $\frac{\sigma_h^2}{\sigma_n^2}$, respectively.

The receiver firstly converts the received channel output vector $\boldsymbol{y}_v^c$ to real values. The abandoned channels are padded with zeros, generating $\boldsymbol{Y}_v \in \mathbb{R}^{l_v \times d}$. Then, $\boldsymbol{Y}_v$ are fed into the JSC decoder $f_d(\cdot;\boldsymbol{\phi})$ parameterized by $\boldsymbol{\phi}$ to reconstruct video semantics $\hat{\boldsymbol{G}}_v \in \mathbb{R}^{l_v \times d}$. Finally, the multimodal fuser $u(\cdot;\boldsymbol{\nu})$ parameterized by $\boldsymbol{\nu}$ is used to fuse video and text information and predict the corresponding answer $a^\star$ to the question $\boldsymbol{S}_q$. The process at the receiver is formulated as:

$$a^\star = u(f_d(\mathrm{Pad}(\mathrm{Unflatten}(\mathrm{C2R}(\boldsymbol{y}_v^c)));\boldsymbol{\phi}), \boldsymbol{S}_q;\boldsymbol{\nu}), \tag{9.5}$$

where Pad$(\cdot)$, Unflatten$(\cdot)$ and C2R$(\cdot)$ denote zero-padding, unflattening and complex-to-real operations, respectively.

The optimization problem of VideoQA-SC is to maximize the average VideoQA accuracy while satisfying bandwidth constraints $B$, which is formulated as:

$$\begin{aligned} \max_{\boldsymbol{\zeta},\boldsymbol{\theta},\boldsymbol{\epsilon},\boldsymbol{\phi},\boldsymbol{\nu}} \quad & \mathbb{E}[\mathrm{ACC}(a^\star, a_{\mathrm{label}})] \\ \text{s.t.} \quad & R \leq B, \end{aligned} \tag{9.6}$$

where ACC$(\cdot,\cdot)$ is an indicator (1 if $a^\star = a_{\mathrm{label}}$ and 0 otherwise).

VideoQA-SC has two customized modules for efficient video semantic extraction and robust video transmission, respectively. For video semantic extraction, Transformer and graph neural network are adopted to model the spatiotemporal correlations in videos and extract compact semantics. For video semantic transmission, cross-attention-based Transformer with learnable rate predictors is adopted to achieve bandwidth-adaptive semantic transmission.

The process of video semantic extraction is based on spatial and temporal modeling of correlations in videos. Firstly, sparse sampling is employed on original videos at a uniform interval, which can reduce transmission and computation costs. Each sampled video includes $l_v$ key frames, which are further divided into $l_c$ clips of length $l_f$ frames. Considering that the computational complexity increases as more frames are involved in the extraction of spatiotemporal relations, the semantic encoder, $g(\cdot;\boldsymbol{\zeta})$, captures spatiotemporal relations of frames in one clip, thereby generating semantics for each clip.

Since adjacent frames in a video often share the same background that contributes little to understanding the video content, significant redundancies exist in videos for VideoQA tasks. Therefore, unlike most video semantic extraction operates directly at the pixel level, VideoQA-SC suggests extract coarse-grained video semantics at the object level, which is more efficient for video modeling and bandwidth savings. Without loss of generality, we take the semantic extraction of $i$-th clip $\boldsymbol{S}_i = \{\boldsymbol{S}_{i,1}, \boldsymbol{S}_{i,2}, \cdots, \boldsymbol{S}_{i,l_f}\}$ as an example to explain the process of spatial modeling and temporal modeling, where $\boldsymbol{S}_{i,j}$ denotes the $j$-th frame of $i$-th clip. Firstly, a pre-trained object detector is applied to extract visual objects in $\boldsymbol{S}_i$, generating object-level feature $\boldsymbol{O}_i \in \mathbb{R}^{l_f \times r \times m}$. $r$ denotes the number of detected objects and $m$ denotes channel dimension.

Temporal modeling is based on Transformer architecture, which is widely used in natural language processing originally. The self-attention mechanism in Transformers effectively captures the relations between each token in the sequence and facilitates feature aggregation. After aggregation, each token becomes a fused representation of all tokens in the sequence. Drawing on this idea, temporal modeling treats the features of each object as a token, with the same object across different frames forming a sequence. Through the self-attention mechanism, each object fuses information from different time steps, thereby capturing the temporal variations of each object. The temporal modeling is for each object, which is formulated as:

$$\sum_{p=1}^{l_f} \alpha_{i,k,j,p} \boldsymbol{O}_{i,p,k} \rightarrow \boldsymbol{O}_{i,j,k}, \tag{9.7}$$

where $\alpha_{i,k,j,p}$ is the attention score. Equation (9.7) indicates that each object $\boldsymbol{O}_{i,j,k} \in \mathbb{R}^m$ $(k \in [1, r])$ is the fusion of the same object across different frames $(\boldsymbol{O}_{i,p,k}, p \in [1, l_f])$ along the time dimension.

Spatial modeling is based on graph neural network architecture, which focuses on the spatial interaction between objects in each frame. A graph $\mathcal{G}_j$ is constructed for every $(\boldsymbol{O}_{i,j}, j \in [1, l_f])$, where each node and each edge represent one object $(\boldsymbol{O}_{i,j,k})$ and the spatial relations between two objects, respectively. Then, each $\boldsymbol{O}_{i,j}$ is fed into a graph neural network to aggregate object features based on $\mathcal{G}_j$. Through the update of nodes, each object becomes a fusion of different objects within the same frame, thereby capturing the spatial relations between these objects.

By sequential temporal and spatial modeling, the object features are effectively aggregated, facilitating the generation of compact semantics for VideoQA. The aggregated $\boldsymbol{O}_i$ is pooled along the object dimension $(\mathbb{R}^{l_f,r,m} \rightarrow \mathbb{R}^{l_f,m})$. In addition, a pre-trained frame-level feature extractor, such as ResNet, extracts frame-level features $\boldsymbol{F}_i \in \mathbb{R}^{l_f \times m}$, which are further integrated with object-level features $\boldsymbol{O}_i$ to produce more comprehensive semantics $\boldsymbol{G}_v \in \mathbb{R}^{l_v \times d}$ by a linear projection:

$$\boldsymbol{G}_v = \text{Linear}(\boldsymbol{F}_i, \boldsymbol{O}_i). \tag{9.8}$$

After video semantic extraction, a cross-attention-based JSC encoder-decoder is employed to ensure the accurate transmission of video semantics, mitigating the effects of wireless channels. Additionally, a semantic-aware bandwidth adaptation method is integrated into the JSC encoder-decoder to further optimize the utilization of limited bandwidth resources.

Transformer serves as the backbone for both the JSC encoder and decoder. A series of rate predictors parameterized by $\boldsymbol{\epsilon}$ is used to generate a learnable mask given the video contents and channel conditions for efficient bandwidth adaptation. Considering that models trained under different rate-distortion (RD) optimizations allocate attention to video frames differently, a learnable tensor $\boldsymbol{G}_{\text{rate}} \in \mathbb{R}^{l_v \times d}$ with the same shape as video semantics $\boldsymbol{G}_v$ is introduced to provide prior information for bandwidth adaptation. Instead of using random initialization, the initial values of $\boldsymbol{G}_{\text{rate}}$ are set as a series of learnable parameters. By training under varying bandwidth constraints, the model acquires distinct initial values for $\boldsymbol{G}_{\text{rate}}$, which serve as prior information reflecting the initial attention distribution for all transmitted frames under specific bandwidth constraints. During inference, the appropriate initial values corresponding to the given bandwidth constraint are selected to initialize $\boldsymbol{G}_{\text{rate}}$.

Both the JSC encoder and decoder consist of a series of cross-attention Transformer block, which enables sufficient interaction between $\boldsymbol{G}_{\text{rate}}$ and $\boldsymbol{G}_v$ in cross-attention form. Each cross-attention Transformer block includes two branches, i.e., *feature* branch and *rate* branch to process $\boldsymbol{G}_v$ and $\boldsymbol{G}_{\text{rate}}$, respectively. Both the branches include a standard Vision Transformer Block. Given the inputs $\boldsymbol{G}_v \in \mathbb{R}^{l_v \times d}$ and $\boldsymbol{G}_{\text{rate}} \in \mathbb{R}^{l_v \times d}$, $\boldsymbol{Q}_v \in \mathbb{R}^{l_v \times d}$, $\boldsymbol{K}_v \in \mathbb{R}^{l_v \times d}$ and $\boldsymbol{V}_v \in \mathbb{R}^{l_v \times d}$ is generated from $\boldsymbol{G}_v$ by learnable tensor $\boldsymbol{W}_v^{QKV}$, and $\boldsymbol{Q}_{\text{rate}} \in \mathbb{R}^{l_v \times d}$, $\boldsymbol{K}_{\text{rate}} \in \mathbb{R}^{l_v \times d}$ and $\boldsymbol{V}_{\text{rate}} \in \mathbb{R}^{l_v \times d}$ is generated from $\boldsymbol{G}_{\text{rate}}$ by learnable tensor $\boldsymbol{W}_{\text{rate}}^{QKV}$. Then, the cross-attention-based interaction between $\boldsymbol{G}_v$ and $\boldsymbol{G}_{\text{rate}}$ is formulated as:

$$\text{CA}(\boldsymbol{G}_{\text{rate}}, \boldsymbol{G}_v) = \text{softmax}\left(\frac{\boldsymbol{Q}_{\text{rate}}\boldsymbol{K}_v^T}{\sqrt{d}}\right)\boldsymbol{V}_v \tag{9.9}$$

and

$$\text{CA}(\boldsymbol{G}_v, \boldsymbol{G}_{\text{rate}}) = \text{softmax}\left(\frac{\boldsymbol{Q}_v\boldsymbol{K}_{\text{rate}}^T}{\sqrt{d}}\right)\boldsymbol{V}_{\text{rate}} \tag{9.10}$$

for rate branch and feature branch, respectively. There are additional feed forward network (FFN) for further nonlinear transformation in both branches. The whole process of cross-attention Transformer block is formulated as:

$$\begin{aligned} \tilde{\boldsymbol{G}}_{\text{rate}}^l &= \text{MHCA}(\text{LN}(\boldsymbol{G}_{\text{rate}}^{l-1}), \text{LN}(\boldsymbol{G}_v^{l-1})) + \boldsymbol{G}_{\text{rate}}^{l-1}, \\ \boldsymbol{G}_{\text{rate}}^l &= \text{FFN}(\text{LN}(\tilde{\boldsymbol{G}}_{\text{rate}}^l)) + \text{LN}(\tilde{\boldsymbol{G}}_{\text{rate}}^l), \end{aligned} \tag{9.11}$$

and

$$\begin{aligned} \tilde{\boldsymbol{G}}_v^l &= \text{MHCA}(\text{LN}(\boldsymbol{G}_v^{l-1}), \text{LN}(\boldsymbol{G}_{\text{rate}}^{l-1})) + \boldsymbol{G}_v^{l-1}, \\ \boldsymbol{G}_v^l &= \text{FFN}(\text{LN}(\tilde{\boldsymbol{G}}_v)) + \text{LN}(\tilde{\boldsymbol{G}}_v^l), \end{aligned} \tag{9.12}$$

for rate branch and feature branch, respectively. $G_{\text{rate}}^{l-1}$ and $G_v^{l-1}$ denote the inputs of the $l$-th cross-attention Transformer block, and $G_{\text{rate}}^l$ and $G_v^l$ denote the outputs of the $l$-th Transformer blocks. MHCA$(\cdot)$ denotes multi-heads cross-attention function CA$(\cdot)$. LN$(\cdot)$ denotes standard layer normalization typically used in Transformer architecture. FFN$(\cdot)$ denotes two fully connected layers with GeLU$(\cdot)$ activation.

The cross-attention Transformer block enables $G_v$ and $G_{\text{rate}}$ to achieve information interaction based on cross-attention process, which facilitates the gradual refinement of each by the other. Moreover, since $G_{\text{rate}}$ is used to scale elements of $G_v$, it can be viewed as the estimation of semantic importance. Therefore, the JSC encoder exploits $G_{\text{rate}}$ to generate a mask for variable-length coding. Specifically, a neural network can take $G_{\text{rate}}$ as input to predict an appropriate mask to achieve bandwidth adaptation. The $i$-th row of the mask $M \in \mathbb{R}^{l_v \times d}$ includes the first $k_i$ ones followed by $(d - k_i)$ zeros ($k_i \in [0, d]$), which indicates that $M$ is used to retain the first $k_i$ channels of $G_{v,i}$ and mask the rest channels.

Instead of using $G_{\text{rate}}$ from the final layer to predict the mask, the JSC encoder utilizes $G_{\text{rate}}$ from all layers to achieve a more informed and reasonable prediction. The network structure of JSC encoder is illustrated in Fig. 9.3. Assuming there are $L$ blocks, $L$ rate predictors are introduced, each of which is located at the end of each block to make a prediction based on the current $G_{\text{rate}}$. Specifically, the $l$-th ($l \in [1, L-1]$) rate predictor uses $l$-th $G_{\text{rate}}$, i.e., $G_{\text{rate}}^l$ as input to give current prediction. Firstly, the $l$-th rate predictor model the local information of $G_{\text{rate}}^l$ by a fully connected layer:

$$Z_{\text{local}}^l = \text{FC}(G_{\text{rate}}^l) \in \mathbb{R}^{l_v \times d}, \tag{9.13}$$

where FC$(\cdot)$ denotes the fully connected layer with GeLU$(\cdot)$ activation. Then, the global information of $G_{\text{rate}}^l$ is given by average pooling $Z_{\text{local}}^l$ along the frame dimension:

$$Z_{\text{global}}^l = \text{Avg}(Z_{\text{local}}^l) \in \mathbb{R}^{1 \times d}, \tag{9.14}$$

where Avg$(\cdot)$ denotes the average pooling operation. $Z_{\text{global}}^l$ is expanded to the same shape of $Z_{\text{local}}^l$ ($\mathbb{R}^{1\times d} \to \mathbb{R}^{l_v \times d}$) and combined with it along the channel dimension to form $Z_{\text{rate}}^l$:

$$Z_{\text{rate}}^l = \text{Concat}(Z_{\text{local}}^l, Z_{\text{global}}^l) \in \mathbb{R}^{l_v \times 2d}, \tag{9.15}$$

where Concat$(\cdot)$ denotes the concatenation operation. Finally, $Z_{\text{rate}}^l$ is inputted into MLP with softmax normalization, obtaining the $l$-th decision score $D_l \in \mathbb{R}^{l_v \times q}$:

$$D_l = \text{Softmax}(\text{MLP}(Z_{\text{rate}}^l)), \tag{9.16}$$

where MLP$(\cdot)$ is MLP with GeLU$(\cdot)$ activation. Softmax$(\cdot)$ denotes the softmax operation. The last rate predictor uses corresponding $Z_{\text{rate}}^L$ and all previous decisions $D_l$ ($l \in [1, L-1]$) as inputs to predict the final decision $D \in \mathbb{R}^{l_v \times q}$.

$$D = \text{Softmax}(\text{MLP}(\text{concat}(Z_{\text{rate}}^L, \beta(D_1 \dots D_{L-1})))), \tag{9.17}$$

where $\beta(D_1 \dots D_{L-1})$ is the fusion of the previous decisions. For convenience, the average operation is adopted to fuse these decisions:

$$\beta(D_1 \dots D_{L-1}) = \frac{1}{L}\sum_{i=1}^{L-1} D_i, \tag{9.18}$$

where $D$ denotes the probability that each $G_{v,i}$ is classified into $q$ different fixed bandwidths. Here, $q$ is a predefined value, which is set to $q = \log_2 d$. The retained channels $k_i$ for each $G_{v,i}$ is selected from the predefined set $\{2, 4, 8, \cdots, d\}$. Note one needs to sample a specific bandwidth for each $G_{v,i}$ given $D$, which is non-differentiable. Thus, the Gumbel-Softmax method described in Sect. 7.3.2 is used to address this problem.

The above bandwidth allocation is only based on the video contents, which can further integrate SNR for more reasonable bandwidth adaptation. SNR is used as additional input of all rate predictors. Therefore, Each $D_l$ without the last one is given by:

$$D_l = \text{MLP}(Z_{\text{rate}}^l, \text{SNR}), \tag{9.19}$$

and the final decision $D$ is given by:

$$D = \text{MLP}(\text{Concat}(Z_{\text{rate}}^L, \beta(D_1 \dots D_{L-1}), \text{SNR})). \tag{9.20}$$

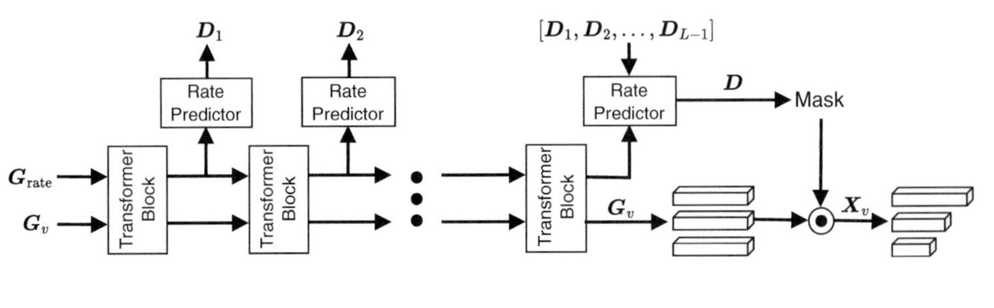


**Fig. 9.3** The network structure of JSC encoder

By replicating the SNR across all rate predictors, VideoQA-SC enhances robustness against noisy channels.

The experiments is conducted based on TGIF-QA dataset, which is one of the popular datasets for VideoQA. TGIF-QA dataset designs four unique task types, i.e., repetition count, repeating action, state transition and frame QA for evaluating the reasoning capabilities of models across different aspects. The comparison methods include VideoQA-SC and other reconstruction-based methods that adopt different video transmission schemes combined with a trained VideoQA model for inference. For traditional video transmission schemes based on separate source-channel coding, H264 and H265 is used for source coding and capacity-achievable channel codes are used for channel coding. For video transmission schemes based on DJSCC, advanced semantic video transmission method DVST is adopted. The upper bound "Error-free" scheme adopts lossless video at the receiver for inference and the lower bound "Text-only" scheme only uses text (questions and answers) at the receiver for inference without video transmission.

The performance comparison of different methods under different CBRs on repeating action task is shown in Fig. 9.4. One basic observation is that VideoQA-SC outperforms other methods across all CBRs and different SNRs. The performance gain is more pronounced under low SNRs, e.g., 0 dB. As reconstruction-based methods need to reconstruct original pixel-level videos at the receiver, they all have a turning point with the necessary transmission bandwidth and the corresponding minimum CBR $R_{\min}$. Once the available bandwidth $R < R_{\min}$, the performance of reconstruction-based methods is degraded rapidly because they lack the necessary information for video reconstruction. As SNR decreases, the necessary bandwidth $R_{\min}$ increases to resist the worse channel conditions. Although DJSCC-based semantic communication system DVST adopts an efficient entropy-aware bandwidth adaptation method, the gain is slight compared with traditional schemes ("H264/H265 + Channel capacity") for VideoQA, which indicates reconstruction-based methods are not efficient for VideoQA tasks. When $R < R_{\min}$ cannot be satisfied, DVST still experiences a rapid performance degradation.

Unlike reconstruction-based methods that perform video semantic extraction and reconstruction at the pixel level, VideoQA-SC processes video at object level and performs inference based on reconstructed semantics directly at the receiver, which significantly reduces bandwidth usage. Therefore, VideoQA-SC achieves promising performance across all CBRs, indicating the efficiency of communications. By adopting bandwidth adaptation ("VideoQA-SC with BA"), VideoQA-SC can make full use of the limited bandwidth and further improve the performance of the system. Particularly, VideoQA-SC achieves 5.17% accuracy gain compared with the DJSCC-based semantic communication system while using only 5% of the bandwidth when SNR is 0 dB.

## 9.3 Semantic Communication for HSI Super-Resolution Reconstruction

With the rapid advancement of AI and DL technologies, multi-source data fusion has demonstrated significant potential in enhancing information transmission and reconstruction quality, becoming a critical technical support for intelligent systems in perception and decision-making. In the field of autonomous driving, the real-time fusion of LiDAR point clouds and camera images provides vehicles with more comprehensive environmental perception capabilities, enabling high-precision obstacle detection and path planning. In medical imaging diagnostics, the collaborative processing of multimodal medical imaging data such as CT and MRI offers more accurate pathological analysis for telemedicine, providing doctors with comprehensive pathological information, thereby significantly optimizing image transmission and diagnostic efficiency in remote medical scenarios. In hyperspectral imaging, the fusion of HR-RGB and LR-HSI data to reconstruct HR-HSI has notably improved the accuracy of environmental monitoring and resource exploration.

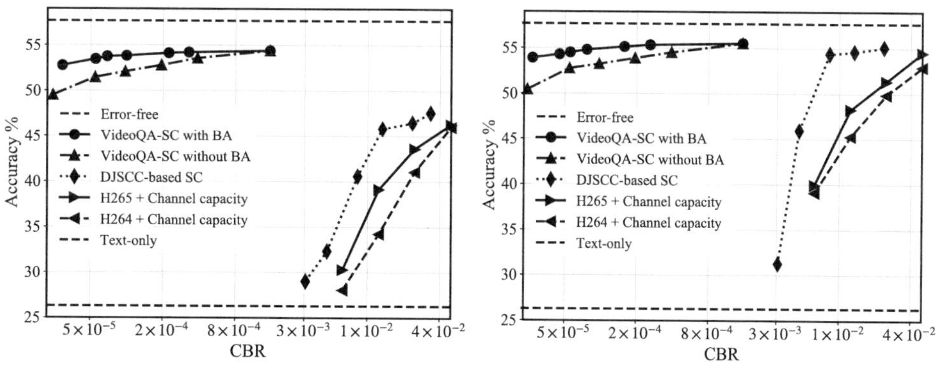


(a) SNR=0dB (b) SNR=10dB

**Fig. 9.4** Accuracy performance of different transmission schemes versus CBR over the AWGN channel

However, the heterogeneity and high-dimensional characteristics of multi-source data in these application scenarios impose severe bandwidth pressure on communication systems. Taking HR-HSI reconstruction as an example, the high-dimensional features after fusion significantly increase the communication load, further exacerbating bandwidth pressure. The HR-HSI reconstruction task requires the fusion of HR-RGB and LR-HSI to generate a target image with both high spatial resolution and high spectral resolution. In this process, the fused data must retain precise spatial details (such as object edges and texture information) while maintaining spectral continuity and ensuring the integrity of global structural information (such as scene semantics and object distribution). This multi-dimensional, multi-granularity feature demand leads to a substantial increase in load during multi-source data fusion. In bandwidth-constrained scenarios such as satellite remote sensing and disaster monitoring, existing communication systems struggle to achieve high-quality transmission with limited resources, further restricting practical application value.

Recent advances in semantic communication research have developed from single-modal data processing to multi-modal fusion approaches, driven by the expanding diversity of AI applications and data types. This evolution addresses the growing need for more sophisticated communication systems capable of handling complex, heterogeneous data streams.

In the domain of cross-modal semantic communication, researchers have developed innovative encoding methods that operate at the semantic level, enabling efficient adaptation to diverse user task requirements [6]. A novel approach involves dynamic adjustment of transmission features based on multi-modal channel conditions, implementing cross-modal semantic fusion for end-to-end data transmission.

The field has also seen the emergence of distributed multimodal semantic communication systems specifically designed for audio-visual parsing tasks. These systems employ distributed terminals to capture audio and video information, which is then processed through multi-task analysis at remote servers. By effectively fusing multi-modal data, these systems achieve notable improvement in parsing performance. Task-oriented communication systems based on multi-source have also made substantial progress, particularly in applications like VQA [7, 8]. These systems exploit the correlations between multimodal data across multiple users, employing deep neural network-based architectures with co-designed and co-optimized transmitters and receivers.

As illustrated in Fig. 9.5, semantic communication systems based on multi-source data fusion typically incur higher communication loads and bandwidth requirements compared to single-modal semantic communication approaches. In HSI super-resolution tasks, researchers have employed various advanced techniques to optimize feature extraction and fusion processes for spatial and spectral characteristics.

Recent studies have demonstrated significant progress in this domain. Convolutional neural networks with multiple kernels have been effectively utilized to extract and fuse spatial details with spectral features. The introduction of Transformer architectures has further enhanced super-resolution performance by establishing long-range dependencies along the spectral dimension during the fusion process. More recently, graph convolutional networks have been implemented to construct long-range fusion associations across both spatial and spectral dimensions, achieving superior reconstruction quality.

To address the challenges of limited bandwidth resources and dynamic channel conditions in multimodal data fusion, this chapter uses a method of multi-source data fusion for

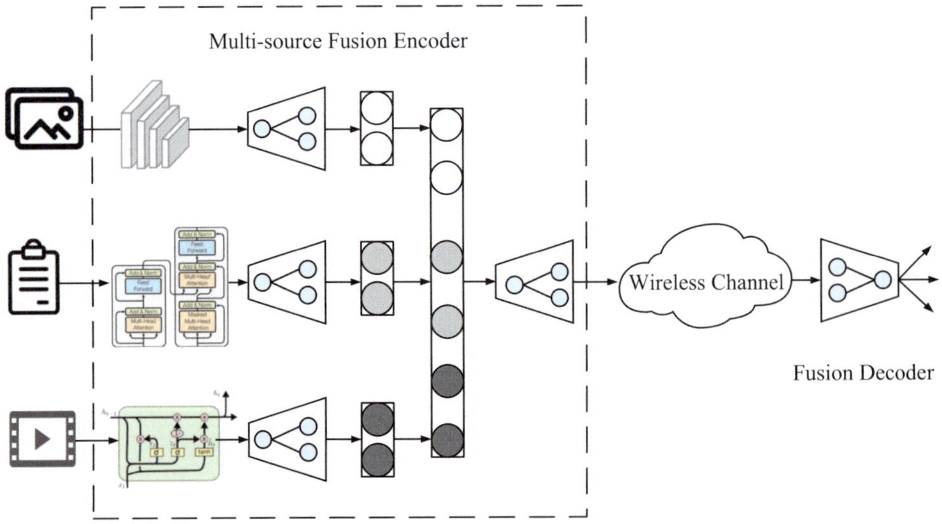


**Fig. 9.5** The impact of multi-source data fusion on bandwidth requirements

semantic communication system. The system incorporates hierarchical perception and channel-adaptive data fusion mechanisms, demonstrating superior performance under bandwidth-constrained conditions. The framework extracts spatial detail features from high-resolution RGB images and spectral characteristics from low-resolution HSI data respectively, subsequently fusing them to capture global structural features.

A key innovation is the implementation of a Transformer-based channel-adaptive attention mechanism. This enables hierarchical feature extraction and multi-scale fusion strategies, where critical features are dynamically selected for transmission based on real-time SNR conditions. By transmitting only the most essential features from each modality, the system achieves effective adaptive operation under strict bandwidth constraints while maintaining reconstruction quality.

Figure 9.6 illustrates the bandwidth-constrained semantic communication architecture based on multi-source data fusion. The system consists of a multi-source data semantic encoder at the transmitter and a semantic decoder at the receiver, forming an efficient multi-source data fusion semantic communication system. To achieve efficient information transmission under bandwidth constraints, the system incorporates a hierarchical perception and channel-adaptive fusion mechanism. This mechanism extracts and transmits key semantic information from multi-source data, effectively capturing complementary semantic features while compressing redundancy among multimodal data, thereby enabling efficient semantic communication under bandwidth-limited conditions. Moreover, the framework demonstrates broad application potential in fields such as remote sensing image processing, medical image analysis, and intelligent transportation systems.

In this system, $\boldsymbol{S}_1 \in \mathbb{R}^{w \times h \times L}$ and $\boldsymbol{S}_2 \in \mathbb{R}^{W \times H \times 3}$ represent LR-HSI and HR-RGB data respectively, both sampled at the same transmitter. $\boldsymbol{S}_3 \in \mathbb{R}^{W \times H \times L}$ denotes the corresponding HR-HSI reconstructed at the receiver. Here, $w \ll W$ and $h \ll H$ indicate that LR-HSI has significantly lower spatial resolution than HR-HSI, while $3 \ll L$ shows HR-HSI possesses higher spectral dimensionality. Therefore, HR-HSI reconstruction requires combining rich spectral and spatial information.

For efficient feature extraction, the transmitter employs dedicated spatial and spectral feature extraction modules to extract detail features from LR-HSI and HR-RGB respectively, with a fusion mechanism further extracting global features. The multi-source semantic encoder comprises three components: fusion feature extraction module $S_f^e(\cdot; \boldsymbol{\alpha}_f)$, spectral feature extraction module $S_s^e(\cdot; \boldsymbol{\theta}_s)$, and spatial feature extraction module $S_p^e(\cdot; \boldsymbol{\delta}_p)$, where $\boldsymbol{\alpha}_f$, $\boldsymbol{\theta}_s$, and $\boldsymbol{\delta}_p$ represent their respective trainable parameters.

The feature extraction process at the transmitter can be formulated as:

$$\begin{aligned} \boldsymbol{X}_s &= S_s^e\left(\boldsymbol{S}_1; \boldsymbol{\theta}_s\right), \\ \boldsymbol{X}_p &= S_p^e\left(\boldsymbol{S}_2; \boldsymbol{\delta}_p\right), \\ \boldsymbol{X}_f &= S_f^e\left(\boldsymbol{X}_s, \boldsymbol{X}_p; \boldsymbol{\alpha}_f\right), \end{aligned} \tag{9.21}$$

where $\boldsymbol{X}_s, \boldsymbol{X}_p, \boldsymbol{X}_f \in \mathbb{R}^{w_1 \times h_1 \times l}$ denote spectral features, spatial features, and fused features respectively. The compressed channel dimension ($l/L \leq 1/2$) reduces bandwidth consumption during transmission.

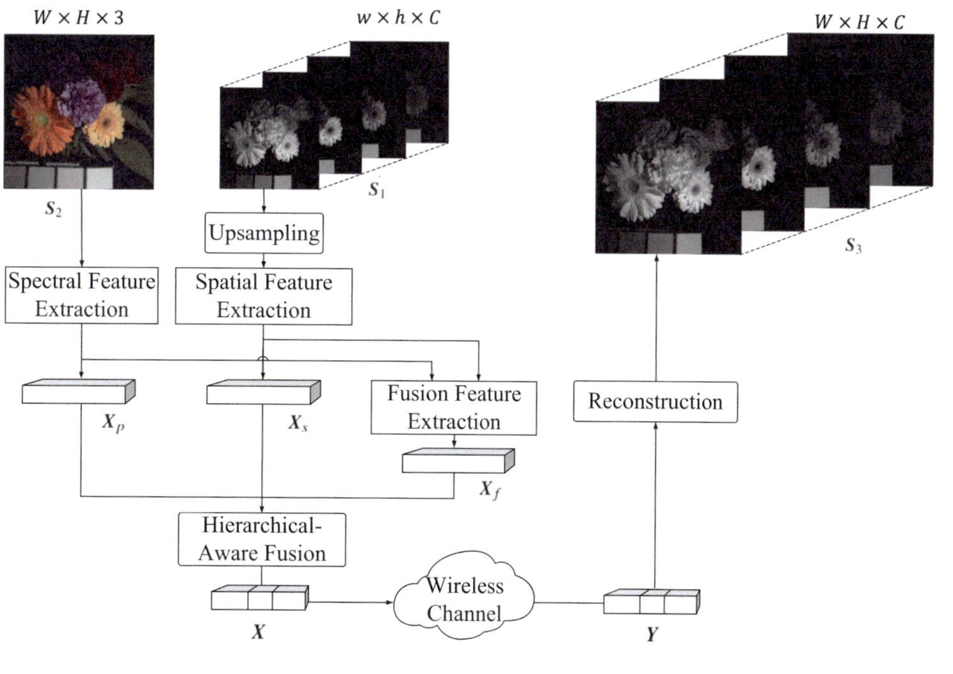


**Fig. 9.6** The bandwidth-constrained semantic communication system based on multi-source data fusion

For HSI super-resolution, both global structural features and spectral/spatial details are crucial. Conventional approaches that fuse deep global features with shallow details during decoding would impose excessive bandwidth demands in wireless scenarios. To address this, a hierarchical perception fusion module is used to adaptively integrates features without increasing bandwidth requirements:

$$\boldsymbol{X} = F\left(\boldsymbol{X}_f, \boldsymbol{X}_s, \boldsymbol{X}_p; \boldsymbol{\beta}\right), \tag{9.22}$$

where $\boldsymbol{X} \in \mathbb{R}^{w_1 \times h_1 \times l}$ maintains dimensional consistency with input features to avoid bandwidth expansion.

In AWGN channels with noise power $\sigma^2$, the received features are:

$$\boldsymbol{Y} = \boldsymbol{X} + \boldsymbol{N}_0, \tag{9.23}$$

where $\boldsymbol{N}_0$ represents channel noise.

The semantic decoder $D(\cdot; \boldsymbol{\rho})$ reconstructs HR-HSI from denoised features:

$$\hat{S}_3 = D\left(\boldsymbol{Y}; \boldsymbol{\rho}\right), \tag{9.24}$$

with $\hat{S}_3 \in \mathbb{R}^{W \times H \times L}$ representing the super-resolved output. The system optimizes reconstruction quality using MSE loss:

$$\mathcal{L}_{\text{MSE}}(S_3, \hat{S}_3; \boldsymbol{\alpha}_f, \boldsymbol{\theta}_s, \boldsymbol{\delta}_p, \boldsymbol{\beta}, \boldsymbol{\rho}) = \left\| S_3 - \hat{S}_3 \right\|_2^2, \tag{9.25}$$

where all parameters $(\boldsymbol{\alpha}_f, \boldsymbol{\theta}_s, \boldsymbol{\delta}_p, \boldsymbol{\beta}, \boldsymbol{\rho})$ are jointly trainable to minimize reconstruction error.

The following content provides a detailed introduction to the bandwidth-constrained semantic communication network architecture based on multi-source data fusion. Comprising transmitter and receiver components, this network is designed to achieve efficient multi-source data fusion and semantic communication under bandwidth-limited conditions.

The transmitter consists of: (1) A spectral feature extraction module (2) A spatial feature extraction module (3) A fusion feature extraction module (4) A hierarchical-aware fusion module

The receiver incorporates a semantic decoder responsible for HR-HSI reconstruction. By adaptively fusing deep structural features with shallow detail features, this framework effectively enhances data fusion performance while ensuring efficient transmission of critical semantic information under bandwidth constraints.

**Spectral and Spatial Feature Extraction Modules**

Shallow detail features are important components in multi-source data fusion, mainly derived from the outputs of the spectral feature extraction module and spatial feature extraction module. These features contain rich local information, such as subtle spectral variations and spatial edge textures, providing an important foundation for subsequent feature fusion and HR-HSI reconstruction.

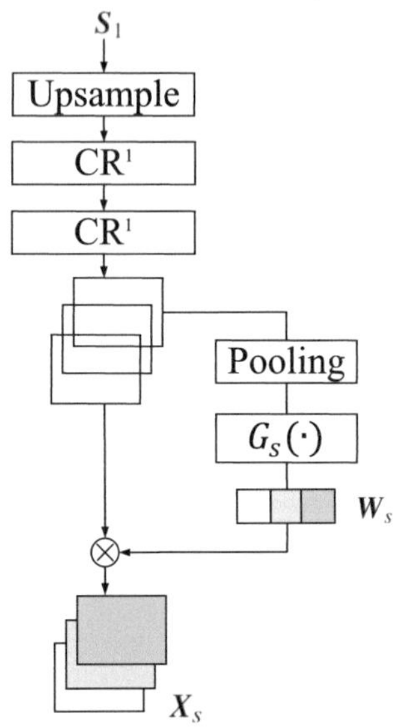


**Fig. 9.7** Structure of the spectral feature extraction module

The spectral feature extraction module extracts detailed spectral channel features from LR-HSI, as shown in Fig. 9.7. Since LR-HSI has high spectral dimensionality, the module adaptively selects important spectral channels through a spectral attention mechanism, thereby reducing redundant information while retaining key spectral features. Specifically, LR-HSI has low spatial resolution, so an upsampling operation is first performed in the spatial dimension to make its spatial resolution consistent with HR-HSI. This process can be expressed as:

$$\boldsymbol{S}_1^{up} = \text{Upsample}\left(\boldsymbol{S}_1\right), \tag{9.26}$$

where Upsample $(\cdot)$ represents the upsampling operation. $\boldsymbol{S}_1^{up}$ denotes the upsampled LR-HSI. To preserve detailed information in the spectral dimension, the spectral feature extraction module adopts a relatively shallow network structure, including two convolutional layers with ReLU as the activation function. Specifically:

$$\boldsymbol{S}_s = F_{cr^1}\left(F_{cr^1}\left(\boldsymbol{S}_1^{up}\right)\right), \tag{9.27}$$

where $F_{cr^1}(\cdot)$ represents a convolutional layer with kernel size 1 and ReLU as the activation function. To enhance the extraction of correlations between spectral channels, the spectral attention mechanism is a key part of the spectral feature extraction module, aiming to dynamically adjust the contribution of each channel by calculating importance weights for each spectral channel, thereby enhancing the expressive capability of spectral features. The specific steps are as follows:

$$\begin{aligned} \boldsymbol{W}_s &= G_s\left(\text{Pooling}\left(\boldsymbol{S}_s\right)\right), \\ \boldsymbol{X}_s &= \boldsymbol{W}_s \otimes \boldsymbol{S}_s, \end{aligned} \tag{9.28}$$

where Pooling $(\cdot)$ represents the pooling operation. $G_s(\cdot)$ represents the spectral factor prediction network.

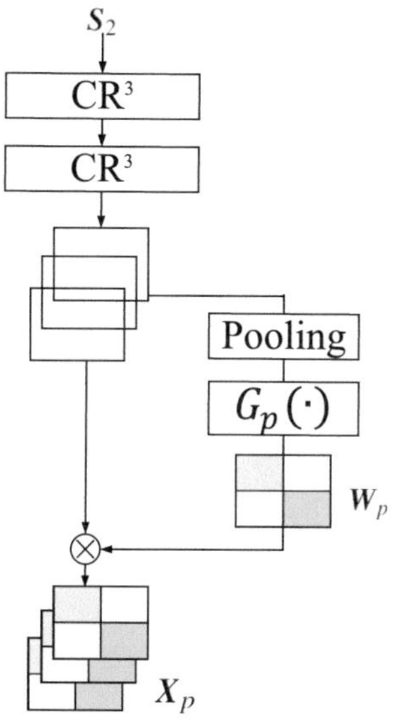


**Fig. 9.8** The structure of spatial feature extraction module

The spatial feature extraction module focuses on extracting spatial dimensional detail features from HR-RGB, as shown in Fig. 9.8. Similar to the spectral feature extraction module, the spatial feature extraction module also adopts a relatively shallow network structure, but its convolution kernel size is 3, to better capture local information in the spatial dimension, such as edges, textures, and other spatial structural features. The detailed process of the spatial feature extraction module:

$$\begin{aligned} S_p &= F_{cr^3}\left(F_{cr^3}\left(S_2\right)\right), \\ W_p &= G_p\left(\text{Pooling}\left(S_p\right)\right), \\ X_p &= W_p \otimes S_p, \end{aligned} \tag{9.29}$$

where $F_{cr^3}(\cdot)$ represents a convolutional layer with kernel size 3 and ReLU as the activation function. $G_p(\cdot)$ represents the spatial factor prediction network.

The spectral feature extraction module and spatial feature extraction module extract key spectral and spatial features from LR-HSI and HR-RGB, respectively. Through the spectral and spatial attention mechanisms, these two modules can adaptively enhance the contribution of important regions (spectral channels or spatial positions), thereby preserving spectral and spatial detail information while reducing redundant information. This design provides high-quality spectral and spatial features for subsequent multi-source data fusion and HR-HSI reconstruction.

**Fusion Feature Extraction Module**

The fusion feature extraction module is responsible for extracting global structural information from shallow detail features. Deep global features can capture the overall structure of the image, thereby improving the reconstruction quality of HR-HSI. The fusion feature extraction module

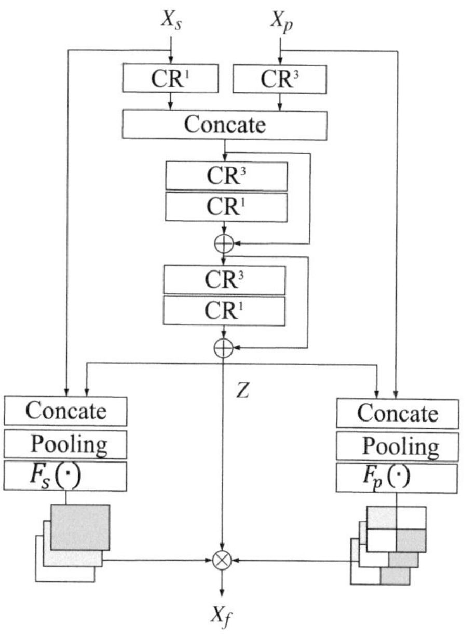


**Fig. 9.9** The structure of the fused feature extraction module at the transmitter

integrates spectral features $X_s$ and spatial features $X_p$, performs preliminary fusion of spectral and spatial features, and further extracts global features through multiple residual blocks as shown in Fig. 9.9. This process is mathematically represented as follows:

$$Z = F_f\left(F_f\left(\left[F_{cr^1}\left(X_s\right), F_{cr^3}\left(X_p\right)\right]\right)\right), \tag{9.30}$$

where $[\cdot]$ denotes the concatenation operation, and $Z$ represents the initial fused features. $F_f(\cdot)$ denotes the residual block operation, defined as:

$$F_f(x) = x + F_{cr^3}\left(F_{cr^1}(x)\right). \tag{9.31}$$

Additionally, the fusion feature extraction module introduces a dual attention mechanism in both spectral and spatial dimensions. After concatenating the preliminary fused features with shallow spectral and spatial features, the module further weights the fused features in different dimensions through spectral and spatial feature factor prediction networks, thereby enhancing the expressive capability of the fused features. Through this hierarchical feature extraction and fusion mechanism, the system can effectively capture complementary semantic features in multi-source data while compressing information redundancy between multimodal data, thus achieving efficient semantic communication under bandwidth-constrained conditions. This process is mathematically represented as follows:

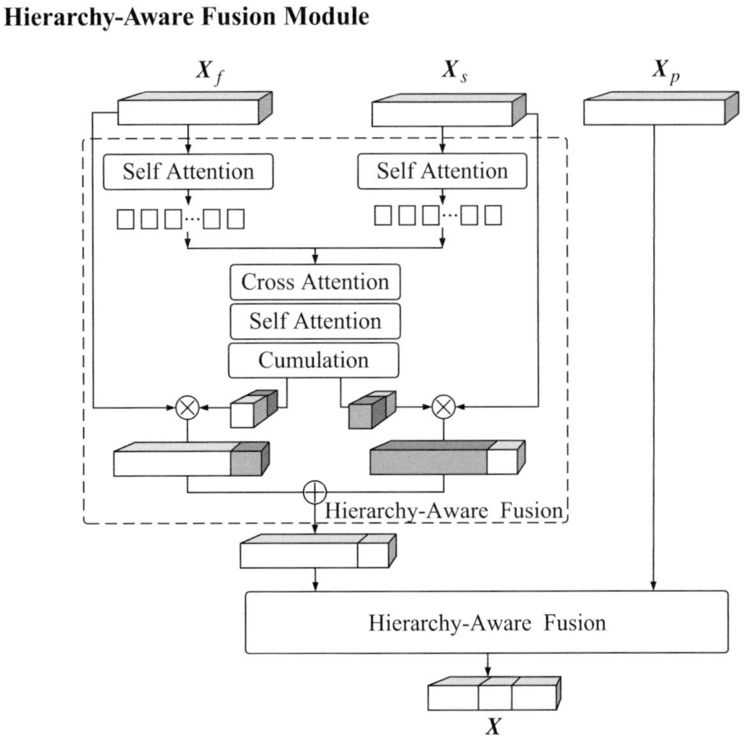


**Fig. 9.10** The structure of the hierarchy-aware and channel-adaptive fusion module [9]

$$
\begin{aligned}
W_f^p &= F_p\left(\text{Pooling}\left([X_p, Z]\right)\right), \\
W_f^s &= F_s\left(\text{Pooling}\left([X_s, Z]\right)\right), \\
X_f &= Z \otimes W_f^s \otimes W_f^p,
\end{aligned}
\tag{9.32}
$$

where $\otimes$ denotes the element-wise multiplication operation. $F_s(\cdot)$ and $F_p(\cdot)$ represent the enhancement factor extraction modules for the spectral and spatial domains, respectively.

**Hierarchy-Aware Fusion Module**

The hierarchy-aware data fusion module integrates shallow detail features with deep global features through a channel-adaptive mechanism. Specifically, the module first performs channel-wise weight allocation for both shallow and deep features, calculating the importance of each channel. Then, it combines the shallow detail features with deep global features through weighted summation. This process can be expressed as.

The channel-adaptive hierarchy-aware data fusion module is a key technology in the bandwidth-constrained semantic communication system, aiming to adaptively integrate deep and shallow features using advanced Transformer architecture without additional bandwidth consumption. Deep features are used to extract global structural information, while shallow features preserve spectral and spatial details. As shown in Fig. 9.10, to simultaneously transmit both global structural features and detail features under bandwidth constraints, a hierarchy-aware fusion module is introduced. By dynamically generating guidance features, it guides the adaptive combination of deep and shallow features, achieving efficient feature fusion and optimization.

The module structure consists of two channel-adaptive fusion blocks that effectively integrate key features by combining Self-Attention (SA) [10] and Cross-Attention (CA) [11]. The Transformer-based channel-adaptive attention module computes dynamic guidance features $N_{(f,p)}, N_{((f,p),s)} \in \mathbb{R}^{w_1 \times h_1 \times l}$ through an SA-CA-SA sequence with softmax function. These guidance features are then processed through an accumulation function to generate mask features $M_{(f,p)}, M_{((f,p),s)} \in \mathbb{R}^{w_1 \times h_1 \times l}$, which dynamically guide the fusion of deep and shallow features. Here, the subscript $(f, p)$ denotes the integration of features $X_f$ and $X_p$, while $((f, p), s)$ represents the further integration of feature $X_s$ after combining $X_f$ and $X_p$.

Within the channel-adaptive fusion block, features $X_f$, $X_p$ and $X_s$ are divided into patches and fed into the SA module. The CA module efficiently extracts interactive information from input feature pairs by combining their complementarity, thereby enhancing the expressive capability of the output features. The features processed by the CA module undergo further SA processing combined with softmax activation to generate guidance features $N_{(f,p)}$. The mask features are obtained through the cumulative calculation of the guidance features:

$$M_{(f,p)}^{(i)} = \begin{cases} N_{(f,p)}^{(i)} & \text{for } i = 1, \\ M_{(f,p)}^{(i-1)} + N_{(f,p)}^{(i)} & \text{for } i = 2, \ldots, l, \end{cases} \quad (9.33)$$

The computation of mask feature $\boldsymbol{M}_{((f,p),s)}$ follows the same approach as $\boldsymbol{M}_{(f,p)}$. These mask features $\boldsymbol{M}_{(f,p)}$ and $\boldsymbol{M}_{((f,p),s)}$ dynamically guide the feature fusion process, effectively combining global structural feature $\boldsymbol{X}_f$ while preserving detail features $\boldsymbol{X}_p$ and $\boldsymbol{X}_s$. This achieves efficient integration of multiple features, where the fused features maintain both global structural information and the spectral/spatial details required for super-resolution reconstruction. The mathematical formulation is:

$$\begin{aligned} \boldsymbol{X}_{(f,p)} &= \boldsymbol{X}_f \times \boldsymbol{M}_{(f,p)} + \boldsymbol{X}_p \times \left(\mathbf{1} - \boldsymbol{M}_{(f,p)}\right), \\ \boldsymbol{X} &= \boldsymbol{X}_{(f,p)} \times \left(\mathbf{1} - \boldsymbol{M}_{((f,p),s)}\right) + \boldsymbol{X}_s \times \boldsymbol{M}_{((f,p),s)}. \end{aligned} \quad (9.34)$$

where $\mathbf{1}$ denotes a matrix with all elements equal to 1.

**Reconstruction Module**

At the receiver, the semantic decoder aims to reconstruct HR-HSI from the received noisy fused features. To address the bandwidth-constrained conditions in semantic communication, the decoder incorporates fusion masks to enhance the reconstruction capability of detail features and optimize the overall reconstruction quality of HSI. Assuming the fusion masks are known at the receiver, these masks occupy minimal bandwidth, which is negligible compared to the bandwidth required for transmitting complete reconstructed features. During the decoding phase, the fusion masks guide the decoder to dynamically extract both global structural features and detail features, effectively improving the quality and resolution of the reconstructed HSI.

As shown in Fig. 9.11, the core structure of the fusion decoder consists of multiple deconvolution layers (Convolution Transpose Layer, CTL), each combined with a ReLU activation function. These CTLs progressively expand the spatial dimensions of the input features while adaptively adjusting the parsing process of the received fused features using the spectral and spatial guidance information provided by the fusion masks. Through this approach, the fusion decoder can better recover the spectral and spatial information in the fused features, ensuring consistency in both detail and structure of the reconstructed HSI, thereby achieving high-quality HR-HSI reconstruction. This process can be expressed as:

$$\begin{aligned} \hat{S}_3 = D\left(\boldsymbol{Y}; \boldsymbol{\rho}\right) &= F_{ctl^3}\left(\ldots F_{ctl^3}\left(\boldsymbol{Y}\right)\right) \\ &+ F_{ctl^3}\left(\boldsymbol{Y} \times \boldsymbol{M}_{(f,p)}\right) + F_{ctl^3}\left(\boldsymbol{Y} \times \left(\mathbf{1} - \boldsymbol{M}_{((f,p),s)}\right)\right), \end{aligned} \quad (9.35)$$

where $F_{ctl^3}(\cdot)$ represents the deconvolution layer combined with ReLU activation function.

To verify the effectiveness of the used method, a performance evaluation was conducted for the semantic communication approach labeled as "HACA-SC", in comparison with single-source semantic communication methods. The experiments were carried out under varying values of $l$ (which is proportional to the required communication bandwidth) and SNR conditions ranging from –5 to 9 dB for both training and testing.

As shown in Fig. 9.12, the experimental results demonstrate that HACA-SC significantly outperforms single-source semantic communication methods across the entire tested SNR range. This validates the effectiveness of multi-source

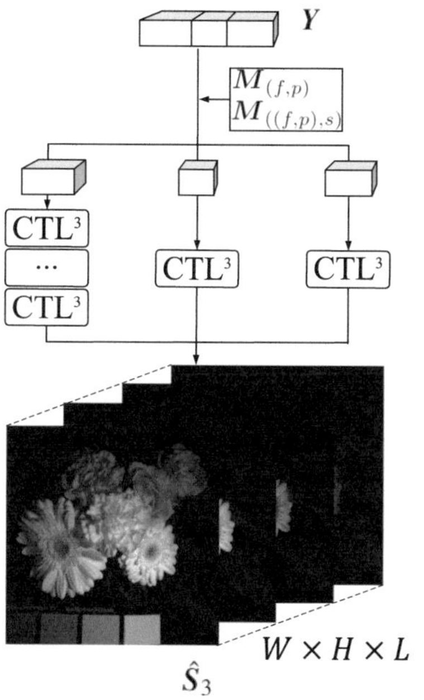


**Fig. 9.11** The structure of the reconstruction module [9]

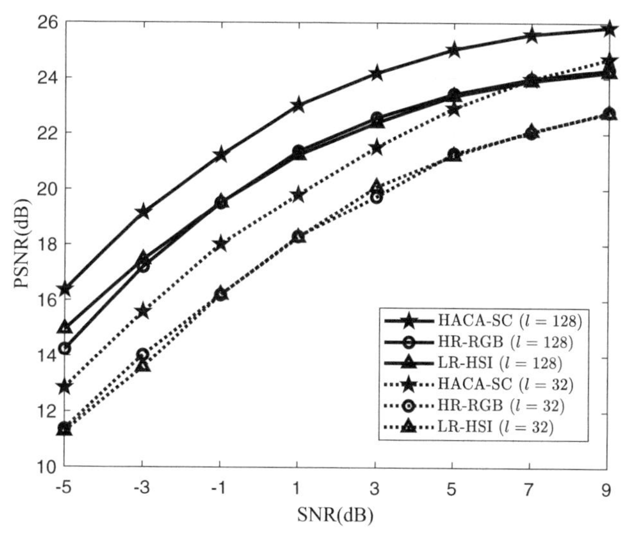


**Fig. 9.12** PSNR comparisons between single-source and HACA-SC for different values of $l$ [9]

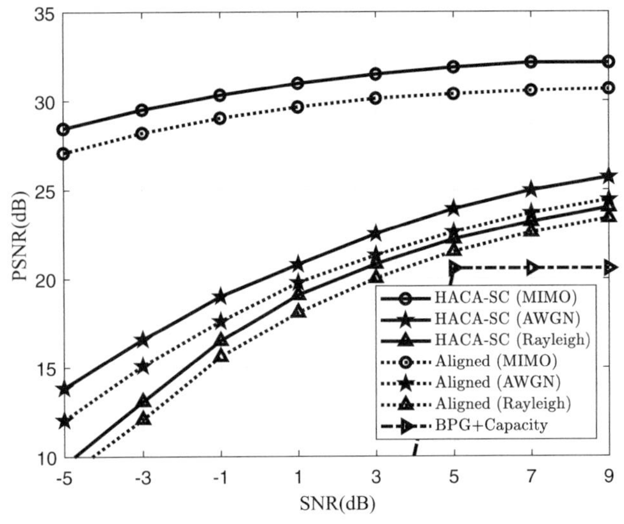


**Fig. 9.13** Performance comparisons of HACA-SC on the CAVE dataset [9]

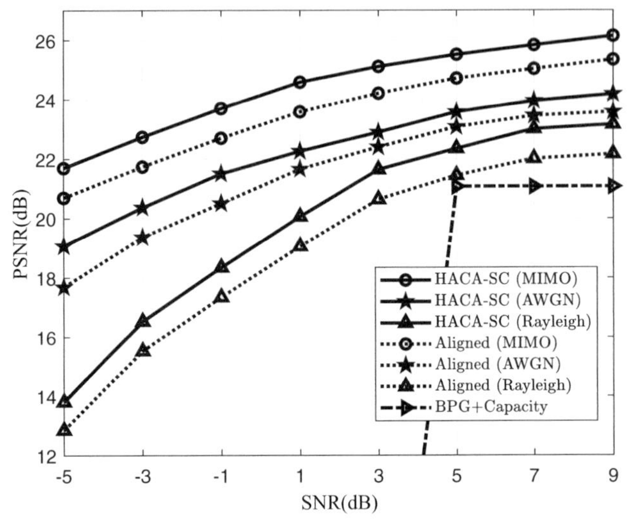


**Fig. 9.14** Performance comparison of HACA-SC on the DC Mall dataset [9]

data fusion based semantic communication in improving HR-HSI reconstruction quality.

As shown in Figs. 9.13 and 9.14, the experimental results demonstrate the PSNR performance of HACA-SC on the CAVE and DC Mall datasets, with comparative analysis against the alignment-based multi-source data fusion semantic communication framework [12, 13] labeled as "Aligned" and the traditional "BPG+Capacity" scheme. The experiments encompass various channel conditions, including AWGN channels, Rayleigh channels, and MIMO channels, to evaluate the robustness and adaptability of the method under different channel types. To ensure fair comparison, the impact of the diffusion model-based adaptive denoising module was similarly excluded when comparing with the "Aligned" method.

The experimental results indicate that HACA-SC outperforms both the alignment-based fusion framework and the "BPG+Capacity" scheme across all tested channel conditions, demonstrating superior reconstruction quality and stronger resistance to channel noise. Particularly on the DC Mall dataset with richer spectral information, HACA-SC achieves significantly higher PSNR values than the comparison methods under all channel conditions. For instance, in low SNR AWGN channels, HACA-SC improves PSNR by approximately 1.5 dB compared to the alignment-based framework and far exceeds the "BPG+Capacity" scheme. Similarly, in Rayleigh and MIMO channels, HACA-SC maintains performance advantages, validating its superiority in handling more complex multi-source data fusion tasks.

## 9.4 Generative Semantic Communication

The evolution from traditional communication to semantic communication can be viewed as a transition from transmitting raw bits to conveying meaningful symbols. In traditional communication systems, information is typically transmitted as a sequence of bits, where the focus is on the accurate and efficient delivery of data. However, in semantic communications, the emphasis shifts toward transmitting not just the data, but the meaning of the data itself.

To draw an analogy, consider your experience with generative AI tools such as ChatGPT, Stable Diffusion, or DALL·E. When you provide a brief prompt to these systems, they generate rich, meaningful outputs—whether in the form of coherent text or vivid images. For example, DALL·E 3 can be used to produce image datasets [14]. Despite receiving minimal input, these tools produce detailed, contextually relevant responses, which is akin to the core idea of semantic communication: conveying meaning rather than simply transmitting data.

When we combine generative models with semantic communication principles, we enter the domain of generative semantic communication (GSC). In contrast to traditional semantic communication, which focuses on the recovery of information (e.g., reconstructing an image or video from lossy data), GSC prioritizes the regeneration of information. This involves creating new content that preserves the essential meaning of the original, but allows for variability in form and structure.

In short, GSC represents a paradigm shift in wireless communication by leveraging advanced generative AI models to enhance the transmission and reconstruction of semantic information. Unlike traditional communication systems that focus on bit-level accuracy, GSC prioritizes the perceptual quality and contextual meaning of the transmitted data. This chapter explores the integration of generative models, such as diffusion models, into semantic communication frameworks, highlighting their advantages and applications.

**The Potential of Generative Models**

Generative models revolutionize semantic communication by enabling task-aware data regeneration and semantic fidelity preservation, transcending traditional bit-level accuracy. Their unique capabilities include:

- Semantic-Guided Generation: Generative models (e.g., diffusion models, GANs, VAEs) decode compressed semantic features (e.g., object contours, text prompts) into contextually meaningful data. For instance, (1)Diffusion models reconstruct photorealistic scenes from noisy semantic maps, preserving critical task-specific details (e.g., vehicle positions in traffic monitoring) while discarding irrelevant pixel-level noise. (2) GANs synthesize high-fidelity multimedia (e.g., speech, video) from low-dimensional semantic vectors, which highly reduces the bandwidth cost
- Robustness to Channel Imperfections: Generative models mitigate channel noise by leveraging data-driven semantic priors rather than purely mathematical redundancy. Shifting from error correction to generation, generative models establish an a prior understanding of "meaning", so that even if the symbols are distorted by channel noise, the model can still reason about the most reasonable interpretation based on the semantic structure rather than mechanically reducing the symbols. For example, (1) Conditional denoising diffusion probabilistic models (DDPMs) iteratively refine noisy inputs, recovering semantic integrity even under severe channel distortions. (2) VAEs map corrupted signals to structured latent spaces, enabling reliable reconstruction for medical imaging or IoT sensor data.
- Multimodal Semantic Extraction and Fusion: Generative models unify diverse data modalities (text, audio, 3D point clouds) into coherent semantic representations: (1) Text-to-image diffusion models (e.g., Stable Diffusion) generate scenes from textual descriptions, enabling energy-efficient metaverse applications. (2)Hybrid encoders prioritize task-relevant semantics (e.g., pedestrian locations over weather details), optimizing bandwidth usage.

**Deep Generative Models in Semantic Communication**

While deep generative models (DGMs) have demonstrated remarkable success in fields like computer vision and natural language processing, their application within communication frameworks remains significantly underexplored. These models have the potential to revolutionize semantic communication frameworks, where the objective is not merely to recover transmitted messages as encoded sequences of bits, but to regenerate semantically consistent content at the receiver.

Transceivers in semantic communication systems can be significantly enhanced by optimizing generative models for information comprehension, transmission and management. Essentially, the purpose of a semantic transceiver is to extract semantics at the sender's end and to restore the semantics at the receiver's end with minimal semantic error under different channel conditions [15].

DGMs focus on understanding the intrinsic structure and patterns within the data, capable of generating high-dimensional content from low-dimensional noise, making it suitable to regenerate the compressed or corrupted source information.

In the point-to-point JSCC transmission system structure, the transmitter processes image $\boldsymbol{s}$ to obtain the symbol $\boldsymbol{x}$ through a joint encoder to realize this encoding process. Subsequently, $\boldsymbol{x}$ is transmitted through a noisy channel, which introduces uncertainty into the communication system, denoted as $\boldsymbol{y}$. The receiver, in turn, works to reconstruct the image $\boldsymbol{s}$ through a joint decoder.

Typical DGMs are GANs, diffusion models, Transformers and VAEs. GANs are adversarial in nature and involve a game between two deep learning sub-models called the generator and discriminator. Usually, the discriminator is used for train supervising ,and the generator is actually the model to create contents. Diffusion models have two diffusion processes. The forward diffusion process is a process of turning an image into noise, and the reverse diffusion process is supposed to turn that noise into the image. But the forward diffusion process usually plays a guidance role to help the reverse diffusion process learn how to eliminate noise. Transformer can be viewed as a backbone network to process features. Although it is widely used in NLP, Transformer models for other modalities are spring up, making them highly versatile. Traditionally, VAEs rely on two neural networks called encoder and decoder, working to compress and reconstruct data, which is useful for semantic tasks. The dual networks work in tandem, so they could directly replace the encoder and decoder in semantic communication systems.

In essence, DGMs possess inherent architectural compatibility to serve as encoders, decoders, or dual-function modules within semantic communication frameworks, leveraging their hierarchical representation learning capabilities to bridge semantic and physical layers.

**a. GAN-based Semantic Communication**

A deep conditional generative semantic communication (DCG-SC) system for image transmission [16] explores the

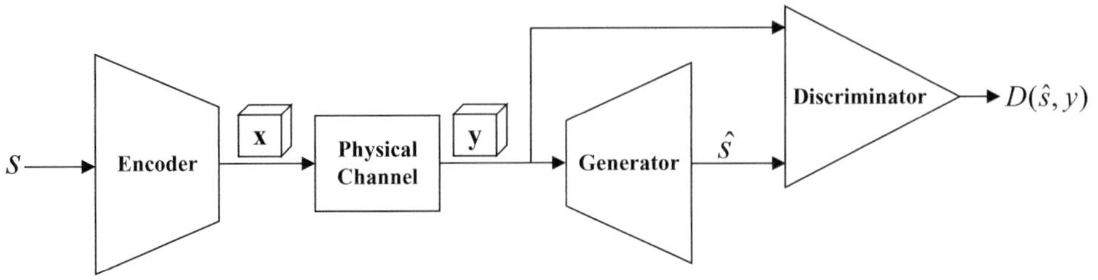


**Fig. 9.15** Framework of DCG-SC in [16]

application of generative models in semantic communication, which establishes a trade-off among distortion, perceptual quality, and transmission rate. DCG-SC consists of three main components: an encoder $E$, a generator $G$, and a discriminator $D$, as shown in Fig. 9.15. The decoder in classical autoencoder framework is replaced by the generator of conditional GAN. This design effectively integrates the principles of both autoencoders and conditional GANs.

The input image $\boldsymbol{s}$ is first processed by $E$, yielding a semantic feature vector $\boldsymbol{x}$. The feature vector plays two key roles: it encapsulates the semantic properties derived from $\boldsymbol{s}$ and also acts as conditional input for $G$ and $D$ subsequently. As the feature vector $\boldsymbol{x}$ is communicated via the physical channel, it encounters channel noise $\boldsymbol{n}$, which alters it to the received feature representation $\boldsymbol{y}$ that may be distorted. The function of $E$ involves two primary aspects. First, it targets the efficient encoding of images, thereby minimizing the symbols needed for representation. Second, it improves robustness by minimizing the effects of channel noise. The transmission rate, represented by $r(\boldsymbol{x})$, measures the cost of encoding $\boldsymbol{x}$, reflecting its length as a semantic feature vector. At the receiver, the semantic feature vector $\boldsymbol{y}$ received acts as the conditional input for $G$, which then reconstructs an image $\hat{\boldsymbol{s}}$. At the same time, the discriminator $D$ assesses the genuineness of the generated image, steering $G$ toward creating outputs that are visually convincing and consistent with human perception.

To assess the quality of the reconstructed image $\hat{\boldsymbol{s}}$, two primary distortion metrics are usually employed. MSE function is used to measure the pixel-wise reconstruction error between the original image $\boldsymbol{s}$ and its reconstructed counterpart $\hat{\boldsymbol{s}}$, denoted as $d_M(\boldsymbol{s}, \hat{\boldsymbol{s}})$. LPIPS evaluates perceptual differences in the feature space of a pre-trained deep neural network (e.g., VGG), capturing high-level structural and semantic discrepancies. The perception distance is denoted as $d_P(\boldsymbol{s}, \hat{\boldsymbol{s}})$. The training objective consists of two loss functions. The first loss function is formulated as:

$$\mathcal{L}_1 = \mathbb{E}_{s\sim p_S}[k_r \underbrace{r(\boldsymbol{x})}_{\text{Rate}} + k_m \underbrace{d_M(\boldsymbol{s}, \hat{\boldsymbol{s}})}_{\text{Distortion}} + k_p \underbrace{d_P(\boldsymbol{s}, \hat{\boldsymbol{s}})}_{\text{Perception}}], \quad (9.36)$$

where $k_r, k_m, k_p$ are hyper-parameters that balance the trade-off among transmission rate, distortion, and perceptual quality. $r(\boldsymbol{x})$, $d_M(\boldsymbol{s}, \hat{\boldsymbol{s}})$, $d_P(\boldsymbol{s}, \hat{\boldsymbol{s}})$ represent the transmission rate, distortion, and perception distance, respectively. $L_1$ loss indicates the capabilities of $E$ and $G$ in image encoding and generation. Minimizing $L_1$ loss can improve the capabilities of $E$ and $G$. The second loss function $L_2$ follows the standard conditional GAN loss formulation:

$$\mathcal{L}_2 = \mathbb{E}_{s\sim p_S}[-\log(1 - D(\hat{\boldsymbol{s}}, \boldsymbol{y}))] + \mathbb{E}_{s\sim p_S}[-\log(D(\boldsymbol{s}, \boldsymbol{y}))], \quad (9.37)$$

where the discriminator $D$ learns to distinguish real images $\boldsymbol{s}$ from generated images $\hat{\boldsymbol{s}}$ conditioned on the received feature vector $\boldsymbol{y}$. Likewise, minimizing $L_2$ loss can improve the capabilities of $E$, $G$, and $D$ in generating high-quality images that align with the original data distribution.

DCG-SC regards the extracted semantic features as supplementary information or conditioning cues that aid the receiver in reconstructing data effectively. This approach aligns with the standard procedure of DJSCC but introduces a significant change: a conditional GAN generator replaces the conventional joint semantic-channel decoder. The received channel output $\boldsymbol{y}$ is utilized in two distinct manners: (1) it serves as the input for the generator $G$, resulting in the reconstructed image $\hat{\boldsymbol{s}} = G(\boldsymbol{y})$, and (2) it acts as an additional input for the discriminator $D$, which assesses both actual and generated images through $D(\boldsymbol{s}, \boldsymbol{y})$ and $D(\hat{\boldsymbol{s}}, \boldsymbol{y})$.

**b. Diffusion Model-based Semantic Communication**

Diffusion models, particularly DDPM, have demonstrated remarkable success in generating high quality, semantically meaningful data. Their iterative denoising process makes them well-suited for semantic communication, where the objective is to reconstruct scenes or objects from compressed semantic representations.

In conventional communication systems, the transmitter sends all data through the physical channel, leading to high bandwidth consumption while disregarding the semantic significance of the transmitted information. However, in many applications—such as video transmission—the receiver often requires only key semantic features to accomplish specific tasks. For instance, in video sequences, significant redundancy exists between consecutive frames due to repeated content, resulting in unnecessary transmission overhead. To address the challenge of limited bandwidth in video transmission, one approach is to transmit only the changes in the target semantic graph. These changes are then used to guide the gen-

**Fig. 9.16** Semantic change driven generative semantic communication framework in [17]

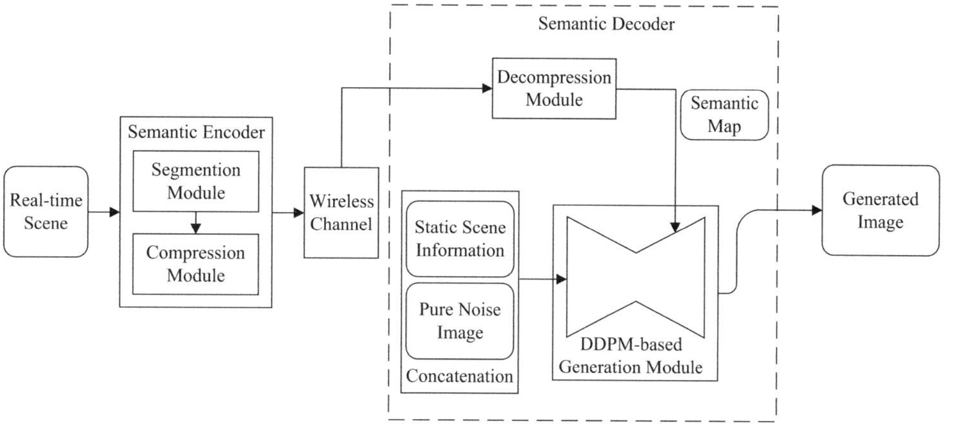


erative model at the receiver to reconstruct the corresponding video frames [17].

The semantic change driven generative semantic communication (SCDGSC) framework [17] employs a conditional DDPM as the core component of the semantic decoder. The model leverages two essential inputs:

- Semantic map: A compact representation of task-relevant information (e.g., object locations and contours) extracted by the semantic encoder.
- Static scene information: A reference image of the remote scene (e.g., background or environmental context), which is updated only when significant changes occur.

As depicted in Fig. 9.16, the encoder contains a segmentation module that isolates changed objects, followed by a compression module that encodes the extracted information into a semantic map. At the receiver, a conditional denoising diffusion probabilistic model acts as the semantic decoder, reconstructing a near-realistic frame image based on the received semantic map and static scene information. The generation process is conditioned on both the static scene $\mathbf{r}$ and the real-time semantic map $\mathbf{s}_t$, formulated as a latent variable model $p_\theta(\mathbf{x}_0) := \int p_\theta(\mathbf{x}_{0:N} | \mathbf{r}, \mathbf{s}_t)\, d\mathbf{x}_{0:N}$, where $\mathbf{x}_0, ..., \mathbf{x}_T$ are latent variables sharing the same dimensionality as the generated data, with $\mathbf{x}_0 \sim q(\mathbf{x}_0)$. The reverse process is defined as: $p_\theta(\mathbf{x}_{0:N} | \mathbf{r}, \mathbf{s}_t) = p(\mathbf{x}_N) \prod_{n=1}^{N} p_\theta(\mathbf{x}_{n-1} | \mathbf{x}_n, \mathbf{r}, \mathbf{s}_t)$.

The conditional DDPM is trained to reconstruct the remote scene by progressively denoising a latent variable while being guided by the semantic map and static scene information. The training objective minimizes the variational bound on negative log-likelihood, ensuring that the generated output accurately reflects the intended semantics:

$$\mathcal{L}_d = \mathbb{E}_{n,\mathbf{x}_n,\epsilon}\left[\left\|\epsilon - \epsilon_\theta\left(\sqrt{\bar{\alpha}_n}\mathbf{x}_n + \sqrt{1-\bar{\alpha}_n}\epsilon, \mathbf{r}, \mathbf{s}_t, n\right)\right\|_2\right], \quad (9.38)$$

**c. Transformer-based Semantic Communication**

The Transformer architecture [18], revolutionized sequential data processing through self-attention mechanisms. Initially designed for natural language processing (NLP), its ability to model long-range dependencies inspired adaptations in computer vision. Vision Transformer (ViT) [19] extended this concept to images by treating them as sequences of pixel patches, enabling applications in object detection, segmentation, and generative modeling. However, ViT faces challenges in dense vision tasks and high-resolution scenarios due to its quadratic computational complexity and low-resolution feature maps. These limitations make it less suitable for DJSCC in semantic communication, potentially leading to suboptimal performance in high CBR scenarios or high-resolution image transmission. Swin Transformer [20] addresses ViT's shortcomings through a hierarchical architecture and shifted-window self-attention. By progressively merging non-overlapping patches and expanding feature dimensions, it efficiently processes high-resolution images while maintaining global context awareness. This capability makes Swin Transformer particularly suited for semantic communication. The architecture of Swin Transformer has served as the backbone network in many semantic communication frameworks, such as GSC system in [21] and SwinJSCC in [22].

The GSC framework combines Transformer and diffusion models, comprising four primary modules:

- Semantic feature encoder module: Utilizes the Swin Transformer to extract and compress semantic features efficiently, enabling the transmitter to generate low-dimensional semantic symbols suitable for wireless transmission.
- Physical wireless channel module: Simulates the effects of noise and channel impairments on the transmitted semantic symbols.

- Semantic feature decoder module: Recovers the semantic representation at the receiver.
- Semantic fine-tuning module: Incorporates a diffusion model to refine and reconstruct high-quality semantic information, effectively mitigating the impact of channel noise in low SNR conditions.

In particular, the diffusion model within the semantic fine-tuning module plays a crucial role at the receiver by reconstructing high-quality images from degraded signals, thereby enhancing perceptual details. This synergistic approach—leveraging the Swin Transformer for efficient feature extraction and the diffusion model for robust reconstruction—enables high-fidelity semantic communication even under challenging channel conditions.

**d. VAE-based Semantic Communication**

VAE is a class of generative models that learn latent representations of data by optimizing a variational lower bound on the marginal likelihood. They consist of an encoder that maps input data to a latent space and a decoder that reconstructs the data from latent variables. The probabilistic nature of VAEs makes them particularly well-suited for semantic communication, where robustness to noise and efficient data compression are critical.

Hierarchical VAE architectures have demonstrated remarkable performance in image compression, with quantization-aware resNet VAE (QARV) [23] being a representative example. Through a bottom-up and top-down design, QARV achieves high-speed decoding and superior rate-distortion performance. Additionally, an adaptive normalization operation was introduced to enable variable-rate compression.

QARV is a lossy image compression framework that employs a ResNet VAE as the block backbone, as illustrated in Fig. 9.17. On the encoding (transmitter) side, a CNN (left-to-right path in the figure) extracts a hierarchy of four feature levels, each at different spatial resolutions (e.g., 8×, 16×, 32×, and 64× downsampled relative to the original image). These extracted features are then fed into the latent variable blocks of the decoding path (right-to-left path) to generate the compressed bitstream. Each latent variable $z_i$ is conditionally dependent on all preceding variables $z_{<i}$.

On the decoding (receiver) side, only the right-to-left path is required to reconstruct the image. The latent variable blocks and upsampling layers operate sequentially, ultimately producing the reconstructed image $\hat{\boldsymbol{x}}$. Notably, since the top-down path is shared between encoding and decoding, most computations occur during encoding, significantly accelerating the decoding process.

QARV supports both fixed-rate and variable-rate compression by conditioning residual blocks on a Lagrange multiplier $\lambda$ which governs the trade-off between rate and distortion. Variable-rate compression is achieved by conditioning all residual blocks in the model on the Rate-Distortion Lagrange multiplier $\lambda$. This is done by two modules: $\lambda$ embedding network and adaptive normalization operation. More details can be found in [23]. When $\lambda$ is fixed, the loss function is given by:

$$\mathcal{L}_{\lambda} = \mathbb{E}_{\mathbf{x}\sim p_{\mathbf{x}},\mathbf{z}\sim q_{\mathbf{z}|\mathbf{x}}} \left[ \sum_{l=1}^{L} -\log p_{\mathbf{z}_l|\mathbf{z}_{<l}}(\mathbf{z}_l \mid \mathbf{z}_{<l}) + \lambda \cdot d(\mathbf{x}, \hat{\mathbf{x}}) \right], \tag{9.39}$$

For variable-rate compression, the loss function is:

$$\mathcal{L}_{\mathcal{VR}} = \mathbb{E}_{\mathbf{x}\sim p_{\mathbf{x}},\mathbf{z}\sim q_{\mathbf{z}|\mathbf{x}},\Lambda} \left[ \sum_{l=1}^{L} -\log p_{\mathbf{z}_l|\mathbf{z}_{<l}}(\mathbf{z}_l \mid \mathbf{z}_{<l}, \Lambda) + \Lambda \cdot d(\mathbf{x}, \hat{\mathbf{x}}) \right], \tag{9.40}$$

In summary, QARV is a novel neural network model that outperforms most existing methods (including fixed-rate ones) with its streamlined design (eliminating context models), enhanced flexibility (rate-variable and hierarchical fea-

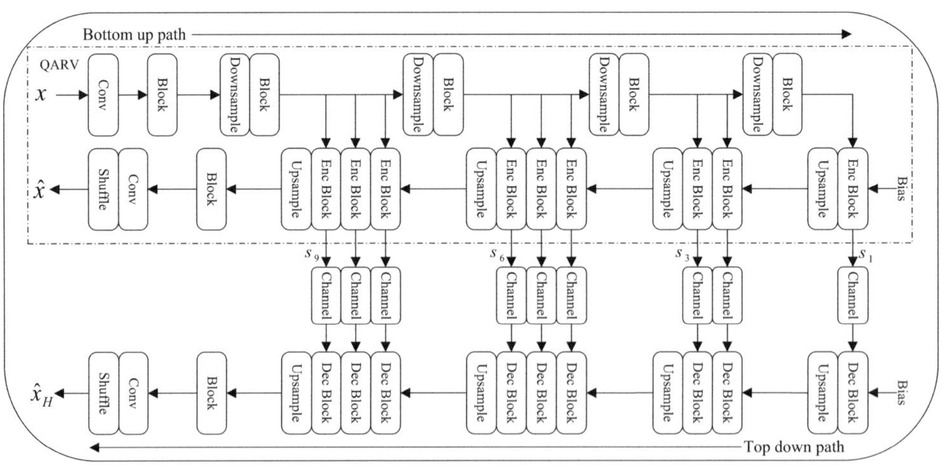


**Fig. 9.17** Framework of QARV and PLIT, where QARV architecture is part of PLIT

tures), improved BD-rate efficiency, and faster CPU decoding [23].

Building upon the foundation of QARV and NTSCC [24] (introduced in Chap. 7), the progressive learned image transmission (PLIT) framework [25] represents an advanced application of hierarchical VAEs in semantic communication, capturing coarse-to-fine structures in source images. This innovative architecture specifically addresses the challenges of progressive image transmission through the dual-path design:

- Bottom-up path: Performs multi-scale feature extraction from input images using a cascade of convolutional blocks, progressively capturing finer visual semantics.
- Top-down path: Generates hierarchical representations autoregressively, enabling progressive refinement of the transmitted image, namely, coarse initial reconstruction from early-layer features and iterative refinement through subsequent layers.

On the encoding side, the transmitter extracts features using the bottom-up path and generates hierarchical representations via the top-down path. These representations are then mapped to channel symbols through a JSCC encoder. Transmission occurs progressively, allowing the receiver to iteratively refine image quality. On the decoding side, the receiver reconstructs the image using received symbols, with missing representations replaced by predicted mean values to maintain consistency with the prior distribution.

The objective loss function for PLIT is:

$$\mathcal{L} = \mathbb{E}_{\mathbf{x}\sim p_{\mathbf{x}},\mathbf{z}\sim q_{\mathbf{z}|\mathbf{x}}}\left[\sum_{l=1}^{L} -\alpha \log p_{\mathbf{z}_l|\mathbf{z}_{<l}}(\mathbf{z}_l \mid \mathbf{z}_{<l}) + \lambda \cdot \left(d(\mathbf{x}, \hat{\mathbf{x}}) + d(\mathbf{x}, \hat{\mathbf{x}}_{\mathrm{H}})\right)\right], \tag{9.41}$$

where the scaling parameter $\alpha$ regulates the balance between the entropy of the latent variable $z_l$ and the transmission rate for $s_l$.

From a successive refinement perspective, encoded representations are transmitted sequentially, ensuring that the receiver can iteratively enhance the image quality using both newly received and previously transmitted data. The primary goal is to achieve a high-fidelity image estimate even when only the initial transmission phases are available.

## 9.5 Summary

In this chapter, task-oriented semantic communication is introduced, a paradigm that prioritizes successful task completion over mere data fidelity by transmitting only essential, context-aware information. This approach leverages AI/ML and aims for efficiency in resource-constrained 6G networks through principles like task-first design and end-to-end optimization. An example of this is the VideoQA-SC system, designed for video question answering, which integrates spatiotemporal video semantic extraction with bandwidth-adaptive joint source-channel coding for efficient task performance. The chapter also delves into methods for complex data scenarios, such as multi-source data fusion and reconstruction.

A channel-adaptive hierarchical fusion method is designed to address the bandwidth demands of multi-source data fusion and reconstruction. By dynamically selecting and transmitting key features, the proposed approach enhances transmission efficiency and reconstruction quality while maintaining the same communication load to optimize resource utilization for bandwidth-constrained scenarios.

GSC leverages generative models like GANs, diffusion models, Transformers, and VAEs to regenerate information while preserving its essential meaning. It enables task-aware data regeneration and semantic fidelity preservation and offers robustness to channel imperfections and multimodal semantic extraction and fusion, thus enhancing the transmission and reconstruction of semantic information in wireless communication.

## Problems

9.1 What is the primary optimization objective in task-oriented semantic communication?

A. Minimizing the loss of signal reconstruction at the receiver.
B. Minimizing task-specific loss such as classification or detection error.
C. Resisting channel noise to ensure symbol-level fidelity.

9.2 Why is traditional communication methods suboptimal for task-oriented communication?

A. It requires specialized hardware for encoding and decoding.
B. It cannot be used for image or video tasks.
C. It optimizes for bit-level accuracy rather than task success.

9.3 What is the biggest challenge in semantic communication for video question answering (VideoQA)?

A. Ensuring all pixels in the video are transmitted with no loss.
B. Replacing neural networks with traditional compression techniques.
C. Efficiently extracting and transmitting only the necessary video semantics for answering questions.

9.4 How does semantic communication optimize video transmission for VideoQA tasks?

A. By selectively transmitting essential video features required for accurate answers.
B. By compressing all video frames equally.
C. By increasing bandwidth usage to accommodate full-resolution video.

9.5 In the semantic communication system for video question answering, what is the main role of the encoder?

A. Transmit complete video frames to preserve pixel-level fidelity.
B. Extract and compress task-relevant video features for transmission.
C. Perform question answering directly without transmitting any data.

9.6 In VideoQA-SC, at which level is semantic encoding performed for improved task efficiency?

A. Object-level.
B. Scene-level.
C. Pixel-level.

9.7 Which of the following is not a typical generative model?

A. Variational autoencoder (VAE).
B. Generative adversarial network (GAN).
C. denoising diffusion probabilistic model (DDPM).
D. Support vector machine (SVM).

9.8 Which of the following descriptions of the basic architecture of a GAN is correct?

A. The generator is used to create copies of real samples, and the discriminator is used to improve the quality of the generated samples.
B. The generator is used to create new, unseen samples, and the discriminator is used to distinguish between generated and real samples.
C. Both generators and discriminators are used to generate new samples, but they use different techniques.
D. Discriminators are used to create new samples, while generators are used to decide whether these samples are real or not.

9.9 Which of the following properties does the LPIPS metric serve to quantify?

A. The bit error rate of transmitted symbols.
B. Perceptual similarity of the generated image to the original image.
C. The spectral efficiency of channel coding.
D. Compression ratio of semantic features.

9.10 What is the main purpose of Conditional DDPM to introduce "static scene information" in the SCDGSC framework?

A. To reduce the transmission bandwidth of semantic graphs.
B. To provide a priori constraints on the scene of the generation process.
C. To speed up the number of iterations of reverse denoising.
D. Optimize the convergence speed of the channel equalizer.

## References


1. D. Gündüz, Z. Qin, I.E. Aguerri, H.S. Dhillon, Z. Yang, A. Yener, K.K. Wong, C.-B. Chae, Beyond transmitting bits: context, semantics, and task-oriented communications. IEEE J. Sel. Areas Commun. **41**(1), 5–41 (2023)
2. Y. Shi, Y. Zhou, D. Wen, Y. Wu, C. Jiang, K.B. Letaief, Task-oriented communications for 6g: vision, principles, and technologies. IEEE Wirel. Commun. **30**(3), 78–85 (2023)
3. J. Shao, Y. Mao, J. Zhang, Task-oriented communication for multi-device cooperative edge inference. IEEE Trans. Wireless Commun. **22**(1), 73–87 (2023)
4. Y.E. Sagduyu, S. Ulukus, A. Yener, Task-oriented communications for nextg: end-to-end deep learning and ai security aspects. IEEE Wirel. Commun. **30**(3), 52–60 (2023)
5. J. Guo, W. Chen, Y. Sun, J. Xu, B. Ai, Videoqa-sc: adaptive semantic communication for video question answering. IEEE J. Select. Areas Commun. 1–1 (2025)
6. G. Zhang, Q. Hu, Z. Qin, Y. Cai, G. Yu, X. Tao, A unified multi-task semantic communication system for multimodal data. IEEE Trans. Commun. **72**(7), 4101–4116 (2024)
7. H. Xie, Z. Qin, G.Y. Li, Task-oriented multi-user semantic communications for VQA. IEEE Wirel. Commun. Lett. **11**(3), 553–557 (2022)
8. H. Xie, Z. Qin, X. Tao, K.B. Letaief, Task-oriented multi-user semantic communications. IEEE J. Sel. Areas Commun. **40**(9), 2584–2597 (2022)
9. L. Guo, W. Chen, Y. Sun, B. Ai, N. Pappas, T.Q.S. Quek, Hierarchy-aware and channel-adaptive semantic communication for bandwidth-limited data fusion. IEEE Wirel. Commun. Lett. **14**(6), 1748–1752 (2025)
10. A. Vaswani, N. Shazeer, N. Parmar, J. Uszkoreit, L. Jones, A.N. Gomez, L. Kaiser, I. Polosukhin, Attention is all you need, in *Neural Information Processing Systems (NIPS)* (2017)
11. A. Jaegle, F. Gimeno, A. Brock, O. Vinyals, A. Zisserman, J. Carreira, Perceiver: general perception with iterative attention, in *International Conference on Machine Learning* (PMLR, 2021), pp. 4651–4664
12. M. Chen, M. Liu, W. Wang, H. Dou, L. Wang, Cross-modal semantic communications in 6G, in *IEEE/CIC International Conference on Communications in China (ICCC)* (2023)
13. W. Wang, M. Liu, M. Chen, Ca deepsc: cross-modal alignment for multi-modal semantic communications, in *GLOBECOM 2023-2023 IEEE Global Communications Conference* (2023), pp. 5871–5876

14. D. Cao, J. Wu, A.K. Bashir, Multimodal large language models driven privacy-preserving wireless semantic communication in 6g, in *2024 IEEE International Conference on Communications Workshops (ICC Workshops)* (IEEE, 2024), pp. 171–176
15. C. Liang, H. Du, Y. Sun, D. Niyato, J. Kang, D. Zhao, M.A. Imran, Generative ai-driven semantic communication networks: architecture, technologies, and applications. IEEE Trans. Cogn. Commun. Netw. **11**(1), 27–47 (2025)
16. G. Xin, P. Fan, K.B. Letaief, C. Peng, Deep conditional generative semantic communication for image transmission, in *2024 IEEE International Conference on Communications Workshops (ICC Workshops)* (IEEE, 2024), pp. 1073–1078
17. W. Yang, Z. Xiong, H. Du, Y. Yuan, T.Q. Quek, Semantic change driven generative semantic communication framework, in *2024 IEEE Wireless Communications and Networking Conference (WCNC)* (2024), pp. 01–06
18. A. Vaswani, N. Shazeer, N. Parmar, J. Uszkoreit, L. Jones, A.N. Gomez, Ł. Kaiser, I. Polosukhin, Attention is all you need, in *Advances in Neural Information Processing Systems*, vol. 30 (2017)
19. A. Dosovitskiy, L. Beyer, A. Kolesnikov, D. Weissenborn, X. Zhai, T. Unterthiner, M. Dehghani, M. Minderer, G. Heigold, S. Gelly, et al., An image is worth 16x16 words: transformers for image recognition at scale (2020). arXiv:2010.11929
20. Z. Liu, Y. Lin, Y. Cao, H. Hu, Y. Wei, Z. Zhang, S. Lin, B. Guo, Swin transformer: hierarchical vision transformer using shifted windows, in *Proceedings of the IEEE/CVF International Conference on Computer Vision* (2021), pp. 10012–10022
21. K. Zhang, L. Li, W. Lin, Y. Yan, R. Li, W. Cheng, Z. Han, Semantic successive refinement: a generative ai-aided semantic communication framework. IEEE Trans. Cogn. Commun. Netw. 1–1 (2025)
22. K. Yang, S. Wang, J. Dai, X. Qin, K. Niu, P. Zhang, Swinjscc: taming swin transformer for deep joint source-channel coding. IEEE Trans. Cogn. Commun. Netw. **11**(1), 90–104 (2025)
23. Z. Duan, M. Lu, J. Ma, Y. Huang, Z. Ma, F. Zhu, Qarv: quantization-aware resnet vae for lossy image compression. IEEE Trans. Pattern Anal. Mach. Intell. (2023)
24. J. Dai, S. Wang, K. Tan, Z. Si, X. Qin, K. Niu, P. Zhang, Nonlinear transform source-channel coding for semantic communications. IEEE J. Sel. Areas Commun. **40**(8), 2300–2316 (2022)
25. G. Zhang, H. Li, Y. Cai, Q. Hu, G. Yu, Z. Qin, Progressive learned image transmission for semantic communication using hierarchical vae. IEEE Trans. Cogn. Commun. Netw. 1–1 (2025)

# Part III

# Compatibility with Conventional Communication System

# 10 Semantic-Aware Source Coding

**Abstract**

This chapter focuses on the semantic-aware source compression coding methods, covering AI-driven encoding techniques for speech, image, and video modalities in greater depth. For AI speech codecs, this chapter primarily introduces the VQ-VAE-based end-to-end AI speech codec, the LPC-enhanced AI speech codec, and the text-based AI speech codec, while also explaining their respective strengths, design motivations, and potential application scenarios. For AI image codecs, this chapter highlights how data-driven designs in transformation coding and entropy modeling replace handcrafted components to enable end-to-end rate-distortion optimization, and it further discusses how innovations such as hyperpriors, context models, and Transformer-based architectures capture both local and global dependencies, thereby improving compression efficiency, perceptual fidelity, adaptability, and decoding speed. For AI video codecs, this chapter explains the process of traditional video compression standards and the metrics of video quality, and based on these backgrounds, it introduces the architecture, design principles, and emerging trends of the existing AI video codec.

**Keywords**

Semantic-aware compression • AI speech codec • VQ-VAE • AI image codec • Entropy modeling • AI video codec • Hybrid coding

## 10.1 AI Codec for Speech Compression

Deep learning-based speech compression coding has significantly improved both quality and compression efficiency compared to traditional methods. The mainstream research directions can be divided into end-to-end architectures and hybrid architectures. The former relies entirely on neural networks and data-driven approaches, while the latter combines neural networks with signal processing models. In addition to these two approaches, there is another research direction for ultra-low bitrate speech coding, i.e., text-based speech codecs using speech recognition and synthesis techniques. This section first introduces the fundamentals of speech compression techniques and then discusses these three AI-based approaches.

### 10.1.1 Introduction to Speech Codec

Speech compression is an essential technology in information transmission and communication systems. Mobile communication has the following requirements for voice compression coding:

- Low coding rate with high voice quality;
- Strong robustness with good resistance to noise interference and error correction performance;
- Low encoding and decoding delay;
- Low complexity of the codec, making it suitable for large-scale integration.

To ensure connectivity in different digital communication systems, international organizations such as ITU-T have developed a series of speech coding standards as shown in Table 10.1.

Traditional speech compression coding methods can be categorized into waveform coding, parametric coding, and hybrid coding.

- **Waveform coding**: These techniques aim to reconstruct the speech waveform, resulting in high speech quality but higher coding rates, such as the ADPCM technique.
- **Parametric coding**: These techniques, also known as analysis-by-synthesis coding, extract feature parameters from the source signal in the transform domain and recon-


W. Chen and Z. Qin, *Semantic Communications*, Textbooks in Telecommunication Engineering,
https://doi.org/10.1007/978-3-032-11005-3_10

**Table 10.1** Speech coding standards

| Standard | Coding technique | Bitrate |
|---|---|---|
| G.711 | A/$\mu$ law nonlinear companding | 64 Kbps |
| G.721/3/6 | Adaptive differential pulse code modulation (ADPCM) | 32/24/16 Kbps |
| G.722 | Sub-band adaptive differential pulse code modulation (SB-ADPCM) | 48 Kbps |
| G.723.1 | Algebraic code excited linear prediction (ACELP) | 5.3 Kbps |
| G.728 | Low-delay code excited linear prediction (LD-CELP), vector quantization | 16 Kbps |
| G.729 | Conjugate-structure algebraic code excited linear prediction (CS-ACELP) | 8 Kbps |
| AMR-NB | ACELP | 4.75–12.2 Kbps |

struct the speech at the receiver end, such as the linear predictive coding (LPC) vocoder.
- **Hybrid coding**: These techniques combine the advantages of both approaches and have been widely applied. For example, technologies like ACELP belong to the category of hybrid coding.

The popular audio codecs Opus [1] and EVS [2] use a divide-and-conquer approach by classifying signals into speech and audio. For speech, EVS uses an improved ACELP (Algebraic Code-Excited Linear Prediction) method. At low bitrates, Opus uses the SILK encoder, which is also based on LPC.

Speech compression coding can be categorized into **lossy coding** and **lossless coding**, depending on whether the speech can be perfectly reconstructed. In practice, lossless coding forms the foundation of lossy coding. Unless explicitly stated otherwise, the coding methods discussed in the following sections refer to lossy coding.

Traditional speech codecs typically rely on modeling the human speech production mechanism. As shown in Fig. 10.1, the model consists of three components: an excitation source $U(z)$, a vocal tract model $V(z)$, and a radiation model $R(z)$, which correspond to the human vocal organs—the larynx, vocal tract, and mouth. The speech signal generation model can be represented by these three sub-models and is expressed by Eq. 10.1.

$$H(z) = U(z) \cdot V(z) \cdot R(z). \tag{10.1}$$

The speech coding techniques discussed above rely on elaborate models and features, which depend heavily on time-consuming and costly listening tests. However, standard speech and audio codecs typically consist of multiple interacting heterogeneous coding blocks, which complicates the tuning process. From the use of white noise and periodic pulses to the use of codebooks for excitation in CELP technology, manual modeling of excitation signals is undoubtedly complex. These manually designed excitation signals also face bottlenecks in sound quality at low bit rates.

With the development of AI technology, intelligent speech processing technologies such as automatic speech recognition (ASR) and text-to-speech (TTS) are becoming more and more

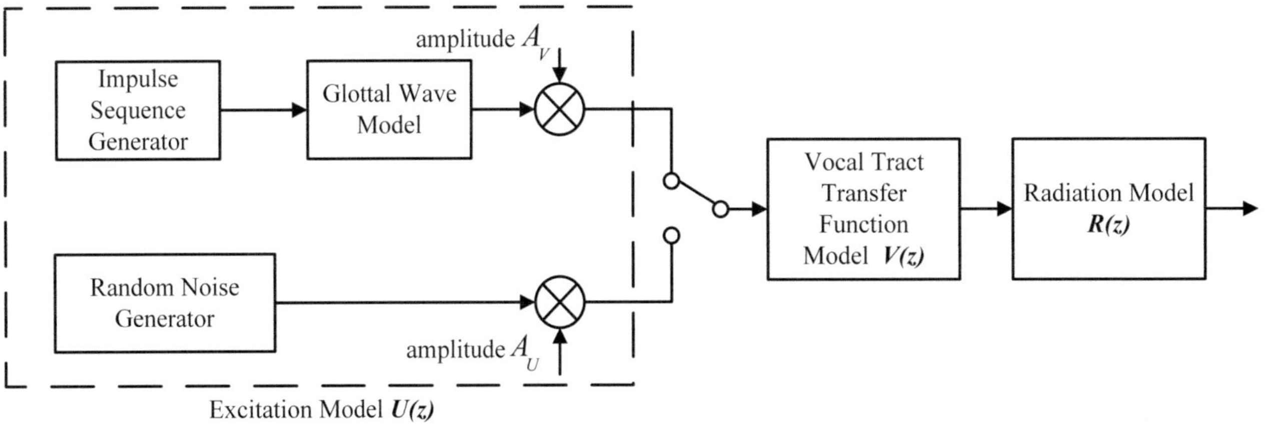


**Fig. 10.1** Voice generation model

**Fig. 10.2** End-to-end AI speech codec

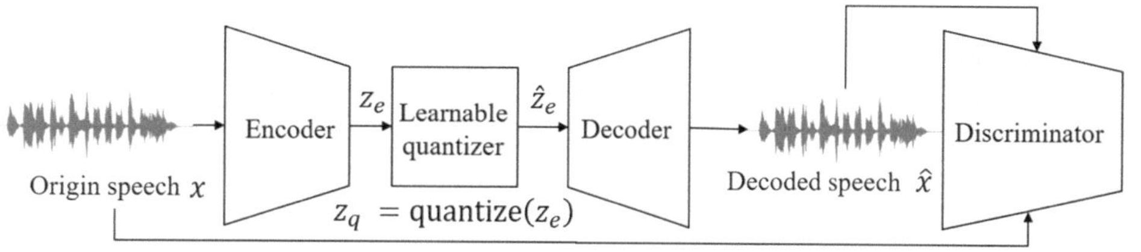


mature. By automatically learning semantic features in speech from large-scale data, AI-based codecs also show the potential to break even lower bit rates and restore higher sound quality.

In the beginning, neural networks were only used to replace traditional audio decoders. For the encoder, traditional coding methods were still used. Later, end-to-end AI speech codecs became mainstream. Recent speech codecs make extensive use of end-to-end deep neural networks and vector quantization techniques. Such codecs, which are not based on sound modeling, have shown potential in dealing with external noise interference and multi-source signal excitation, and have outperformed conventional codecs in terms of sound quality at many bit rates. The following section describes the architecture and ideas for an end-to-end AI speech codec.

### 10.1.2 End-to-End AI Speech Codec

End-to-end neural speech compression, along with neural compression methods for images and videos, all belong to the broader category of neural data compression techniques. These neural data compression methods across different modalities are heavily inspired by deep generative models. Representative works in deep generative modeling include GANs [3], VAEs [4], flow-based models [5], and autoregressive models [6].

#### VQ-VAE Architecture

Currently, the mainstream end-to-end neural speech compression methods are primarily based on the vector quantized variational autoencoder (VQ-VAE) architecture [7]. As shown in Fig. 10.2, an end-to-end AI speech codec is generally composed of an encoder, a learnable quantizer, a decoder, and a discriminator. The VQ-VAE architecture is a variant of the VAE architecture. Let's review the VAE as background.

The variational autoencoder architecture consists of several key components: an encoder that learns the posterior distribution $q(z \mid x)$, mapping speech signals to low-dimensional latent variables; a latent space where these variables $z$ capture essential features like pitch and timbre; and a decoder that generates reconstructed speech $\hat{x}$ from this latent representation, modeled as $p(x \mid z)$. The system's optimization objective combines reconstruction loss (ensuring output fidelity to the original input) and KL divergence regularization (guiding the posterior distribution $q(z \mid x)$ toward a standard normal distribution $p(z)$), with the total loss function being a weighted sum of these components.

$$\begin{aligned}\mathcal{L} &= \mathcal{L}_{\text{recon}} + \beta \cdot \mathcal{L}_{\text{KL}} \\ &= -\mathbb{E}_{q(z|x)}[\log p(x \mid z)] + \beta \cdot D_{\text{KL}}(q(z \mid x) \| p(z)).\end{aligned} \tag{10.2}$$

The latent space in VAE is continuous, but communication systems often require transmitting discrete bits. VQ-VAE adopts a discrete latent space based on the VAE framework. Specifically, it performs vector quantization on the encoder's output $z_e(x)$. The most intuitive approach is to use nearest neighbor search in the discrete latent variable space $\mathbf{e} \in \mathbb{R}^{K \times D}$.

$$z_q(x) = \text{Quantize}\,(z_e(x)) = e_k, \quad \text{where } k = \arg\min_j \|z_e(x) - \mathbf{e}_j\|_2 \tag{10.3}$$

By defining the prior distribution $p(z)$ as uniform, we can calculate that the KL divergence is a constant.

In VAE, the posterior distribution $q(z \mid x)$ is a Gaussian distribution, whereas in VQ-VAE, the posterior distribution $q(z \mid x)$ is a **categorical distribution** and represented in a one-hot manner as follows:

$$q(z = k \mid x) = \begin{cases} 1 & \text{for } k = \arg\min_j \|z_e(x) - \mathbf{e}_j\|_2 \\ 0 & \text{otherwise} \end{cases} \tag{10.4}$$

The input to the decoder is the discrete latent representation $z_q(x)$, which generates the reconstructed data $\hat{x}$. The entire process of VQ-VAE can be expressed as follows:

$$x \to z_e(x) = \text{Enc}(x) \to z_q(x) = \text{Quantize}\,(z_e(x)) \to \hat{x} = \text{Dec}\left(z_q(x)\right). \tag{10.5}$$

Let us revisit the training objective of VAE in Eq. 10.2. Since the KL divergence in the second term is a constant and does not affect the training objective, it can be ignored. This implies that the latent distribution is no longer explicitly constrained to match the prior distribution, but this also brings new problems with quantization process and embedding space learning in VQ-VAE.

A key challenge in the VQ-VAE training process is how to backpropagate gradients through the non-differentiable quantization operation. Since the nearest neighbor lookup is a non-differentiable function, it cannot directly perform backpropagation. To address this, VQ-VAE employs a straight-through estimator that approximates the gradient for backpropagation:

$$\text{quantize}(z_e) = z_e + (\text{quantize}(z_e) - z_e) \cdot \text{detach}(). \quad (10.6)$$

The detach() operation ensures that the quantization part is computed during the forward pass, but during backpropagation, the gradient is directly copied back to the input $z_e$, enabling the transmission of pseudo-gradients.

Another important aspect of VQ-VAE is how the embedding space (codebook) participates in the learning process. The authors use one of the simplest dictionary learning algorithms, which directly computes the mean square error between the encoder output $z_e(x)$ and the quantized vector $e_k$:

$$\min_{e_k} \|\text{sg}[z_e(x)] - e_k\|_2^2. \quad (10.7)$$

Here, sg refers to the stop gradient operation, which means that this mean square error loss will only update the embedding space and will not propagate to the encoder. It is worth noting that there are more efficient methods for learning embedded Spaces, such as exponential moving smoothing (EMA).

### SoundStream Framework

In the previous section, we provided a detailed explanation of the fundamental architecture in end-to-end neural speech compression. We highlighted the basic nearest neighbor search quantizer and addressed the issues of gradient propagation and codebook learning. However, further details about the encoder, decoder, and other components were not discussed. In this section, we take SoundStream [8], a classic architecture, as an example to further explore the current methods of end-to-end neural speech compression.

In SoundStream, dilated causal convolutions are used to implement both the encoder and decoder. During encoding, the number of channels produced by the convolution block doubles at each convolutional layer. The number of convolutional blocks $B_{\text{enc}}$ and the striding of each convolution determine the feature sampling ratio. For example, if $B_{\text{enc}} = 4$ and the striding of the encoder is (2, 4, 5, 8), then every $M = 2 \cdot 4 \cdot 5 \cdot 8 = 320$ sampled audio points correspond to one embedded feature. For each frame of single-channel audio $x \in \mathbb{R}^T$, the output of the encoder is $z_e = \text{enc}(x) \in \mathbb{R}^{S \times D}$, where $D$ is the embedding dimension (set to 256 in the original experiment), and $S$ is the number of embedded features in each audio frame. The decoder performs exactly the reverse process.

The residual vector quantization approach is employed to address challenges in direct quantization. The vector quantization of a signal takes every $K$ consecutive points in the signal sequence as a single sample, forming a vector in a $K$-dimensional Euclidean space that is then quantized. Direct vector quantization for audio compression requires an enormous vector codebook. The rate of the vector quantizer (measured in bits per sample) can be calculated as follows: $r = \frac{B}{k} = \frac{\log_2 N^k}{k} = \log_2 N$, where $N$ represents the size of the codebook, and $B$ is the number of bits used to encode each codeword.

It can be observed that for a sampling rate of 24 kHz, if the total downsampling ratio of the encoder is set to $M = 320$ (as configured in the previous section), the number of features output per second of speech is $S = 75$. Encoding this at a rate of $r = 6$ kbps results in each codeword having $B = \frac{6000}{75} = 80$ bits. At this point, the codebook size would be $N = 2^{80}$, which is practically infeasible.

Residual Vector Quantization (RVQ) adopts a step-by-step quantization approach, performing hierarchical quantization through multiple quantizers. RVQ is essentially a multi-stage vector quantizer. Given the output of the encoder $z_e = \text{enc}(x)$ and vector quantizers $Q_i$ for $i = 1 \ldots N_q$, the algorithm can be formulated as follows:

$$z_q^{(0)} = 0 \quad (10.8)$$

$$r^{(0)} = z_e \quad (10.9)$$

$$z_q^{(i)} = z_q^{(i-1)} + Q_i(r^{(i-1)}), \quad \text{for } i = 1, \ldots, N_q \quad (10.10)$$

$$r^{(i)} = r^{(i-1)} - Q_i(r^{(i-1)}), \quad \text{for } i = 1, \ldots, N_q \quad (10.11)$$

The final quantized output is $z_q = z_q^{(N_q)}$.

Feature reconstruction is a critical aspect of end-to-end neural speech compression, where designing standards that closely align with subjective evaluation becomes essential. Commonly used metrics include mean absolute error and mean square error losses in the time domain, as well as mean absolute error and mean square error losses of the Mel spectrogram in the frequency domain. To account for features at different scales, multi-scale spectral reconstruction is generally adopted. For example, in EnCodec [9], the reconstruction loss is defined as:

$$\ell_f(x, \hat{x}) = \frac{1}{|\alpha| \cdot |s|} \sum_{\alpha_i \in \alpha} \sum_{i \in e} \left\| S_i(x) - S_i(\hat{x}) \right\|_1 + \alpha_i \left\| S_i(x) - S_i(\hat{x}) \right\|_2, \quad (10.12)$$

where $S_i$ is a 64-bin mel-spectrogram using a normalized STFT with a window size of $2^i$ and hop length of $2^i/4$, $e = 5, \ldots, 11$. The parameter $\alpha$ is used to balance the mean absolute error and mean square error losses.

Another important advancement in neural speech compression is the incorporation of generative adversarial training. In SoundStream, a carefully designed discriminator is introduced, while the end-to-end encoder-decoder acts as the generator. This approach enables the neural codec to better mimic the characteristics of real human speech through fine-tuning. Two main types of discriminators have proven effective in this context. Wave-based discriminators take the time-domain waveform as input and operate directly on it. The multi-scale discriminator (MSD) can discriminate waveforms at different levels of downsampling, with each individual discriminator functioning as a Markov window-based discriminator. Another architecture, the multi-period discriminator (MPD), segments the waveform based on fixed periods to extract periodic signal features. Alternatively, STFT-based discriminators (STFTD) take the real and imaginary parts of the Short-Time Fourier Transform as input, with multi-scale discrimination possible using different STFT window sizes.

Recent developments have further enhanced end-to-end speech compression techniques. Notable works include EnCodec [9], DAC [10], and HiFi-Codec [11], which follow a trajectory similar to SoundStream [8]. These methods introduce various improvements in vector quantization and discriminator design, achieving remarkable audio quality at around 1.2 kbps. A significant breakthrough came with SemantiCodec [12], the first work to use a diffusion model as a decoder in neural audio coding. This approach achieves high-quality speech reconstruction at extremely low bit rates (0.31 kbps), though at the cost of increased model size and computational complexity.

### 10.1.3 LPC-Enhanced AI Speech Codec

The pursuit of low bitrate, high quality, and minimized complexity is the eternal goal of speech codecs. Combining AI with traditional model-based methods has the potential to maintain high quality while reducing complexity, offering a new perspective. Among these approaches combining traditional signal processing with neural networks, linear predictive analytics-enhanced AI speech codecs are promising.

#### Linear Predictive Coding

One important type of redundancy in speech signal compression is the temporal correlation between adjacent samples of the speech signal. In short-time speech segments, the current speech signal $x(n)$ can be approximately represented as a linear combination of the previous $p$ samples:

$$x(n) = e(n) + \sum_{k=1}^{p} a_k x(n-k), \tag{10.13}$$

where $a_k$ are the linear predictive coefficients used to describe the characteristics of the vocal tract, and $p$ is the prediction order, which is typically related to the spectral complexity of the speech signal.

The above relationship can be transformed into a filter representation, where the residual signal is treated as the excitation source (source), and the original speech can be reconstructed by passing it through an all-pole filter (filter) defined by the LPC coefficients:

$$H(z) = \frac{X(z)}{E(z)} = \frac{1}{1 + \sum_{k=1}^{p} a_k z^{-k}}. \tag{10.14}$$

$H(z)$ represents the transfer function of the vocal tract. The formants of the vocal tract correspond to the pole locations of $H(z)$, which are determined by the linear predictive coefficients $a_k$. By adjusting $a_k$, the frequency modulation effect of the vocal tract on the speech signal can be accurately modeled.

As shown in Fig. 10.3, Voiced/Unvoiced Decision and Pitch Determination use DSP techniques such as pattern matching and short-term autocorrelation to provide the voiced/unvoiced parameter $U/V$ and the pitch period $P$ for the current frame. The LPC Analyzer calculates the prediction coefficients $a_k$ using the autocorrelation method and the Levinson-Durbin recursion algorithm. After encoding the above parameters, the excitation source can be reconstructed at the receiver, and speech is synthesized in the LPC Synthesizer through the all-pole filter. It is worth noting that the above process is performed frame by frame.

The original LPC Vocoder achieves a coding rate of 2.4 kbps, but the synthesized speech quality is low. The main reason lies in the insufficient modeling of complex signals by the excitation model. Subsequent research introduced high-quality hybrid coding algorithms, such as the Mixed Excitation Linear Prediction vocoder (MELP) and the Code-Excited Linear Prediction vocoder (CELP).

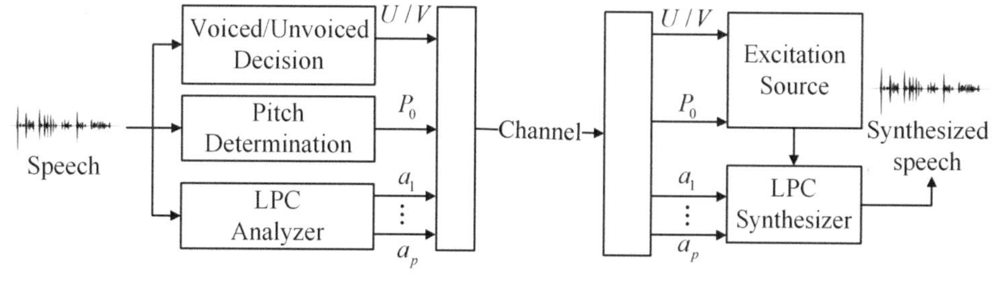


**Fig. 10.3** Basic LPC vocoder

**LPCNet Architecture**

From single impulse signals and periodic signal excitation to codebook excitation and sinusoidal excitation, the modeling of excitation signals has always been a key focus of the source-filter framework. On the other hand, the modeling of the vocal tract filter has been proven over time to be simple and efficient. The idea behind LPCNet [13] is to delegate the modeling of the excitation signal to a neural network, while the linear part is handled using DSP techniques.

As shown in Fig. 10.4a, at the receiver side of LPCNet, the LPC Analyzer calculates the LPC coefficients based on the received bitstream. The frame-rate network, built on a convolutional architecture, generates conditional features $f$ once per frame. The sample-rate network, an improved version of WaveRNN, produces the excitation signal $\hat{e}_t$ at each sampling point. The excitation signal is then added to the prediction signal (which is the process of LPC synthesis) to produce the speech signal. This can be expressed mathematically as follows:

$$\begin{aligned} \boldsymbol{f} &\leftarrow \mathcal{D}_{\text{frame}}(\boldsymbol{h}) \\ u_t, a_1, \ldots, a_p &\leftarrow \mathcal{D}_{\text{LPC}}\left(\boldsymbol{h}, \left[\hat{x}_{t-p}, \ldots, \hat{x}_{t-1}\right]\right) \\ \hat{e}_t &\leftarrow \mathcal{D}_{\text{sample}}\left(\hat{e}_{t-1}, \hat{x}_{t-1}, u_t, \boldsymbol{f}\right) \\ \hat{x}_t &= \hat{e}_t + u_t \end{aligned} \tag{10.15}$$

Models represented by LPCNet combine neural networks with the source-filter model, offering simpler modeling and lower computational complexity compared to other waveform generation models. The overall approach involves using the generative model in neural networks to model the excitation signal. The basic process is summarized in Fig. 10.4b. This is an intuitive yet challenging approach.

### 10.1.4 Text-Based AI Speech Codec

From the perspective of information theory, the theoretical limit of speech compression can be as low as 50 bps. To approach this limit, parameterized speech coding based on the "recognition-synthesis" paradigm has been proposed. Early works [14] relied on hidden Markov models (HMM) for speech recognition, which not only involved complex designs but also suffered from burst errors caused by inaccurate recognition. In recent years, advancements in deep learning for speech recognition and synthesis have shown promise in overcoming these limitations. This type of speech codec is called as text-based AI speech codec.

Deep semantic communication for speech transmission (DeepSC-ST) [15] extracts speech into text-related information for transmission, significantly reducing the transmission bandwidth. Its architecture is shown in the figure below. It primarily employs CNN, gated recurrent units (GRU), and fully connected layers for semantic encoding. The fully connected layers serve as both channel encoder and decoder, while a

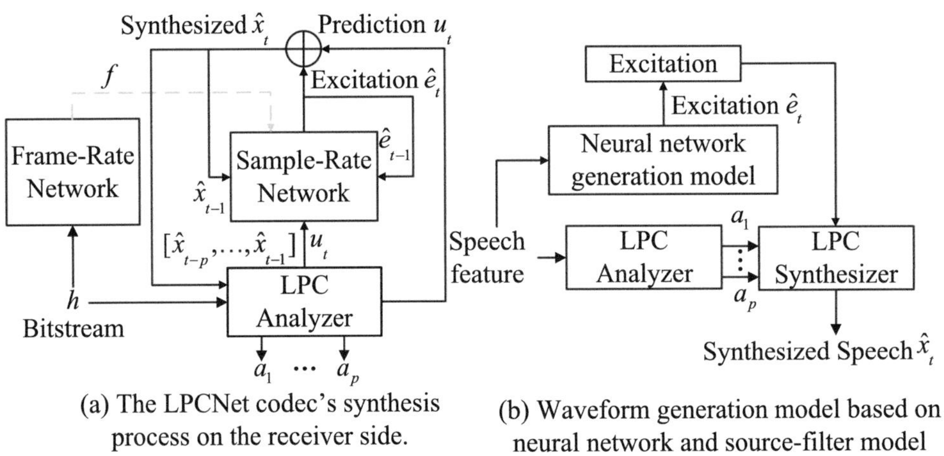


**Fig. 10.4** LPCNet and the basic process of combining AI with source-filter model

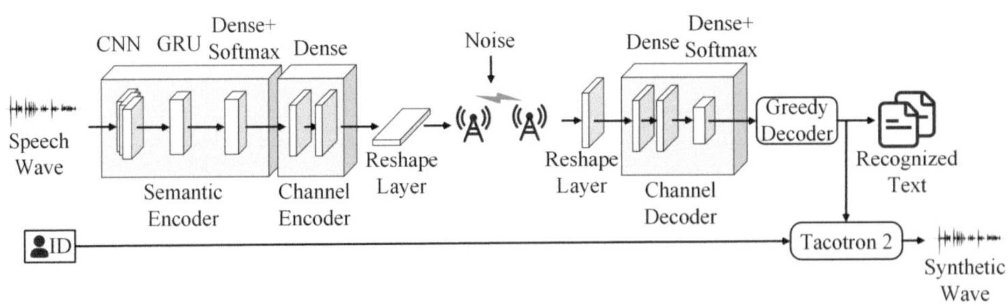


**Fig. 10.5** The architecture of DeepSC-ST

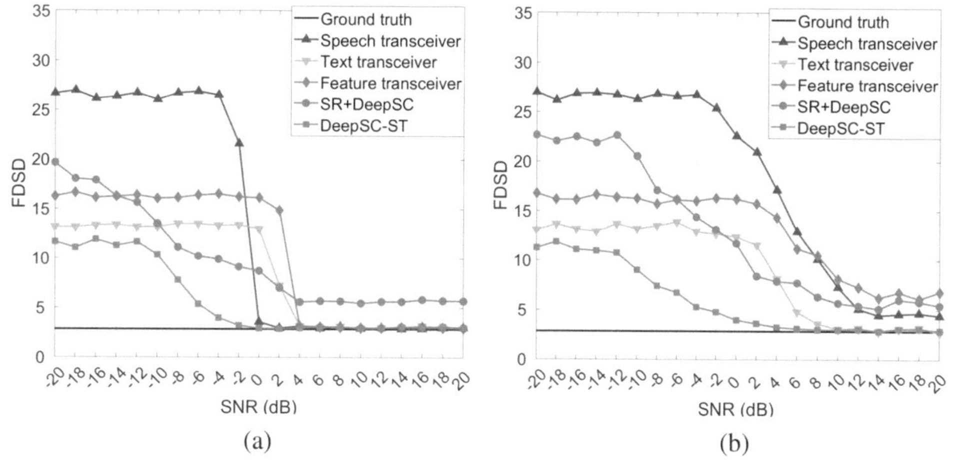


**Fig. 10.6** FDSD score versus SNR for the speech transceiver, the text transceiver, the feature transceiver, the SR+DeepSC, and the proposed DeepSC-ST

greedy decoder is used for speech recognition, and Tacotron2 is adopted for speech synthesis (Fig. 10.5).

In DeepSC-ST, the Frechet Deep Speech Distance (FDSD) metric is used to measure the similarity between the transmitted and received speech. Traditional speech transceivers, traditional text transceivers, feature transceivers, and the combination of speech recognition + DeepSC (SR+DeepSC) are adopted as baseline models. The speech synthesis performance of these methods is compared under different channel environments, highlighting the advantages of DeepSC-ST in terms of speech quality and robustness. The figure below illustrates the speech synthesis performance over AWGN and Rayleigh channels. The data is sourced from the original paper (Fig. 10.6).

FDSD becomes smaller as the speech quality improves. It is generally considered that when FDSD is less than 10, the speech quality is acceptable. It can be observed that, in the DeepSC-ST method, the threshold for usable speech quality is SNR $\geq -10$ dB, which is better than other baseline methods. This demonstrates robustness under low signal-to-noise ratio (SNR) conditions. Moreover, since the thresholds are identical across different fading channel environments, DeepSC-ST also exhibits robustness to various channel conditions. In terms of effectiveness, the original paper shows that the compressed symbol rate of DeepSC-ST is 520 symbols/s.

Text-based speech codecs hold the promise of breaking through to lower bit rates, but there are also new bottlenecks. For example, how to transmit other representations such as speaker characteristics and voice style. Also, as AI codecs develop, computational complexity will become an increasingly important consideration in addition to sound quality and bit rate. But in general, with the increasing abundance of computational resources, AI speech codecs will undoubtedly be used in more scenarios.

## 10.2 AI Codec for Image Compression

### 10.2.1 Introduction to AI Image Codec

Image compression is essential for reducing data size, minimizing storage requirements, and reducing bandwidth usage. It is categorized into lossless and lossy methods. Lossless compression preserves all original data, enabling exact image reconstruction, while lossy compression allows for higher compression ratios by discarding less critical information.

According to information theory, the minimum number of bits required to represent an image depends on its entropy, which relies on accurate probability distribution assessment. However, high-dimensional joint distributions among pixel points make obtaining an exact distribution nearly impossible. Lossless compression approximates this entropy by exploiting redundancies like spatial correlations.

Lossy compression leverages rate-distortion theory and human visual perception limitations. By omitting subtle changes in color or texture that are imperceptible to humans, lossy algorithms achieve significant file size reductions without compromising perceived quality. This makes them ideal for applications where perfect fidelity is not crucial, such as web graphics and streaming services. Evaluating the performance of image compression algorithms is essential for comparing different techniques and optimizing compression parameters. The optimization of image compression is denoted as

$$\mathcal{L} = R + \lambda \cdot D, \tag{10.16}$$

where $R$ is the bit rate, $D$ is the image quality, and $\lambda$ is a parameter that balances the trade-off between rate and distortion. The goal of image compression is to minimize the rate-distortion function. The choice of $\lambda$ determines the trade-off between compression ratio and image quality. A higher $\lambda$ value prioritizes image quality over compression ratio, while a lower $\lambda$ value favors higher compression ratios at the expense of image quality.

For compression rate, the most common metric is the bit rate, which measures the number of bits required to represent the compressed image. The bit rate is typically expressed in bits per pixel (bpp), depending on the color space used. In this book, we mainly use bpp for measuring compression rate, which is given by

$$R = \text{bpp} = \frac{\text{Total bits}}{\text{Image size}}. \tag{10.17}$$

The typical pipeline for image compression consists of three main stages:

- **Transformation Encoding:** Converts the image from the spatial domain to a frequency-domain representation. Common transformations include discrete cosine transform (DCT) and more recently, learned neural networks.
- **Quantization:** Reduces the precision of the transformed coefficients to decrease the amount of data. Quantization levels can be adjusted to balance between compression ratio and image quality. The quantized coefficients are integers representing the compressed data.
- **Entropy Coding:** Compresses the quantized coefficients using statistical methods to minimize the bit representation. Techniques include Huffman coding, arithmetic coding, and more advanced methods like context-adaptive binary arithmetic coding (CABAC). The final compressed bitstream is produced, ready for storage or transmission.

For instance, let $s$ denote the input image and $\hat{s}$ denote the output feature. The transformation encoding step converts the image from the spatial domain to a frequency-domain representation using transformations such as:

$$\hat{s} = \lfloor \mathbf{DAC}s \rceil, \tag{10.18}$$

where $\mathbf{C}$ represents a pixel-wise color transform, $\mathbf{A}$ denotes DCT, $\mathbf{D}$ is a diagonal matrix scaling the coefficients, and $\lfloor\cdot\rceil$ denotes rounding to the nearest integer.

After transformation and quantization, entropy coding compresses the quantized coefficients using statistical methods to minimize the bit representation. Let $E(\cdot)$ denote the entropy coding function. The final compressed bitstream can be represented as:

$$\mathbf{x} = E(\hat{s}), \tag{10.19}$$

where $x$ is the resulting compressed bitstream ready for storage or transmission. Techniques for entropy coding include Huffman coding, arithmetic coding, and more advanced methods like CABAC.

Industry-standard formats like JPEG and MPEG are widely supported and optimized for real-time use, making lossy compression practical and efficient. Recent advancements in neural image compression, using models like VAEs, have further improved both compression efficiency and image quality, addressing longstanding challenges and paving the way for future innovations.

By exploring these topics, this book aims to provide a comprehensive understanding of image compression technologies, highlighting their significance in today's digital world and their potential for further development. Through detailed examination of transformation encoding, quantization, and entropy coding, readers will gain insights into how modern image compression techniques work and their potential for future innovation.

### 10.2.2 Transformation Coding and Quantization for AI Image Codec

Traditional image compression methods such as JPEG, JPEG2000, and BPG rely on a well-defined pipeline consisting of transformation coding, quantization, and entropy coding. In these standards, a linear transform—such as DCT in JPEG or the discrete wavelet transform (DWT) in JPEG2000—is applied to decorrelate pixel values and concentrate signal energy into a small number of coefficients. These coefficients are then quantized and entropy-coded, typically using Huffman or arithmetic coding. Despite their effectiveness, these methods rely on fixed, handcrafted components that are not easily adapted to different image distributions or task-specific objectives.

Neural image compression (NIC) methods aim to learn these components—particularly the transformation and quantization processes—directly from data using deep learning. However, despite significant progress in representation learning, neural compression models still depend on classical entropy coding techniques. In particular, most state-of-the-art NIC systems adopt Asymmetric Numeral Systems [16] (ANS) for entropy coding, which, although efficient, is not integrated into the learning process. The entropy model is learned to predict the probability distribution of quantized latent variables, but the actual encoding step is carried out using an external, fixed algorithm. As such, entropy coding is not yet fully differentiable nor trainable end-to-end, limiting the degree to which compression performance can be jointly optimized across all stages.

Because of this limitation, current research on neural image compression focuses heavily on the transformation and quantization modules, which are trainable and have the greatest impact on compression efficiency. In this book, we follow this emphasis by exploring the principles, architectures, and training methodologies that underpin transformation coding and quantization in neural compression systems.

## Transformation Coding

Considering the joint distribution among the pixels of an image, $p(\mathbf{x}) = p(x_1, x_2, \ldots, x_N)$, with the image denoted as $\mathbf{x}$, we are unable to accurately estimate this high-dimensional distribution. Therefore, we aim to approximate the high-dimensional joint distribution using the product of marginal distributions $q(\mathbf{x}) = \prod_{i=1}^{N} p(x_i)$, that is,

$$\text{minimize} \quad D_{\mathrm{KL}}(p(\mathbf{x}) \| q(\mathbf{x})). \tag{10.20}$$

The theoretical limit for compressing $\mathbf{x}$, denoted as $R_{\min}$, corresponds to its self-information, which can be expressed as $R_{\min} = -\log p(\mathbf{x})$. This can also be written considering the approximation error as $R_{\min} = -\log q(\mathbf{x}) + D_{\mathrm{KL}}(q(\mathbf{x}) \| p(\mathbf{x}))$. The coding rate achieved in practice is denoted by $R = -\log \mathrm{q}(\mathrm{x})$. Therefore,

$$R = R_{\min} + D_{\mathrm{KL}}(q(\mathbf{x}) \| p(\mathbf{x})). \tag{10.21}$$

Considering that $D_{\mathrm{KL}}(q(\mathbf{x}) \| p(\mathbf{x}))$ is always positive, it follows that the practical coding rate $R$ will always be greater than the theoretical minimum rate. So the goal of Transformation Coding is to minimize the $\mathrm{D}_{KL}(q(\mathbf{x}) \| p(\mathbf{x}))$.

A foundational framework in this area was introduced by Ballé et al. [17]. As illustrated in Fig. 10.7, the model comprises an encoder network $g_a$ that transforms the input image $x$ into a latent representation $y$, which is then quantized to $\hat{y}$ or approximated as $\tilde{y}$ using additive uniform noise during training. A decoder network $g_s$ reconstructs the image from the quantized representation. To improve entropy modeling, an auxiliary hyperprior network is introduced: $h_a$ maps $y$ to side information $z$, which is quantized or approximated as $\hat{z}$ or $\tilde{z}$, and a synthesis transform $h_s$ predicts the distribution parameters of $y$ from $z$.

The transformation networks in this framework are constructed using CNNs augmented with generalized divisive normalization (GDN) [18]. Each convolutional layer uses a $5 \times 5$ kernel with a stride of 2, enabling progressive spatial downsampling and compact latent representations. GDN is a biologically inspired activation function designed to normalize each channel based on a weighted combination of all channels at the same spatial location. Specifically, the GDN activation is defined as:

$$u_i = \frac{y_i}{\sqrt{\beta_i + \sum_j \gamma_{ij} y_j^2}}, \tag{10.22}$$

where $\beta_i$ and $\gamma_{ij}$ are learnable, non-negative parameters. This operation encourages statistical independence among the latent activations and transforms their distribution to be more Gaussian-like—a property beneficial for subsequent entropy coding. Compared to conventional nonlinearities like ReLU, GDN leads to better rate-distortion performance in learned image compression models by facilitating more efficient and decorrelated latent representations.

## Quantization

In end-to-end optimized image compression, one of the key challenges is the non-differentiability of the quantization operation, which hinders the application of standard gradient-based optimization methods. The quantization process is typically defined as:

$$\hat{y}_i = q_i = \mathrm{round}(y_i), \tag{10.23}$$

where $i$ indexes all elements in the latent representation, spanning spatial dimensions and channels. The rounding operation introduces discontinuities in the function, whose derivatives are zero almost everywhere, thus impeding backpropagation.

To enable gradient-based training, Ballé et al. [19] proposed a differentiable approximation by injecting additive uniform noise during training. Instead of applying a non-differentiable quantizer directly, the latent variable $y$ is perturbed with i.i.d. uniform noise:

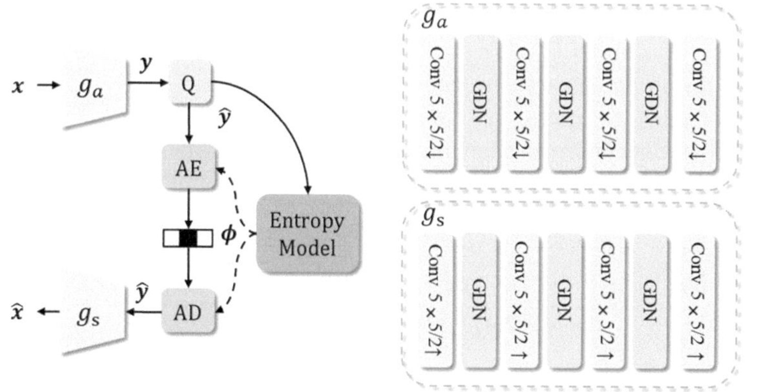


**Fig. 10.7** Transformation coding and quantization for image compression

$$\tilde{y} = y + \mathcal{U}(-0.5, 0.5), \tag{10.24}$$

where $\mathcal{U}(-0.5, 0.5)$ denotes the uniform distribution. This relaxation simulates the quantization effect while preserving differentiability:

$$\frac{\partial \tilde{y}}{\partial y} = 1. \tag{10.25}$$

Assuming the quantization bins are of unit width and centered on integers, the probability mass function of the quantized variable $q_i$ can be expressed in terms of the continuous density $p_{y_i}(t)$ of the pre-quantized value:

$$P_{q_i}(n) = \int_{n-\frac{1}{2}}^{n+\frac{1}{2}} p_{y_i}(t)\, dt, \quad \text{for all } n \in \mathbb{Z}. \tag{10.26}$$

With this approximation, the density of the noisy latent variable $\tilde{y}$ becomes a continuous proxy for the mass function of the quantized variable:

$$p_{\tilde{\boldsymbol{y}}}(\boldsymbol{n}) = P_{\boldsymbol{q}}(\boldsymbol{n}), \quad \text{for all } \boldsymbol{n} \in \mathbb{Z}^M, \tag{10.27}$$

allowing the differential entropy of $\tilde{\boldsymbol{y}}$ to serve as an estimate of the discrete entropy of $\boldsymbol{q}$.

### Optimization

The overall objective in neural image compression follows the classical rate-distortion trade-off, balancing bit rate ($R$) and reconstruction fidelity ($D$). This can be expressed as the following loss function:

$$\mathcal{L} = R + \lambda \cdot D, \tag{10.28}$$

where $\lambda$ controls the trade-off between compression rate and distortion. Interestingly, this formulation closely parallels the loss function used in VAEs:

$$\mathcal{L}_{\text{VAE}} = \mathbb{E}_{\mathbf{z}\sim q_\phi(\mathbf{z}|\mathbf{x})}[-\log p_\theta(\mathbf{x}|\mathbf{z})] + \beta \cdot D_{\text{KL}}(q_\phi(\mathbf{z}|\mathbf{x}) \| p(\mathbf{z})), \tag{10.29}$$

where $q_\phi(\mathbf{z}|\mathbf{x})$ is the approximate posterior (encoder), $p_\theta(\mathbf{x}|\mathbf{z})$ is the decoder likelihood, and $p(\mathbf{z})$ is the prior distribution. The first term corresponds to the reconstruction loss, analogous to distortion $D$, while the KL divergence term measures the divergence between the learned latent distribution and the prior, playing the role of rate $R$.

To further understand this, assume that the true distribution of quantized features is $p$, and the estimated entropy model is $q$. The KL divergence between them is:

$$\begin{aligned} D_{\text{KL}}[p\|q] &= \mathbb{E}_{y\sim p}[-\log q(y)] - \mathbb{E}_{y\sim p}[-\log p(y)] \\ &= R - H(p), \end{aligned} \tag{10.30}$$

where $R$ is the estimated rate and $H(p)$ is the entropy of the true distribution, which is constant during optimization. Therefore, minimizing the KL divergence effectively reduces the rate:

$$\mathcal{L} = R + \lambda \cdot D = \mathcal{L}_{\text{VAE}} + \text{constant}. \tag{10.31}$$

This connection reveals that optimizing a neural image compression model under the rate-distortion objective is mathematically equivalent to training a VAE, where the entropy model functions as a learnable prior over the latent space. Importantly, the quality of the learned entropy model $q$ directly affects the rate $R$, highlighting its central role in determining overall compression efficiency.

### Further Improvements

By combining expressive neural transforms, probabilistic modeling, and differentiable approximations, this framework has laid the foundation for modern deep image compression. In the remainder of this book, we focus on transformation coding and quantization—the two most critical components in neural compression pipelines, where data-driven learning offers the most flexibility and potential for advancement.

Following Ballé's variational framework, subsequent research has sought to improve the transformation modules to enhance representation capacity and trainability. Notably, Cheng et al. [20] proposed replacing GDN with residual blocks, arguing that residual learning provides stronger nonlinearity and facilitates optimization without the computational overhead of GDN. Their design integrates multiple residual blocks after each convolutional layer, resulting in improved rate-distortion performance.

Building on this, ELIC [21] further deepens the transformation networks by stacking additional residual blocks in both encoder and decoder paths. This enhances the model's ability to capture complex image structures and leads to better compression efficiency, especially at low bit rates.

More recently, Transformer-based architectures have been introduced to further improve long-range context modeling in the latent space. For example, the Swin Transformer [22] has been adopted in several image compression frameworks to replace or complement convolutional layers, allowing the transformation network to capture hierarchical spatial dependencies more effectively.

Moreover, the LIC-TCM model [23] adopts a hybrid architecture that combines CNNs with Transformer modules for transformation coding. The encoder and decoder incorporate CNNs for local feature extraction and spatial downsampling, while Transformer blocks are employed to capture global dependencies through self-attention over tokenized image patches. This hybrid design enables the model to leverage both local and long-range contextual information, effectively reducing redundancy and improving perceptual reconstruction quality.

Overall, the evolution of transformation coding in neural image compression reflects a clear shift from handcrafted components like GDN to increasingly powerful, learnable architectures. Starting with residual blocks that improve nonlinear modeling and optimization, subsequent advances such as ELIC demonstrate the benefits of deeper networks. More recently, the incorporation of Transformer-based modules—including Swin Transformers and token-based designs like LIC-TCM—has enabled models to capture long-range dependencies and global structure, further enhancing compression efficiency and perceptual quality. These innovations underscore the central role of transformation design in modern learned codecs and lay the foundation for the next generation of adaptive, content-aware compression systems.

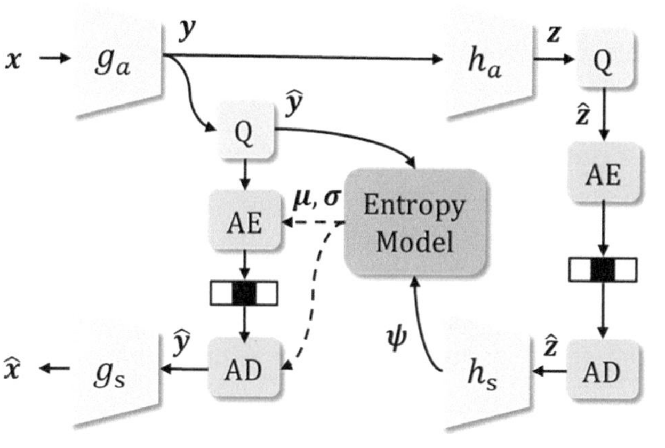


**Fig. 10.8** Hyperprior for image compression

## 10.2.3 Entropy Model for AI Image Codec

One of the defining characteristics that sets neural image compression (NIC) apart from traditional image codecs is the presence of a learnable entropy model. While conventional methods such as JPEG or JPEG2000 rely on fixed statistical assumptions or handcrafted entropy coding rules—typically treating transformed coefficients independently—NIC systems introduce trainable models to predict the probability distribution of latent representations. This allows the compression system to allocate bits more efficiently and adaptively, depending on the image content.

The concept of a learnable entropy model was first introduced by Ballé et al. [19] through the use of a hyperprior network. In their framework, the entropy model conditions the probability distribution of the main latent representation on auxiliary side information, known as the hyperprior. This side information is also learned and transmitted alongside the compressed representation. The use of a hyperprior enables the model to capture spatial variability and uncertainty in the image features, leading to significantly better rate-distortion performance compared to fixed or unconditioned models.

Since its introduction, entropy modeling has become a central focus of research in NIC. Numerous extensions have been proposed to further improve coding efficiency, including autoregressive models, context-adaptive entropy models, and attention-based mechanisms. These advances allow the entropy model to consider both local and global context when estimating symbol probabilities, leading to more accurate bitrate estimation and improved compression quality.

In light of its central role in neural image compression, this book will devote sections to explaining the underlying principles of entropy modeling. We will introduce the foundational concepts, starting from the hyperprior model proposed by Ballé et al. [19], and trace the evolution toward more advanced architectures. Rather than focusing solely on mathematical details, our emphasis will be on the intuition, structure, and design motivations behind these models, aiming to provide readers with a clear and practical understanding of how entropy models contribute to the performance and adaptability of modern learned image codecs.

### Hyperprior

In early NIC models, the latent representation $\tilde{y}$ is typically assumed to follow a fully factorized distribution, implying that each element is statistically independent. However, in practice, there exist significant spatial dependencies among the elements of $\tilde{y}$, as nearby regions in natural images tend to exhibit strong structural correlations. This renders the factorized assumption suboptimal for entropy coding and motivates the need for a more expressive entropy model.

To address this issue, Ballé et al. [19] proposed introducing an additional set of latent variables $\tilde{z}$, referred to as a hyperprior. The core idea is to condition the distribution of $\tilde{y}$ on spatially adaptive parameters—specifically, on a set of predicted standard deviations $\tilde{\sigma}$ that vary across positions. These standard deviations are derived by applying a learned transform $h_s$ to the hyperprior $\tilde{z}$:

$$p_{\tilde{y}|\tilde{z}}(\tilde{y} \mid \tilde{z}) = \prod_i \left( \mathcal{N}(0, \tilde{\sigma}_i^2) * \mathcal{U}\left(-\tfrac{1}{2}, \tfrac{1}{2}\right) \right) (\tilde{y}_i), \quad (10.32)$$

where the standard deviation vector is computed as $\tilde{\sigma} = h_s(\tilde{z})$. This hierarchical structure enables the model to better account for the varying uncertainty and information density across the image, thereby improving entropy estimation and compression performance. This framework is illustrated in Fig. 10.8.

To infer the hyperprior $\tilde{z}$ during encoding, a parametric transformation $h_a$ is applied to the original latent representation $y$, producing a side-channel latent variable $z$. During training, quantized versions of both $y$ and $z$—denoted $\tilde{y}$ and

$\tilde{z}$ respectively—are sampled by adding uniform noise to simulate quantization. The resulting joint variational posterior becomes:

$$q(\tilde{\mathbf{y}}, \tilde{z} \mid \mathbf{x}) = \prod_i \mathcal{U}\left(\tilde{y}_i \mid y_i - \tfrac{1}{2}, y_i + \tfrac{1}{2}\right) \cdot \prod_j \mathcal{U}\left(\tilde{z}_j \mid z_j - \tfrac{1}{2}, z_j + \tfrac{1}{2}\right), \tag{10.33}$$

where $\mathbf{y} = g_a(\mathbf{x})$ and $z = h_a(\mathbf{y})$ are obtained via learnable neural networks. The hyperprior itself is modeled using a fully factorized non-parametric density, which allows for flexibility without making strong assumptions about its form:

$$p_{\tilde{z}}(\tilde{z}) = \prod_i \left(p_{\tilde{z}_i} * \mathcal{U}\left(-\tfrac{1}{2}, \tfrac{1}{2}\right)\right)(\tilde{z}_i). \tag{10.34}$$

Through this design, the hyperprior serves two complementary purposes: it enhances the entropy model's capacity by capturing spatial correlations in the latent space, and it provides side information that is itself compact and compressible. This innovation marked a major advancement in learned image compression and has since become a standard component in most state-of-the-art NIC architectures.

### Context Model

While the hyperprior enables the entropy model to capture spatial variation in uncertainty by conditioning on side information, it does not explicitly model the sequential dependencies among the elements of the latent representation $\tilde{\mathbf{y}}$. In practice, these elements—particularly in images—exhibit strong local correlations. To further exploit these dependencies, autoregressive entropy models have been introduced.

In Sect. 10.2.2 we want to model the joint distribution $p(\mathbf{y})$ as the product of edge distribution $q(\mathbf{y}) = \prod_{i=1}^{N} p(y_i)$. However, this model is not sufficient enough. Using the multiplication formula of probability, each latent variable $\tilde{y}_i$ is conditioned on a subset of previously decoded elements. The model estimates the conditional distribution of each element as:

$$q(\tilde{\mathbf{y}}) = \prod_i p(\tilde{y}_i \mid \tilde{y}_{<i}), \tag{10.35}$$

where $\tilde{y}_{<i}$ denotes all latent elements decoded before position $i$. This is the autoregressive model for image codec, whose formulation allows the model to adaptively assign probabilities based on already-known context, significantly improving coding efficiency.

To implement this in practice, convolutional neural networks with masked kernels are used to ensure that the prediction of $\tilde{y}_i$ does not access future or unavailable information. Popular architectures—such as PixelCNN-style models—have been adapted for this purpose. Minnen et al. [24] introduced a joint autoregressive and hierarchical prior model for learned image compression, combining the advantages

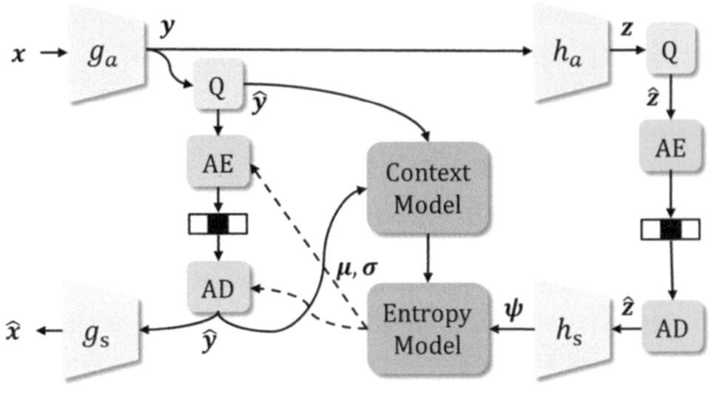


**Fig. 10.9** Context model for image compression

of autoregressive entropy models and hierarchical priors. As shown in Fig. 10.9, the model splits the latent representation into two components: a hierarchical prior that captures dependencies across scales, and an autoregressive model that captures fine-grained spatial dependencies within each scale.

The hierarchical prior is modeled by introducing latent variables at different levels of abstraction, where each level conditionally depends on the one above it. The autoregressive model captures spatial correlations by conditioning each latent variable on previously decoded variables at the same scale. This combination allows for more efficient compression by exploiting both global structure and local texture patterns. Mathematically, the joint entropy model can be expressed as:

$$q(\tilde{\mathbf{y}}) = \prod_i p(\tilde{y}_i \mid \tilde{\mathbf{y}}_{<i}, \tilde{z}), \tag{10.36}$$

where $\tilde{\mathbf{y}}_{<i}$ represents the previously decoded elements, and $\tilde{z}$ denotes the hierarchical latent variables (hyperprior) that model the global context. This approach allows the model to effectively capture both long-range dependencies across different scales and local correlations within each scale.

Figure 10.10 presents a rate-distortion comparison of three representative entropy modeling strategies, simply using PSNR as the quality metric compared on Kodak dataset with RGB color domain. The horizontal axis represents BPP, while the vertical axis corresponds to the PSNR in decibels (dB). The three configurations are plotted using distinguishable line styles and markers to enable interpretation under grayscale rendering:

The results indicate that the hyperprior model achieves substantial improvements in PSNR over the baseline across all bitrate levels. The additional incorporation of an autoregressive context model further improves performance, especially in the low-bitrate region. The solid-line curve (Context + Hyperprior) consistently dominates the others, confirming the effectiveness of joint modeling strategies in reducing entropy and preserving visual fidelity.

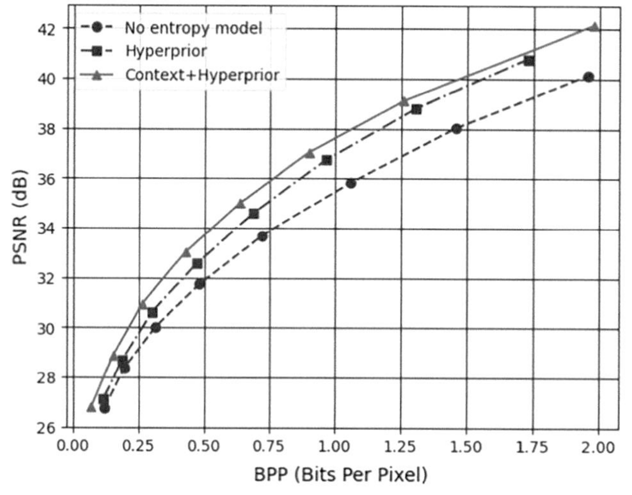


**Fig. 10.10** Rate-distortion comparison using PSNR on Kodak dataset among three entropy models: no entropy model [17], hyperprior [19], and context + hyperprior [24]. The curves are distinguished by marker shapes and line styles: circles with dashed lines for the baseline, squares with dash-dotted lines for the hyperprior model, and triangles with solid lines for the joint context and hyperprior model

**Further Improvements**

After the introduction of Minnen et al.'s joint autoregressive and hierarchical prior model, a key limitation became apparent: the slow decoding speed due to the sequential nature of autoregressive models. To address this issue, several subsequent works have proposed optimizations aimed at improving the decoding speed without compromising compression performance.

The checkerboard context model [25] reduces the decoding bottleneck by dividing the latent representation into two disjoint sets: even and odd positions, akin to a checkerboard pattern. This allows for parallel decoding of these two sets independently, greatly enhancing the speed of the process. The model processes the latent variables in two passes, decoding the "anchor" set first and then using it as context for the "non-anchor" set. This approach significantly improves computational efficiency, enabling faster encoding and decoding without sacrificing rate-distortion performance, which is essential for real-time applications.

The channel-wise context model [26] takes a different approach by modeling the dependencies within each channel independently. By reducing the complexity of entropy coding through channel-wise separation, this model allows for more efficient decoding. The independence between channels enables parallelism during decoding and simplifies the overall entropy coding process, improving the overall efficiency of the model. This method, when combined with latent residual prediction, significantly enhances compression performance and decoding speed.

The space-channel context model (SCCTX) in ELIC [21] combines spatial and channel context models to enhance compression performance and decoding speed. The model reduces redundancy by modeling spatial dependencies within each region and channel dependencies across the image. By applying autoregressive convolutions for spatial context and treating each channel separately for channel context, SCCTX enables efficient entropy coding while maintaining parallelism during decoding. Additionally, the uneven channel grouping technique optimizes the model by prioritizing channels that contain more critical information, allowing for faster decoding without sacrificing compression efficiency. This approach significantly improves computational efficiency, enabling faster encoding and decoding without compromising rate-distortion performance.

In summary, the development of entropy models for image compression reflects a shift from traditional, fixed statistical methods to learnable, adaptive models. The goal of these models is to more efficiently allocate bits by capturing the complex dependencies and variations in image data. By introducing methods such as the hyperprior and context models, entropy models are able to adapt to spatial and channel-wise correlations, improving both compression performance and computational efficiency. These advances allow for faster decoding, enhanced rate-distortion performance, and greater scalability, addressing the key challenge of balancing compression quality with computational speed.

## 10.3 AI Codec for Video Compression

### 10.3.1 Introduction

Video can be simply regarded as a sequence of consecutive images, whose content exhibits continuity over time. Video compression is an extension of image compression that aims to reduce the storage and transmission requirements of video data. The optimization of video compression aligns with that of image compression, both adhering to rate-distortion optimization (RDO). The primary goal is to achieve efficient compression by balancing bit rate and distortion. Unlike static images, video compression not only involves compressing each individual frame, but also considers the temporal correlation between video frames. The core goal of video compression is to minimize the size of the video file by removing redundant data, while preserving visual quality and interframe continuity as much as possible.

Within the process of video compression, two important techniques are intra-frame compression and inter-frame compression. Intra-frame compression, similar to image compression, reduces the redundancy within a single video frame to compress it. On the other hand, inter-frame compression works by analyzing the temporal redundancy between consecutive video frames, particularly in static scenes or slow-motion scenes, where only the differences between consecutive frames are encoded to reduce the data size.

In this book, we mainly focus on lossy video compression techniques. Similar to lossy image compression, lossy video compression achieves higher compression ratios by discarding some detail information, which typically has a minimal perceptual impact on the human eye. This approach is particularly suitable for video streaming and storage applications that require significant data reduction, such as online video platforms, streaming services, and high-definition video transmission.

Current mainstream lossy video compression standards, such as H.264, H.265, and H.266, employ various advanced encoding techniques, including predictive coding, transform coding, motion compensation, and quantization. These techniques enable efficient video data compression while maintaining satisfactory video quality at low bitrates.

### 10.3.2 Architecture of Video Compression Systems

The H.264 standard, as one of the most widely used international video coding standards, provides a modular architecture capable of efficiently handling both spatial and temporal redundancies in video data. Its design laid the technical foundation for subsequent standards such as H.265 and H.266. The architecture, as illustrated in Fig. 10.11, consists of several key components: preprocessing, transform coding, quantization, entropy coding, and predictive coding. These modules work together to eliminate redundancies in video content, achieve efficient data compression, and maintain high visual quality in compressed video.

**Preprocessing**
Preprocessing is the first step in video compression, designed to adjust the raw video data to facilitate more efficient compression. This stage typically involves the following steps:

- **Color Space Transformation**: Video data is converted from the RGB format to the YUV format, where Y represents the luminance component, and U and V represent the chrominance components. This separation aligns with the human visual system's sensitivity to brightness and color, enabling further compression of the chrominance components without significantly affecting perceived video quality.
- **Downsampling**: The resolution of the U and V chrominance components is reduced. For example, in the commonly used YUV420 format, each macroblock contains 256 luminance samples but only 64 U and 64 V samples, where each pair of U and V values is shared by four neighboring luminance samples. Downsampling significantly reduces the data volume, making it a critical step in optimizing the data structure for video compression.

In deep learning video compression systems, the preprocessing module may be omitted as neural networks can directly learn optimal representations from raw data.

**Video Frame Classification**
H.264 organizes video frames into sequences, with each sequence acting as a coding cycle. These sequences typically contain one or more groups of pictures, which include multiple types of video frames, each corresponding to a different coding scheme. Common frame types include:

- **Intra-coded Frame (I-Frame)**: I-frames are video frames that can be independently decoded without relying on other frames. They are similar to static image compression and serve as reference points within the video sequence. Although I-frames have lower compression efficiency, they are the simplest to decode.
- **Predicted Frame (P-Frame)**: P-frames depend on preceding I-frames or P-frames for decoding. They store the changes relative to the previous frame, resulting in higher compression efficiency than I-frames. When decoding a P-frame, the corresponding I-frame or P-frame must first be decoded to reconstruct the current frame using motion vectors and residuals. P-frames effectively reduce temporal redundancies and improve video compression efficiency.
- **Bidirectional Frame (B-Frame)**: B-frames use both preceding and subsequent frames to predict the current frame's content, achieving even higher compression ratios. However, because B-frame decoding requires information from both directions, it is primarily used in video storage scenarios and is less suitable for real-time streaming.

Considering that I-frames resemble static image compression, this book focuses on discussing P-frames, which are the most widely used in various application scenarios.

**Video Frame Partitioning**
Input video frames are first divided into slices, each consisting of an integer number of macroblocks. A macroblock, the basic unit for encoding and decoding in H.264, comprises $16 \times 16$ pixels, with each pixel containing Y, U, and V components. The primary purpose of slices is to control error propagation. H.264 stipulates that data within a slice can only reference data from the same slice and not from other slices. This restriction confines stream errors to a single slice, preventing errors from spreading across the entire frame.

During the encoding process, H.264 further partitions macroblocks based on specific standards to find the optimal partitioning and prediction modes, minimizing the final bitstream size.

**Predictive Coding**

Predictive coding leverages the correlations between discrete signals to predict the next signal based on previous signals. The difference between the actual value and the predicted value, called the residual, is then encoded. Accurate prediction results in smaller residuals, allowing fewer bits to encode the data, thereby achieving compression. Predictive coding addresses both spatial and temporal redundancies in video:

- **Intra Prediction**: Intra prediction uses only information within the current frame to estimate pixel values, reducing spatial redundancies. It relies on adjacent blocks' brightness values to predict a block's content. While the compression ratio of intra-prediction is lower than that of inter prediction, it simplifies decoding.
- **Inter Prediction**: Inter prediction employs motion estimation and motion compensation to predict the current frame's blocks based on reference frames. By finding similar blocks and recording motion vectors, inter prediction reduces the data to be transmitted. It achieves much higher compression ratios than intra-prediction and is a key advantage of video compression.

The residual data generated by predictive coding serves as input for subsequent transform coding, further optimizing the compression process.

**Transform Coding**

Transform coding converts video data from the spatial domain to the frequency domain, reducing spatial redundancies. H.264 employs integer DCT (IDCT), which offers the following advantages:

- **Spatial-to-Frequency Conversion**: DCT transforms pixel data into frequency coefficients, where each coefficient represents a spatial variation mode.
- **Sparsity in Frequency Domain**: Low-frequency components typically capture most of the image's energy, while high-frequency components describe finer details. Transform coding exploits this sparsity to optimize subsequent stages.

**Quantization**

Quantization introduces lossy compression by reducing the precision of frequency coefficients, significantly lowering the data volume:

- **Coefficient Discretization**: DCT coefficients are mapped to discrete quantization levels based on quantization parameters (QP).
- **Trade-off Between Quality and Compression**: Smaller QP values retain more detail but result in lower compression ratios, while larger QP values discard more detail, achieving higher compression ratios. The choice of QP affects the balance between compression efficiency and video quality.

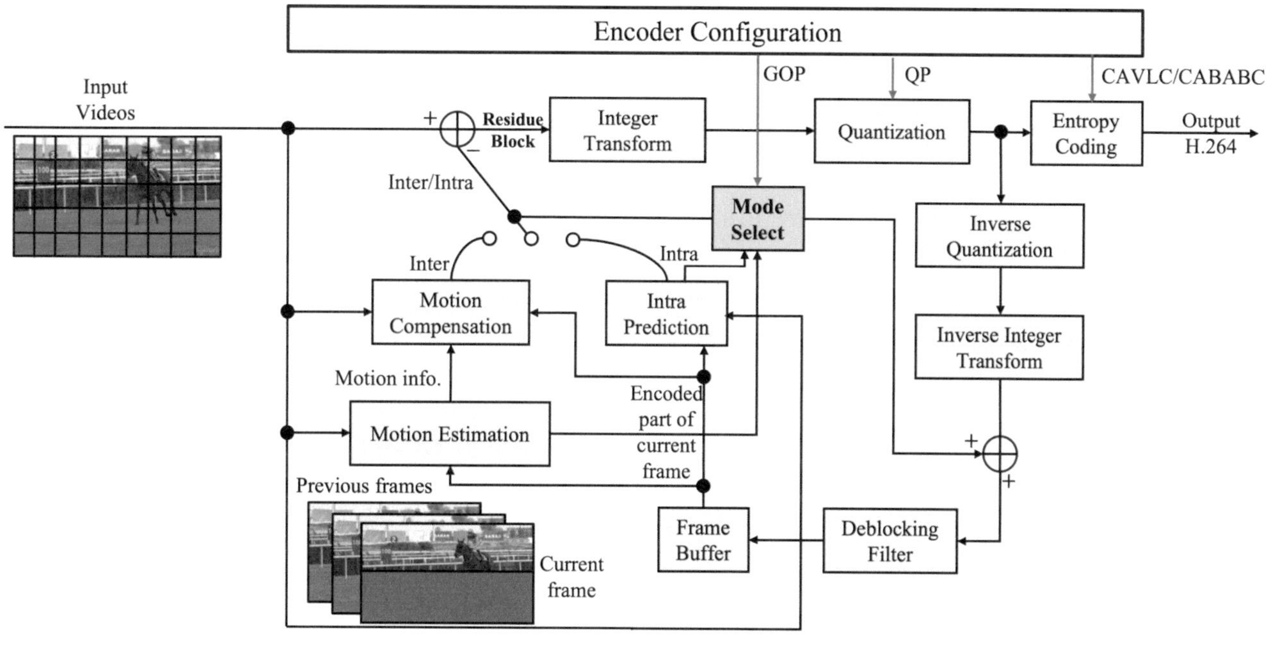


**Fig. 10.11** Architecture of a video compression system

Quantized coefficients have smaller absolute values and a higher proportion of zeros, facilitating further optimization in entropy coding. In deep learning-based systems, quantization parameters are often learned adaptively.

**Entropy Coding**
Entropy coding is the final step in video compression and the only lossless module. It uses statistical models to further compress quantized data. H.264 supports two entropy coding methods:

- **Context-Adaptive Variable Length Coding (CAVLC)**: CAVLC assigns shorter codewords to more frequent symbols, making it suitable for low-complexity scenarios.
- **Context-Adaptive Binary Arithmetic Coding (CABAC)**: CABAC uses probability models to encode symbols more efficiently, achieving higher compression ratios at the cost of increased computational complexity.

Entropy coding converts quantized data into a compact bitstream, ensuring efficient storage and transmission.

**Decoding and Post-processing**
The decoding process reverses the encoding steps:

- **Inverse Quantization**: Restores approximate values of the original coefficients.
- **Inverse Transform**: Converts frequency-domain data back into the spatial domain to reconstruct the video frame.
- **Deblocking Filter**: Smooths block boundaries to reduce artifacts and enhance visual quality.

Post-processing adapts the decoded video for various application scenarios, such as format conversion.

**Summary of the Encoding Process**
The P-frame encoding process begins with preprocessing, slicing, and macroblock partitioning. The encoder performs motion estimation to find the best-matching macroblock in the reference frame, generating motion vectors to describe the spatial displacement. Motion compensation predicts the current macroblock based on reference data, and residuals are calculated by subtracting the prediction from the actual data. These residuals undergo DCT to extract features and are then quantized to reduce data volume. Finally, quantized residuals and motion vectors are entropy-encoded (e.g., using CAVLC or CABAC) to produce the compressed P-frame bitstream. The process is optimized via rate-distortion trade-offs, searching for the best prediction modes, partitioning, and macroblock matches to achieve the highest compression efficiency with minimal distortion.

### 10.3.3 Deep Learning-Based Video Compression Methods

In recent years, iterative advances in video coding standards such as H.264, H.265, and H.266 have achieved remarkable compression performance.However, the increasing complexity of their algorithmic designs, encompassing modules like block partitioning, motion estimation, and transform coding, poses significant challenges. For instance, the number of intra-prediction directions in H.264, H.265, and H.266 is 9, 35, and 65, respectively. Although the addition of prediction directions enhances context extraction and redundancy reduction, it also introduces substantial complexity. Reliance on RDO to search for the optimal coding mode further amplifies computational demands, with encoding a single 1080p (1920×1080) frame potentially taking up to half an hour. This highlights the difficulty and slow pace of achieving breakthroughs in traditional video coding methods.

Currently, AI-driven techniques offer a novel path for video coding. The moving picture, audio, and data coding by artificial intelligence (MPAI) organization has been established to explore AI-based solutions for image, audio, and data coding. MPAI has proposed two technical approaches: (1) enhancing or replacing traditional coding modules with AI, such as the AI-based block partitioning and prediction algorithms introduced in H.266, and (2) constructing end-to-end AI-based video coding models, which leverage the strong spatiotemporal representation capabilities of neural networks to effectively model intra-frame structures and capture inter-frame dependencies. This reduces overall complexity while maintaining high performance. This book focuses on the latter approach, detailing AI-based video coding models for both general and specific scenarios.

**General Scenarios**
Deep video compression (DVC) [27] is the first end-to-end video coding method based on deep neural networks. Its architecture, as shown in Fig. 10.12, retains the principles of the traditional framework while replacing modules such as motion estimation, motion compensation, and residual encoding with CNN-based implementations. DVC performs optical flow estimation in the pixel domain to determine the displacement of each pixel between two consecutive frames, capturing the motion between video frames. The estimated motion vector $m_t$ is locally encoded, quantized, and decoded to produce $\hat{m}_t$. This motion vector, combined with the reference frame $\hat{x}_{t-1}$, is processed by the motion compensation network to generate the context $\bar{x}$. The context $\bar{x}$ is subtracted from the current frame $x_t$ to produce the residual $y_t$, which is subsequently encoded, quantized, and sent to an entropy encoder for conversion into a bitstream.

Both the motion vector and the residual undergo entropy modeling for rate estimation. Quantization, entropy model-

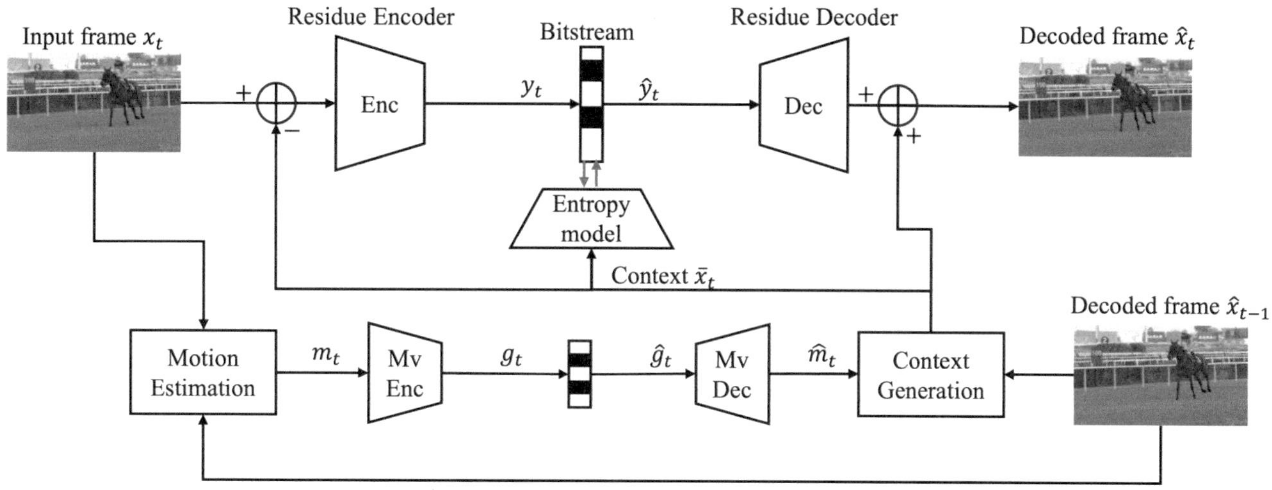


**Fig. 10.12** Architecture of DVC

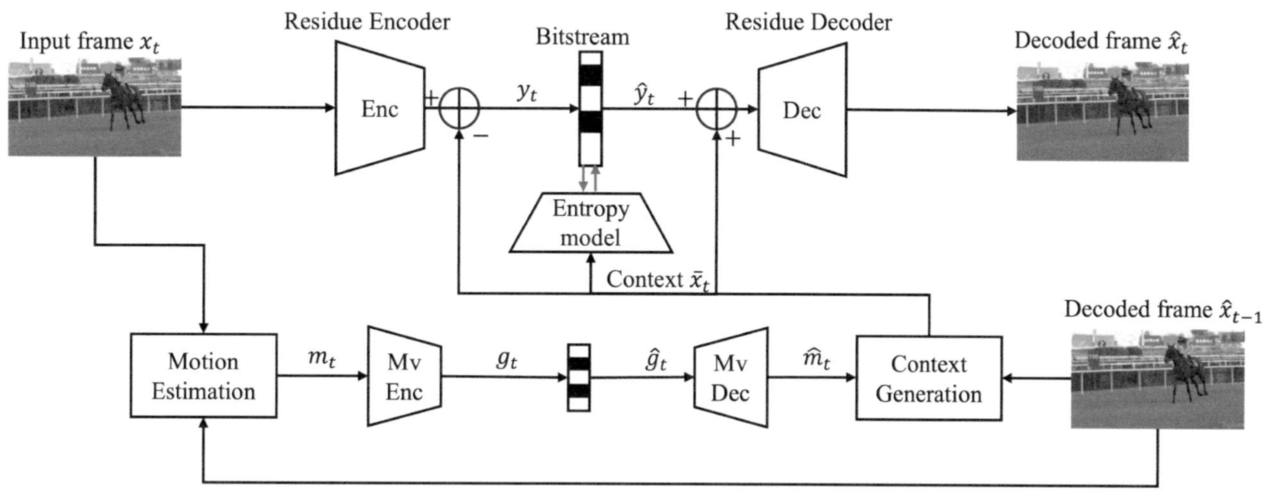


**Fig. 10.13** Architecture of FVC

ing, and hyperprior mechanisms follow principles similar to those discussed in deep learning-based image compression. DVC employs hyperpriors for better estimation of the probability distribution of $y_t$ and $g_t$. The end-to-end optimization loss function aligns with the (10.16) where $R$ represents the bitrate, comprising residuals, residual hyperpriors, motion vectors, and motion vector hyperpriors, and $D$ denotes the reconstruction quality loss. This probability modeling and loss function design have been widely adopted in subsequent works.

Feature space video coding (FVC) [28] provides an alternative approach by performing all compression operations in the feature domain, including motion estimation, motion compression, motion compensation, and residual compression, as shown in Fig. 10.13. This mitigates inaccuracies caused by pixel-domain motion estimation and compensation. Building on FVC, the Coarse-to-Fine (C2F) [29] encoder introduces a two-stage optical flow estimation in the feature domain, employing coarse and fine-grained branches for motion compensation to further reduce temporal redundancy.

Deep contextual video compression (DCVC) [30] represents another milestone with its distinct architecture. Unlike methods that compute explicit residuals by subtracting reference frames or features, DCVC conditions the coding process on context, as shown in Fig. 10.14. As $H(x_t - \bar{x}_t) \geq H(x_t|\bar{x}_t)$, conditional coding achieves a lower entropy, leading to better performance. Building on this architecture, various landmark studies have proposed multi-scale context generation DCVC-TCM [31] and stepwise entropy estimation schemes DCVC-DC [32], outperforming the H.266 standard under comparable settings.

Toward practical real-time neural video compression, DCVC-RT [33] introduces a novel approach that prioritizes operational efficiency while maintaining high compression performance. Unlike prior neural video codecs, which primarily focus on reducing computational complexity,

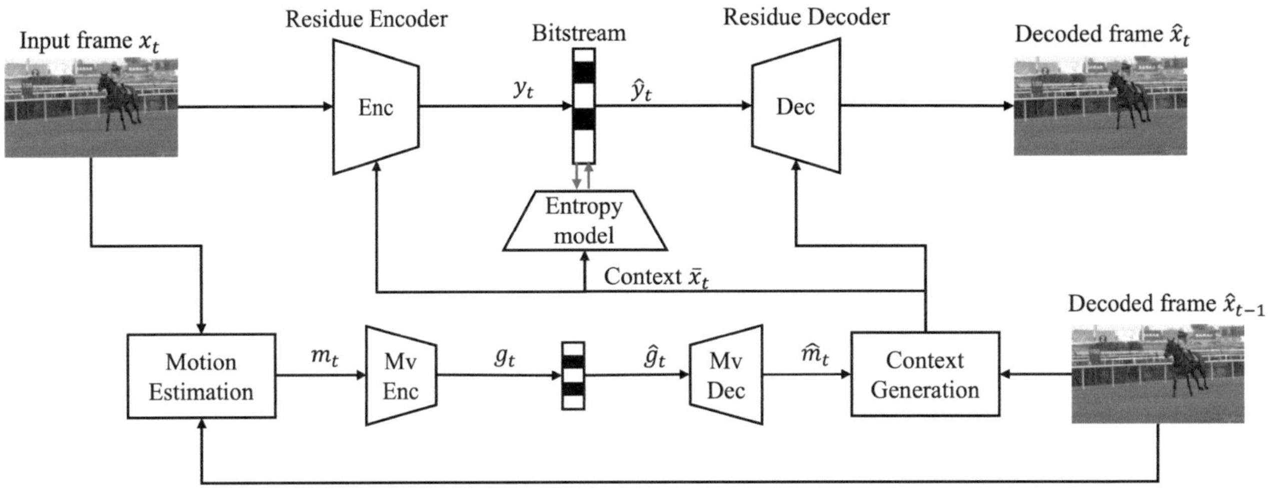


**Fig. 10.14** Architecture of DCVC

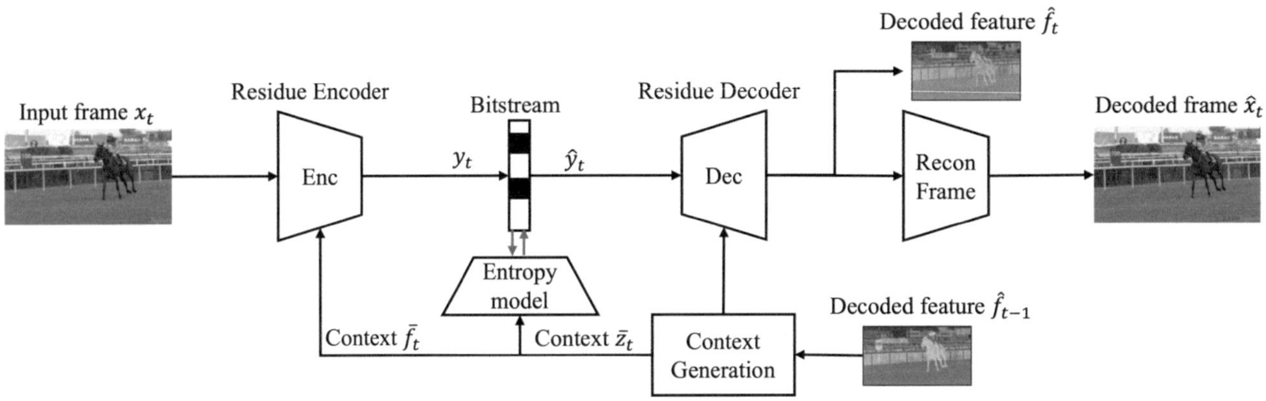


**Fig. 10.15** Architecture of DCVC-RT

DCVC-RT identifies operational complexity—including function call overhead and memory I/O—as the key bottleneck for real-time performance. To address this, DCVC-RT employs several innovative techniques:

- **Implicit Temporal Modeling**: Unlike traditional motion-based methods, such as DCVC-DC, that rely on explicit motion estimation and compensation, DCVC-RT removes these components entirely, as shown in Fig. 10.15, significantly reducing operational overhead while maintaining effective redundancy reduction.
- **Single Low-Resolution Latent Representation**: Instead of utilizing a multi-scale hierarchical approach, DCVC-RT encodes video frames at a single, fixed low-resolution (e.g., 1/8 of the original frame), reducing memory usage and computational requirements.
- **Integerization for Cross-Device Consistency**: Floating-point operations in deep learning-based codecs often result in inconsistencies across hardware platforms. DCVC-RT addresses this by implementing a deterministic integer-based computation framework, ensuring stable cross-device performance.
- **Module-Bank-Based Rate Control**: To improve practical adaptability, DCVC-RT introduces a modular rate control scheme, allowing fine-grained bitrate adjustments while maintaining high-quality compression.

These design choices enable DCVC-RT to achieve real-time performance on consumer hardware, a significant advancement over previous neural video codecs. DCVC-RT achieves an encoding speed of 125.2 fps and a decoding speed of 112.8 fps for 1080p videos, while saving an average of 21% in bitrate compared to H.266/VTM. This makes DCVC-RT the first practical neural video codec capable of real-time 1080p encoding and decoding on widely available GPUs, marking a breakthrough in the deployment of deep learning-based video compression.

**Specific Scenarios**

In this book, we use video conferencing as a case study to discuss how to achieve effective video compression in specific scenarios. Video conferencing typically involves single or small groups of speakers. For single-speaker scenarios, the video primarily focuses on the speaker's head movements and facial expressions, with a relatively static background. In multi-speaker scenarios, while the active speaker may switch, the frame still emphasizes the speaker's actions against a stable background. This characteristic allows compression to focus on dynamic regions (e.g., the speaker's face) while employing more efficient storage or transmission strategies for the background, reducing unnecessary data.

Currently, AI techniques can learn to represent facial keypoints through self-supervised learning, enabling facial generation based on these features. Alternatively, facial representations can be generated using 3D mesh sampling. For video conferencing compression, the P-frame compression can be simplified to encoding the residual information of facial features relative to the reference frame. The decoder reconstructs the video by generating facial expressions based on the reference frame and the residual information, achieving extremely low bitrates while maintaining reasonable subjective quality.

However, challenges remain. Handling large variations in head poses or scaling can lead to distortions or blurring. Additionally, performance saturates at higher bitrates, and processing high-resolution videos remains challenging. Future advancements may include:

- Developing new neural network architectures to enhance generalization and adaptability.
- Integrating audio and video compression. For instance, analyzing emotional features and speech rhythms in audio can assist video compression algorithms in capturing key facial expressions and motions, optimizing frame selection and processing to improve compression efficiency and reconstruction quality.

## 10.4 Summary

This chapter provides an overview of AI codecs across speech, image, and video modalities. For speech codecs, it introduces end-to-end AI speech codecs based on VQ-VAE, LPC-enhanced neural models, and text-driven compression frameworks. For image codecs, it outlines the mathematical formulation and model design principles for building learnable neural codecs. For video codecs, it first outlines the architecture and components of traditional video codecs, and introduces AI video codecs, detailing their design principles, motion modeling strategies, and adaptations for specific scenarios like video conferencing. Overall, this chapter outlines the foundations and advances that enable AI codecs in future intelligent communication systems.

## Problems

10.1 Which of the following is not a requirement for speech compression in mobile communication?

A. Supporting only a single language
B. High speech quality
C. Strong robustness against noise interference
D. Low complexity of the codec

10.2 What is the purpose of vector quantization in VQ-VAE for speech compression?

A. To convert the continuous latent space into discrete space for transmission
B. To reduce the dimensionality of the input speech signal
C. To model the prior distribution of the latent variables
D. To perform error correction in the encoded bitstream

10.3 Which of the following are common metrics for feature reconstruction in end-to-end neural speech compression?

A. Mean squared error (MSE) in the time domain
B. L1 loss of the Mel spectrogram
C. Perceptual evaluation of speech quality (PESQ)
D. Both A and B

10.4 In LPCNet, what part of the speech generation process is handled by the neural network?

A. Calculating LPC coefficients
B. Modeling the excitation signal
C. Voiced/unvoiced decision
D. Pitch determination

10.5 What is the main advantage of text-based AI speech codecs compared to other methods?

A. Higher speech quality
B. Lower computational complexity
C. Robustness against noise and channel errors
D. Extremely low bitrate

10.6 (a) Consider a random variable $X \in \{0, 1\}$. A sample is drawn from $X$ to form a sequence $\{x_n\}$ of length $N$. Assuming $N = 10$, we obtain the following sample sequence: 0010000000. Based on this sequence, it can be calculated that $P(X = 0) = 0.9$ and $P(X = 1) = 0.1$.

Based on these probabilities, calculate the entropy $H(X)$ of the random variable $X$.

(b) Now, process the sample sequence $\{x_n\}$ by replacing every 0 with 01 and every 1 with 10 to get a new sequence $\{y_n\}$ of length $2N$: 01011001010101010101. According to statistical results, it is judged that the random variable $Y$ follows the distribution $P(Y = 0) = 0.5$ and $P(Y = 1) = 0.5$. Calculate the entropy $H(Y)$ of the random variable $Y$. Is $H(Y)$ equal to $H(X)$? Justify your answer.

10.7 Before applying entropy coding methods such as Huffman coding or arithmetic coding, the data sequence to be encoded is considered as a sample sequence $\{x_n\}$ from a random variable $X$. By statistical methods, one can obtain the cumulative distribution function (CDF) of $X$. Specifically, for image compression, where pixel values are integers within $[0, 255]$, it is straightforward to compute the CDF and subsequently estimate the entropy $H(X)$ of an image. This allows for calculating a lower bound on the encoded length based on the CDF. However, this approach introduces redundancy, and current lossless compression efforts continue to surpass the bounds established by these methods. Explain why this compression technique inherently contains redundancy and discuss potential reasons for ongoing advancements that push beyond these traditional limits.

10.8 Suppose the output of the encoder $g_a$ is $\mathbf{y}$. The quantized result using rounding is $\hat{\mathbf{y}} = \text{round}(\mathbf{y})$, and the result using noise-based quantization approximation is $\tilde{\mathbf{y}} = \mathbf{y} + U(-0.5, 0.5)$, where $U(-0.5, 0.5)$ denotes uniform noise. The round operation $\hat{\mathbf{y}} = \text{round}(\mathbf{y})$ is non-differentiable and cannot propagate gradients, hence $\tilde{\mathbf{y}} = \mathbf{y} + U(-0.5, 0.5)$ is used to approximate the quantization process. In fact, the straight-through estimator can also be used to enable gradient propagation through the rounding operation. However, in training NIC models, we still prefer using noise-based quantization. Please explain why.

10.9 Based on the autoregressive formula $q(\tilde{\mathbf{y}}) = \prod_i p(\tilde{y}_i \mid \tilde{\mathbf{y}}_{<i}, \tilde{z})$ and considering the work mentioned in Sect. 10.2.3, how to design an autoregressive model that balances model capacity and encoding time?

10.10 Which of the following video compression techniques is most effective at reducing temporal redundancy between frames in a video?

A. Intra-frame Compression
B. Inter-frame Compression
C. Transform Coding
D. Quantization

10.11 In the H.264 video compression architecture, what is the role of the preprocessing step?

A. It adjusts the video resolution to fit the desired output format.
B. It converts the video color space from RGB to YUV and downsamples chrominance components.
C. It applies motion compensation to video frames.
D. It partitions video frames into macroblocks.

10.12 In the context of Deep Video Compression (DVC), what is the role of the motion compensation network?

A. It quantizes motion vectors for more efficient encoding.
B. It performs optical flow estimation to capture pixel-level displacement between frames.
C. It predicts future frames based on previous frames using deep learning.
D. It generates a reference frame to reduce residuals.

10.13 What distinguishes DCVC-RT (Deep Contextual Video Compression - Real-Time) from traditional neural video codecs in terms of achieving real-time performance?

A. It uses multi-scale motion compensation to enhance frame prediction.
B. It focuses on removing explicit motion compensation to reduce operational overhead.
C. It employs an adaptive rate control scheme to vary the bitrate dynamically.
D. It introduces a higher resolution video encoding method to improve visual quality.

## References

1. J.-M. Valin, K. Vos, T. Terriberry, Definition of the opus audio codec, tech. rep. (2012)
2. M. Dietz, M. Multrus, V. Eksler, V. Malenovsky, E. Norvell, H. Pobloth, L. Miao, Z. Wang, L. Laaksonen, A. Vasilache, et al., Overview of the evs codec architecture, in *2015 IEEE International Conference on Acoustics, Speech and Signal Processing (ICASSP)* (IEEE, 2015), pp. 5698–5702
3. I. Goodfellow, J. Pouget-Abadie, M. Mirza, B. Xu, D. Warde-Farley, S. Ozair, A. Courville, Y. Bengio, Generative adversarial networks. Commun. ACM **63**(11), 139–144 (2020)
4. D.P. Kingma, Auto-encoding variational bayes, arXiv preprint arXiv:1312.6114 (2013)
5. D. Rezende, S. Mohamed, Variational inference with normalizing flows, in *International Conference on Machine Learning* (PMLR, 2015), pp. 1530–1538
6. I. Sutskever, Sequence to sequence learning with neural networks, arXiv preprint arXiv:1409.3215 (2014)
7. A. Van Den Oord, O. Vinyals, et al., Neural discrete representation learning. *Advances in Neural Information Processing Systems*, vol. 30 (2017)

8. N. Zeghidour, A. Luebs, A. Omran, J. Skoglund, M. Tagliasacchi, Soundstream: an end-to-end neural audio codec. IEEE/ACM Trans. Audio, Speech, Lang. Process. **30**, 495–507 (2021)
9. A. Défossez, J. Copet, G. Synnaeve, Y. Adi, High fidelity neural audio compression, arXiv preprint arXiv:2210.13438 (2022)
10. R. Kumar, P. Seetharaman, A. Luebs, I. Kumar, K. Kumar, High-fidelity audio compression with improved rvqgan. *Advances in Neural Information Processing Systems*, vol. 36 (2024)
11. D. Yang, S. Liu, R. Huang, J. Tian, C. Weng, Y. Zou, Hifi-codec: group-residual vector quantization for high fidelity audio codec, arXiv preprint arXiv:2305.02765 (2023)
12. H. Liu, X. Xu, Y. Yuan, M. Wu, W. Wang, M.D. Plumbley, Semanticodec: an ultra low bitrate semantic audio codec for general sound, arXiv preprint arXiv:2405.00233 (2024)
13. J.-M. Valin, J. Skoglund, Lpcnet: improving neural speech synthesis through linear prediction, in *ICASSP 2019-2019 IEEE International Conference on Acoustics, Speech and Signal Processing (ICASSP)* (IEEE, 2019), pp. 5891–5895
14. K. Tokuda, T. Masuko, J. Hiroi, T. Kobayashi, T. Kitamura, A very low bit rate speech coder using hmm-based speech recognition/synthesis techniques, in *Proceedings of the 1998 IEEE International Conference on Acoustics, Speech and Signal Processing, ICASSP'98 (Cat. No. 98CH36181)*, vol. 2 (IEEE, 1998), pp. 609–612
15. Z. Weng, Z. Qin, X. Tao, C. Pan, G. Liu, G.Y. Li, Deep learning enabled semantic communications with speech recognition and synthesis. IEEE Trans. Wireless Commun. **22**(9), 6227–6240 (2023)
16. J. Duda, Asymmetric numeral systems, arXiv preprint arXiv:0902.0271 (2009)
17. J. Ballé, V. Laparra, E.P. Simoncelli, End-to-end optimized image compression, in *International Conference on Learning Representations (ICLR)* (2017)
18. J. Ballé, V. Laparra, E.P. Simoncelli, Density modeling of images using a generalized normalization transformation, in *International Conference on Learning Representations (ICLR)* (2016)
19. J. Ballé, D. Minnen, S. Singh, S.J. Hwang, N. Johnston, Variational image compression with a scale hyperprior, arXiv preprint arXiv:1802.01436 (2018)
20. Z. Cheng, H. Sun, K. Takeuchi, J. Katto, Learned image compression with discretized gaussian mixture likelihoods and attention modules, in *Proceedings of the IEEE/CVF Conference on Computer Vision and Pattern Recognition (CVPR)* (2020), pp. 7939–7948
21. S. He, Z. Liu, S. Li, H. Li, Elic: efficient learned image compression with unevenly grouped space-channel contextual adaptive coding, in *European Conference on Computer Vision (ECCV)* (2022)
22. Z. Liu, Y. Lin, Y. Cao, H. Hu, Y. Wei, Z. Zhang, S. Lin, B. Guo, Swin transformer: hierarchical vision transformer using shifted windows, in *Proceedings of the IEEE/CVF International Conference on Computer Vision (ICCV)* (2021), pp. 10012–10022
23. K. Li, M. Xu, W. Lin, J. Xu, Lic-tcm: a token-based compression model with learnable context modeling for image compression, in *Proceedings of the IEEE/CVF Conference on Computer Vision and Pattern Recognition (CVPR)* (2023)
24. D. Minnen, S. Singh, N. Johnston, J. Ballé, Joint autoregressive and hierarchical priors for learned image compression, in *Advances in Neural Information Processing Systems (NeurIPS)* (2018)
25. D. He, Y. Zheng, B. Sun, Y. Wang, H. Qin, Checkerboard context model for efficient learned image compression, in *Proceedings of the IEEE/CVF Conference on Computer Vision and Pattern Recognition (CVPR)* (2020), pp. 7939–7948
26. D. Minnen, S. Singh, Channel-wise autoregressive entropy models for learned image compression, in *Proceedings of the IEEE International Conference on Image Processing (ICIP)* (2020)
27. G. Lu, W. Ouyang, D. Xu, X. Zhang, C. Cai, Z. Gao, Dvc: an end-to-end deep video compression framework, in *Proceedings of the IEEE/CVF conference on computer vision and pattern recognition* (2019), pp. 11006–11015
28. Z. Hu, G. Lu, D. Xu, Fvc: a new framework towards deep video compression in feature space, in *Proceedings of the IEEE/CVF Conference on Computer Vision and Pattern Recognition* (2021), pp. 1502–1511
29. Z. Hu, G. Lu, J. Guo, S. Liu, W. Jiang, D. Xu, Coarse-to-fine deep video coding with hyperprior-guided mode prediction, in *Proceedings of the IEEE/CVF Conference on Computer Vision and Pattern Recognition* (2022), pp. 5921–5930
30. J. Li, B. Li, Y. Lu, Deep contextual video compression. Adv. Neural. Inf. Process. Syst. **34**, 18114–18125 (2021)
31. X. Sheng, J. Li, B. Li, L. Li, D. Liu, Y. Lu, Temporal context mining for learned video compression. IEEE Trans. Multimedia **25**, 7311–7322 (2022)
32. J. Li, B. Li, Y. Lu, Neural video compression with diverse contexts, in *Proceedings of the IEEE/CVF Conference on Computer Vision and Pattern Recognition* (2023), pp. 22616–22626
33. Z. Jia, B. Li, J. Li, W. Xie, L. Qi, H. Li, Y. Lu, Towards practical real-time neural video compression, arXiv preprint arXiv:2502.20762 (2025)

# Waveform Optimization for Semantic Communication in OFDM System

# 11

**Abstract**

Recent advancements in DJSCC have introduced a new paradigm for semantic communications. DJSCC-based semantic communications have demonstrated notable performance, particularly in scenarios with limited bandwidth and low signal-to-noise ratios. This chapter investigates waveform optimization for semantic communication in OFDM systems. While OFDM is widely adopted for its spectral efficiency and resilience against multipath fading, the integration of neural network-based semantic encoders and decoders introduces new challenges. In particular, DJSCC enables end-to-end optimization and improved adaptability under varying channel conditions, but also leads to significantly higher PAPR due to continuous-valued semantic signals across multiple subcarriers. To address this issue, the chapter reviews both conventional and learning-based PAPR reduction techniques, including clipping, companding, SLM, PTS, PAPR-aware loss design, and clipping with retraining. By applying these methods, the excellent image reconstruction performance of DJSCC can be preserved while keeping PAPR within an acceptable range in the OFDM system. These insights underline the importance of PAPR-aware waveform design in advancing reliable and efficient semantic communication over OFDM systems, particularly for future intelligent wireless networks.

**Keywords**

Semantic communication • Joint semantic-channel coding • OFDM transmission • PAPR reduction • Waveform optimization

## 11.1 Introduction

OFDM is an efficient multicarrier modulation technique extensively employed in both wired and wireless communication systems, including Digital Subscriber Line (DSL) Internet access, Asymmetric Digital Subscriber Line (ADSL), Wireless Local Area Network (WLAN), Worldwide Interoperability for Microwave Access (WiMAX), Digital Video Broadcasting (DVB), Digital Television Broadcasting (DTVB), and fourth-generation Long-Term Evolution (LTE) networks. In recent years, advanced OFDM-based technologies, such as Filtered OFDM (F-OFDM), Universal Filtered OFDM (U-OFDM), and Filter Bank Multicarrier (FBMC), have also been introduced and successfully integrated into fifth-generation (5G) systems.

As a multicarrier modulation technique, OFDM provides high data rates and strong resilience against frequency-selective fading, making it effective at mitigating multipath interference and impulse noise. Compared with alternative multicarrier schemes, OFDM exhibits lower susceptibility to echo and nonlinear distortion, as well as enhanced robustness against narrowband co-channel interference. Consequently, these characteristics collectively contribute to substantially improved system capacity.

OFDM implements modulation and demodulation functions through the Inverse Fast Fourier Transform (IFFT) and Fast Fourier Transform (FFT), respectively, simplifying engineering implementation due to its lower complexity. Moreover, the orthogonality between its subcarriers substantially improves spectral efficiency. Figure 11.1 illustrates the spectral differences between conventional Multicarrier Modulation (MCM) and OFDM. In conventional MCM, each sub-


W. Chen and Z. Qin, *Semantic Communications*, Textbooks in Telecommunication Engineering,
https://doi.org/10.1007/978-3-032-11005-3_11

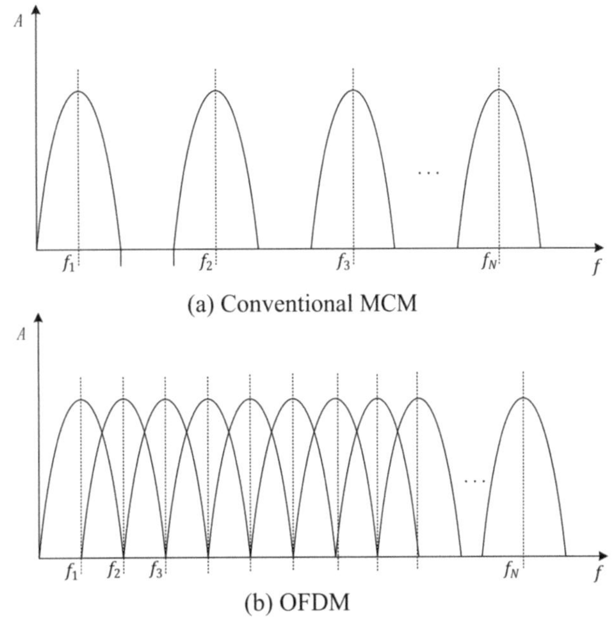


**Fig. 11.1** An Example of **a** Conventional MCM transmission and **b** OFDM transmission

band is completely isolated, and the spacing between sub-bands must exceed the Nyquist bandwidth to avoid inter-carrier interference, resulting in wasted spectral resources. In contrast, orthogonal subcarriers of OFDM allow sub-band overlap, significantly increasing bandwidth utilization.

For high-speed data transmission, OFDM divides a single high-speed serial data stream into multiple lower-speed parallel streams, each modulated onto a distinct subcarrier. This technique effectively mitigates delay spread issues associated with multipath propagation. As a result, in contemporary spectrum-limited wireless communication environments, OFDM remains highly competitive owing to its exceptional spectral efficiency and robust interference resistance.

**Conventional Communication in OFDM System**

In traditional communication systems, OFDM improves spectral efficiency and enhances resistance to multipath interference by dividing a high-speed data stream into multiple lower-speed parallel streams, each assigned to a distinct subcarrier. The core principle of OFDM involves employing a set of mutually orthogonal subcarriers, with each subcarrier occupying its own frequency band. Because these subcarriers remain orthogonal, interference among them is effectively eliminated. To achieve this, an OFDM system typically comprises two main components: a transmitter and a receiver, as illustrated in Fig. 11.2.

Assume the modulated complex symbol sequence being transmitted is $C_k = A_k + B_k i$, where $k \in \{0, 1, 2, \ldots, N-1\}$. For the $k$-th subcarrier, the signal sequence can be decomposed into in-phase (I) and quadrature (Q) components, represented by $\sin(2\pi \Delta f kt)$ and $\cos(2\pi \Delta f kt)$, respectively, where $\Delta f$ denotes the subcarrier spacing. This configuration ensures the orthogonality of different subcarriers, meaning that over a symbol period, all subcarriers are mutually orthogonal, as the integral of their product is zero. The modulated signal can be represented as:

$$s(t) = C_k G_k(t) = \sum_{k=0}^{N-1} \left(A_k \sin(2\pi \Delta f kt) + B_k \cos(2\pi \Delta f kt)\right). \tag{11.1}$$

At the receiver, the corresponding subcarrier signal is extracted, and inner product operations are performed to remove the interference of other subcarriers on the given subcarrier. This process can be expressed in complex form as:

$$s(t) = \sum_{k=0}^{N-1} C_k e^{j2\pi \Delta f kt}. \tag{11.2}$$

This representation is consistent with the Fourier series form. Therefore, the continuous OFDM signal can be represented as:

$$x(t) = \frac{1}{\sqrt{N}} \sum_{k=0}^{N-1} X_k e^{j2\pi \Delta f kt}, \quad (0 \le t < NT), \tag{11.3}$$

where $X_k$ represents the modulated symbol sequence, $N$ is the number of subcarriers, $\Delta f = \frac{1}{NT}$ denotes the subcarrier spacing, and $T$ represents the symbol period.

In discrete time, the modulation process of OFDM can be achieved via the Inverse Discrete Fourier Transform (IDFT), represented by the following formula:

$$x_k = \frac{1}{\sqrt{N}} \sum_{n=0}^{N-1} X_n e^{j2\pi kn/N}, \tag{11.4}$$

where $x_k$ denotes the OFDM signal in the time domain, where $k \in \{0, 1, 2, \ldots, N-1\}$. In practical engineering implementations, this step is usually performed using the FFT and IFFT to improve computational efficiency.

The detailed OFDM implementation process is illustrated in the diagram. At the transmitter, the input data sequence is first modulated, converting it into complex symbols $C_k$. Subsequently, a Serial-to-Parallel (S/P) conversion divides the modulated data sequence into $N$ parallel data streams, each corresponding to a subcarrier. Then, the IFFT is used to transform the frequency-domain signal into the time domain, as shown in Eq. (11.4). This process synthesizes all subcarrier signals into a single time-domain signal, which facilitates parallel transmission over a single path.

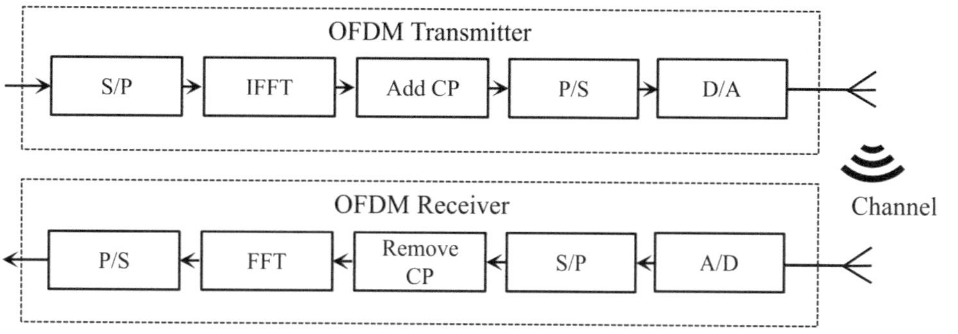


**Fig. 11.2** Implementation of OFDM system

To mitigate the effects of multipath interference, the OFDM system inserts a Cyclic Prefix (CP) into the time-domain signal, with a length generally matched to the maximum delay spread of the channel to ensure the orthogonality of subcarriers. After inserting the CP, the parallel data stream undergoes Parallel-to-Serial (P/S) conversion to form a continuous serial signal. This signal is then converted to an analog signal via Digital-to-Analog (D/A) conversion and transmitted through the wireless channel. During transmission, the signal may be affected by noise and multipath interference.

At the receiver, an Analog-to-Digital (A/D) converter first digitizes the received analog signal, which is then subjected to a Serial-to-Parallel (S/P) conversion to retrieve the parallel data streams corresponding to each subcarrier. The CP is removed to eliminate the effects of multipath interference on orthogonality. Subsequently, an FFT converts the time-domain signal back to the frequency domain, yielding the frequency-domain signals for each subcarrier.

To counteract distortion introduced by phase variations and frequency offset in the channel, the receiver employs phase tracking and Carrier Frequency Offset (CFO) correction modules to rectify the phase and frequency deviations in the received signal. The parallel data streams are then converted back to a serial data stream through Parallel-to-Serial (P/S) conversion, and the demodulation module finally extracts the transmitted data content.

**Semantic Communication in OFDM System**

In a semantic communication system, integrating neural network-based semantic encoders and decoders with OFDM differs markedly from traditional syntax-oriented communication architectures. Consider image transmission as an example: conventional syntax-based systems typically separate source coding (e.g., BPG for image compression) and channel coding (e.g., LDPC for error correction), with source coding first followed by channel coding to mitigate channel noise and interference. Because these two coding processes operate independently, the system exhibits reduced robustness under dynamic channel conditions and entails greater overall complexity. Under high-noise scenarios, the performance of this separated coding paradigm is constrained, making it difficult to satisfy the requirements of efficient image and video transmission.

The neural network-based semantic encoder and decoder combined with OFDM adopts an end-to-end optimization approach [1]. As shown in Fig. 11.3, a point-to-point OFDM system with neural network-based semantic encoding maps the $n$-dimensional input $\boldsymbol{s} \in \mathbb{R}^n$ to a $k$-dimensional complex-valued feature $\boldsymbol{z} \in \mathbb{C}^k$, satisfying the average power constraint $\frac{1}{k}\mathbb{E}(zz^*) \leq 1$, where $z^*$ represents the complex conjugate of $z$. The bandwidth-to-signal ratio is defined as $R = \frac{k}{n}$. The encoding function $f_\theta : \mathbb{R}^n \to \mathbb{C}^k$ is a neural network parameterized by $\theta$, and the encoding process is represented as $z = f_\theta(\boldsymbol{s})$. The encoded symbol sequence $z$ is transformed to a time-domain signal $\boldsymbol{x}$ via the IFFT and transmitted through an additive Gaussian white noise (AWGN) channel. The channel output $\boldsymbol{y} \in \mathbb{C}^k$ can be expressed as:

$$\boldsymbol{y} = \boldsymbol{x} + \boldsymbol{n}, \tag{11.5}$$

where $\boldsymbol{n} \in \mathbb{C}^k$ is an independent, identically distributed Gaussian noise vector with mean 0 and variance $\sigma^2$.

The received signal $\boldsymbol{y}$ is transformed by the FFT into a frequency-domain signal $\hat{z}$, which serves as the input to the decoder. The decoder maps $\hat{z}$ back to $\mathbb{R}^n$ to estimate the original vector $\boldsymbol{s} \in \mathbb{R}^n$. Similar to the encoder, the decoder function $g_\phi : \mathbb{C}^k \to \mathbb{R}^n$ is a neural network parameterized by $\phi$, and the decoding process is represented as:

$$\hat{\boldsymbol{s}} = g_\phi(\hat{z}) = g_\phi\left(FFT(f_\theta(IFFT(\boldsymbol{s}))) + \boldsymbol{n}\right). \tag{11.6}$$

The distortion between the original signal $\boldsymbol{s}$ and the reconstructed signal $\hat{\boldsymbol{s}}$ can be expressed as:

$$d(\boldsymbol{s}, \hat{\boldsymbol{s}}) = \frac{1}{n}\sum_{i=1}^{n}(s_i - \hat{s}_i)^2, \tag{11.7}$$

where $s_i$ and $\hat{s}_i$ represent the $i$-th elements of $\boldsymbol{s}$ and $\hat{\boldsymbol{s}}$, respectively. Given a fixed-bandwidth-to-signal ratio $R$, the neural network-based semantic encoder and decoder aim to mini-

**Fig. 11.3** DJSCC system model with IFFT and FFT

mize the end-to-end distortion by optimizing the parameters $\theta^*$ and $\phi^*$ as follows:

$$(\theta^*, \phi^*) = \arg\min_{\theta,\phi} \mathbb{E}_{p(s,\hat{s})}[d(s, \hat{s})], \tag{11.8}$$

where $p(s, \hat{s})$ is the joint probability distribution of the original and reconstructed signals.

The neural network-based semantic encoder and decoder combined with OFDM significantly improves system adaptability across various channel environments by jointly optimizing source and channel processing while directly considering channel characteristics, such as multipath fading and additive Gaussian white noise. This integrated approach enhances the system's performance stability, even under changing channel conditions, distinguishing it from traditional OFDM architectures, where source and channel coding are independently implemented.

## 11.2 PAPR Reduction Methods in OFDM System

### 11.2.1 PAPR Challenges in OFDM-Based Semantic Communication

In OFDM systems, the Peak-to-Average Power Ratio (PAPR) is an essential metric for evaluating the ratio of the instantaneous peak power to the average power of a signal. The magnitude of PAPR has a direct impact on the system's power amplifier efficiency and its ability to resist nonlinear distortion. Typically, PAPR for both continuous and discrete OFDM signals is defined as follows:

For continuous OFDM signals, PAPR can be expressed as:

$$PAPR(dB) = 10\log_{10}\frac{\max|x(t)|^2}{\frac{1}{NT}\int_0^{NT}|x(t)|^2dt}, \tag{11.9}$$

where $x(t)$ is the amplitude of the OFDM signal at time $t$, and $NT$ is the symbol duration.

For discrete OFDM signals, PAPR is defined as:

$$PAPR(dB) = 10\log_{10}\frac{\max|x_k|^2}{E[|x_k|^2]}, \tag{11.10}$$

where $x_k$ is the amplitude of the discrete OFDM signal, and $E[\cdot]$ denotes the expectation.

An OFDM signal is composed of multiple subcarriers superimposed upon one another, where the amplitude and phase of each subcarrier may vary randomly over time. When the phases of several subcarriers align at a particular time, constructive interference occurs, causing the instantaneous peak of the signal to significantly exceed its average power. Consequently, OFDM signals generally exhibit high PAPR, and this phenomenon becomes more pronounced with an increase in the number of subcarriers. This high PAPR characteristic poses challenges to system design, especially in the power amplifier, where high PAPR can push the amplifier into a nonlinear operating region, resulting in signal distortion and degradation of transmission quality.

In semantic communication systems, the combination of neural network-based semantic encoders and decoders with OFDM exacerbates the PAPR problem. Unlike traditional digital communication that utilizes discrete modulation constellations, neural network-based semantic encoders generate continuous amplitude signals instead of discrete constellation points. The semantic encoder maps semantic information (e.g., images or video content) into a series of continuous values, providing greater freedom in amplitude representation and thus preserving more semantic information in the channel. However, when these continuous amplitude signals are modulated by OFDM, the dynamic nature of their amplitudes further increases the instantaneous peaks of the signal. Figure 11.4 shows the constellation diagram of modulated signals in a neural network-based semantic commu-

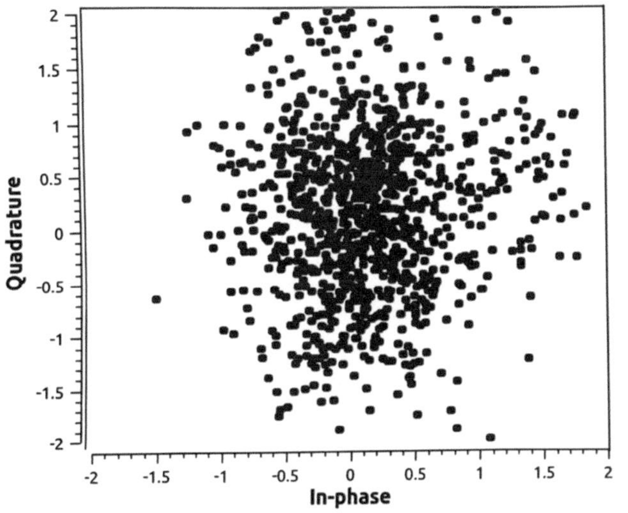


**Fig. 11.4** DJSCC constellation diagram

nication system. It can be observed that the points are distributed continuously, indicating that the signals generated by the semantic encoder exhibit greater freedom in phase and amplitude. This continuous amplitude signal, combined with the superposition of multiple OFDM subcarriers, results in higher instantaneous peaks, thereby significantly increasing the PAPR.

The performance of PAPR is generally characterized by the Complementary Cumulative Distribution Function (CCDF), defined as the probability that the PAPR of a signal exceeds a specific threshold:

$$CCDF(PAPR0) = P\{PAPR > PAPR0\}, \quad (11.11)$$

where $P$ represents probability. A steeper CCDF curve implies a lower PAPR in the system, which is favorable for power amplifier design.

Figure 11.5 presents the PAPR performance for DJSCC and traditional digital modulation (e.g., 16QAM) with different numbers of subcarriers. It can be observed that, with the same number of subcarriers, the PAPR of the DJSCC-based semantic communication system is significantly higher than that of traditional 16QAM modulation. As the number of subcarriers increases, the PAPR for DJSCC further increases. The figure shows that for 16QAM modulation, with 64, 128, and 256 subcarriers, the PAPR values are relatively low, and the CCDF curves are more concentrated. However, for DJSCC-generated signals, as the number of subcarriers increases, the PAPR increases significantly, and the CCDF curve only starts to drop at higher PAPR values, reflecting the higher instantaneous peaks.

These results indicate that the continuous amplitude signals generated by DJSCC in a multi-subcarrier configuration lead to a substantial increase in PAPR. To mitigate the impact

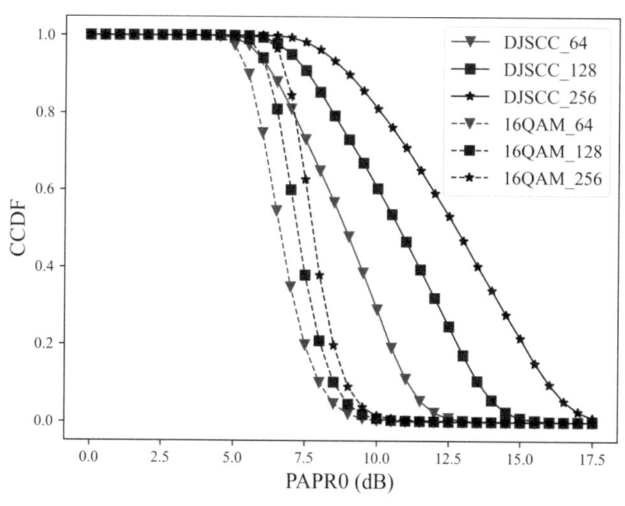


**Fig. 11.5** PAPR performance comparison between DJSCC and 16QAM

of high PAPR on nonlinear distortion in power amplifiers, the system needs to adopt more effective PAPR reduction techniques to ensure the stability of signal transmission and overall energy efficiency. The following introduces the traditional PAPR reduction techniques and deep learning-based methods for PAPR reduction [1].

## 11.2.2 PAPR Reduction Techniques

### 11.2.2.1 Conventional Signal Distortion Techniques

The fundamental principle of signal distortion-based PAPR suppression algorithms involves pre-distorting the OFDM time-domain signal so that its peak-to-average power ratio is already low before passing through the nonlinear hardware amplifier. Although the pre-distorted signals may be either entirely unrecoverable or partially recoverable, these methods generally offer straightforward implementation and low computational complexity.

**Clipping** Clipping is a process that limits the signal amplitude to a predefined threshold, thereby reducing peak values while preserving most low-amplitude samples. The clipped signal can be expressed as:

$$x_{clip}[n] = \begin{cases} x[n]e^{j\phi[n]}, \ x[n] < \rho\sqrt{\bar{P}} \\ \rho\sqrt{\bar{P}}e^{j\phi[n]}, \ x[n] \geq \rho\sqrt{\bar{P}} \end{cases}, \quad (11.12)$$

where $x[n]$ represents the OFDM time-domain sample sequence at phase $\phi[n]$, and the phase remains unchanged before and after clipping. The parameter $\rho$ denotes the clipping rate (CR), defined as the ratio of the clipping threshold amplitude to the root-mean-square (RMS) value of the original signal samples.

**$\mu$-Law Companding** The companding method applies a companding function to compress large-amplitude components of the OFDM time-domain signal and to expand smaller amplitudes, thereby reducing the PAPR. Unlike clipping, companding allows partial recovery of the original signal through an inverse companding function, with the degree of signal restoration depending on the specific companding function employed. The $\mu$-law companding function is expressed as:

$$z_{n,k} = \frac{V x_{n,k}}{\ln(1+\mu)|x_{n,k}|} \ln\left(1 + \frac{\mu}{V}|x_{n,k}|\right), \quad (11.13)$$

where $x_{n,k}$ represents the $k$-th sample of the $n$-th OFDM symbol of the transmitted signal. $V$ denotes the average magnitude of the time-domain OFDM symbols, and $|x_{n,k}|$ is the modulus of the complex sample $x_{n,k}$. The parameter $\mu$ is a constant that determines the degree of compression.

The inverse companding function applied at the receiver is expressed as:

$$x_{n,k} = \frac{V' z_{n,k}'}{\mu |z_{n,k}'|} e^{(\frac{|z_{n,k}'| \ln(1+\mu)}{V'}) - 1}, \quad (11.14)$$

where $V'$ denotes the average amplitude of the received signal $z_{n,k}'$.

#### 11.2.2.2 Conventional Signal Non-distortion Techniques

Signal distortion-free PAPR suppression techniques do not introduce signal distortion when mitigating the PAPR of OFDM signals, thereby preserving receiver performance. However, these algorithms often pose implementation challenges and entail high computational complexity. This section primarily focuses on two distortion-free PAPR suppression methods, namely Selected Mapping (SLM) and Partial Transmit Sequence (PTS). Both approaches generate multiple signal representations by applying different phase factors and then select the representation exhibiting the lowest PAPR for transmission. By reducing the probability of large peak occurrences, these techniques fall under the probabilistic class of PAPR mitigation methods.

**SLM** In the SLM system block diagram (see Fig. 11.6), the frequency-domain signal $Z$ is replicated $V$ times as $\{Z_1, Z_2, Z_3, \ldots, Z_{V-1}, Z_V\}$. Each copy is then multiplied by a phase factor sequence $p_v$, $v \in 1, 2, \ldots, V$, where a single phase factor is defined as $p_v{}^{(k)} = \exp(j\varphi)$, $\varphi \in [0, 2\pi]$. This yields the phase-transformed signal set $Z'$ and, after an IFFT operation, the corresponding time-domain OFDM signal set $X$. A selector chooses the time-domain signal in $X$ that exhibits the lowest PAPR for transmission. The phase factor sequence index associated with the selected signal is then forwarded alongside the transmitted data, enabling the receiver to recover the original signal by applying the correct phase factor sequence. Notably, SLM does not introduce additional distortion to the original signal.

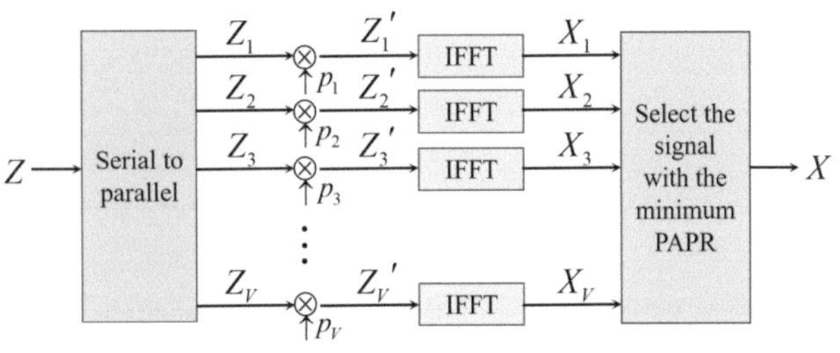


**Fig. 11.6** SLM system framework

**PTS** PTS is a technique closely related to SLM, and its structure is illustrated in Fig. 11.7. A modulated symbol sequence $Z = [Z_1, Z_2, Z_3, \ldots, Z_{N-1}, Z_N]$, where $N$ denotes the number of subcarriers, is partitioned into $V$ segments. Each segment contains $\frac{N}{V}$ valid data symbols, and the remaining positions are filled with zeros to form a data block of length $N$. An $N$-point IFFT is then applied to each data block to generate the time-domain signal. A random sequence of $V$ phase factors, each of length $N$, is generated as $p_v$, where $v \in \{1, 2, \ldots, V\}$. Each phase factor is defined as $p_v(k) = \exp(j\phi)$, with $\phi \in [0, 2\pi]$ and $k \in \{1, 2, \ldots, N\}$. These phase factors are typically selected from the set $\{\pm 1, \ pmj\}$. For each branch, the phase factor sequence is applied to the corresponding time-domain signal, and the resulting signals are weighted and summed. The PAPR is then calculated for each summed signal.

This process is repeated $M$ times, and the phase factor sequence that results in the smallest PAPR is selected as the final phase factor sequence. The corresponding time-domain emission signal, generated using this optimal phase factor sequence, achieves a reduction in PAPR.

#### 11.2.2.3 Deep Learning-Based Techniques

Deep learning can determine the correspondence between inputs and outputs based on the loss function and the training dataset. During the training process, the loss function significantly influences the direction in which the network parameters are updated, thereby achieving the desired outputs. Additionally, incorporating nonlinear functions into the network enables it to learn outputs that conform to specific features [2].

**PAPR Loss** According to [3], incorporating PAPR into the loss function of a neural network facilitates the training of a network model that produces outputs with low PAPR. To minimize the distortion between the network's input and output, the original DJSCC loss function $L_1$ can be expressed as:

$$L_1 = \|\boldsymbol{s} - g(FFT(IFFT(f(\boldsymbol{s}, \theta)) + \boldsymbol{n}), \phi)\|_2. \quad (11.15)$$

To reduce the PAPR, the loss function $L_2$ can be expressed as:

$$L_2 = \mathbb{E}[PAPR\{IFFT(f(\boldsymbol{s}, \theta))\}], \quad (11.16)$$

where $\mathbb{E}$ denotes the mean value.

To minimize both distortion and PAPR, the new DJSCC loss function is formulated by combining the original distortion loss $L_1$ and an additional PAPR-related loss $L_2$. The combined loss function $L$ is expressed as:

$$L = L_1 + \lambda L_2, \quad (11.17)$$

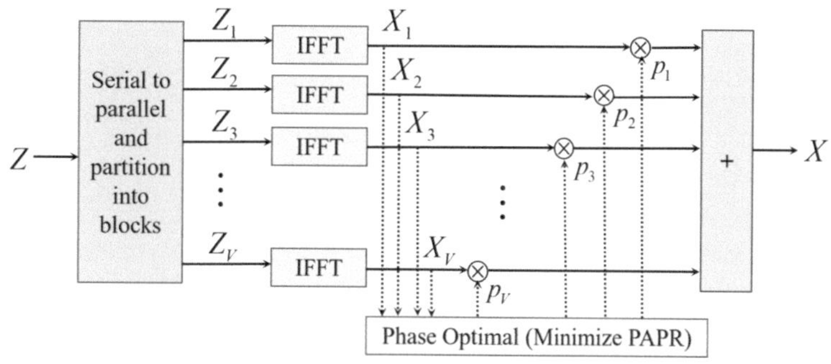


**Fig. 11.7** PTS system framework

where $\lambda$ denotes the weighting factor that balances the importance of minimizing distortion against reducing PAPR. A larger $\lambda$ means that the learning is more biased toward reducing the PAPR, whereas a smaller $\lambda$ places greater emphasis on reducing the distortion between the reconstructed signal and the original signal.

By integrating $L_2$ into the loss function, the neural network is encouraged to produce outputs that not only closely match the input data but also possess lower PAPR values. This dual-objective optimization enhances the efficiency and robustness of the communication system by ensuring high-quality signal reconstruction while mitigating the adverse effects of high PAPR on the transmission hardware.

**Clipping with retraining** Deep learning can enable a network to generate outputs with specific characteristics by incorporating nonlinear functions within its architecture. Consequently, clipping can be integrated as a nonlinear function during the training process to enhance performance further. The clipping function is defined as follows:

$$x_{clip}[n] = x[n]\left(1 - \frac{ReLU\,(|x[n]| - \rho\bar{x}[n])}{|x[n]| + \gamma}\right), \quad (11.18)$$

where $x[n]$ represents the signal obtained after applying the IFFT to the output of the joint encoder, $\rho$ denotes the clipping ratio, $|x[n]|$ signifies the magnitude of the complex symbol, $\bar{x}[n]$ indicates the average amplitude of the signal, and $\gamma$ is a small constant used to prevent the denominator from becoming zero.

## 11.3 Semantic-Aware OFDM Waveform

In recent years, there has been a growing interest in developing semantic-aware end-to-end communication systems through data-driven approaches. Compared with conventional digital communication, deep learning-enabled semantic communications employ DJSCC, extracting semantic-aware and goal-oriented information directly from the source. This approach not only achieves superior performance but also provides robustness against channel variations. To cope with the transmission under multipath fading channels, a DJSCC scheme for wireless image transmission with OFDM is introduced in [4]. However, semantic communications with DJSCC exhibit a large PAPR in the transmitted signal, as illustrated in Fig. 11.5. In this section, the performance of various PAPR reduction techniques for OFDM in DJSCC systems is evaluated.

In the experiment, the ImageNet dataset is utilized to train the encoder and decoder of DJSCC, with the Adam optimizer used at a learning rate of $10^{-4}$ and a batch size of 128. During training, the SNR is sampled from a uniform distribution within the range [0, 20] dB. For simplicity, the neural network is trained under an AWGN channel, as the extension to a fading channel is straightforward. The Kodak dataset, comprising 24 images of size $768 \times 512$, is employed to evaluate the performance of DJSCC. Each image in the Kodak dataset is transmitted 10 times with different noise realizations. Across all experiments, the number of OFDM subcarriers is fixed at 64, and the DJSCC bandwidth-to-signal ratio is set to 1/6.

Here we provide a performance comparison of different PAPR reduction methods given in Sect. 11.2, including clipping, $\mu$-law companding, SLM, PTS, PAPR loss, and clipping with retraining [1].

- **Clipping**. Figure 11.8 illustrates the PAPR and image peak signal-to-noise ratio (PSNR) performance of DJSCC after conventional clipping. A significant improvement in PAPR performance is observed following clipping, with greater improvements as the CR decreases. However, the signal distortion introduced by clipping results in a degradation of PSNR performance, with the magnitude of the degradation increasing as CR decreases.
- **$\mu$-law Companding** The performance of DJSCC using the companding method is presented in Fig. 11.8. A significant enhancement in PAPR performance is observed after companding, but this comes at the cost of reduced PSNR performance. Additionally, the improvement in PAPR becomes more pronounced with increasing $\mu$, while the quality of image reconstruction decreases correspondingly with higher $\mu$.

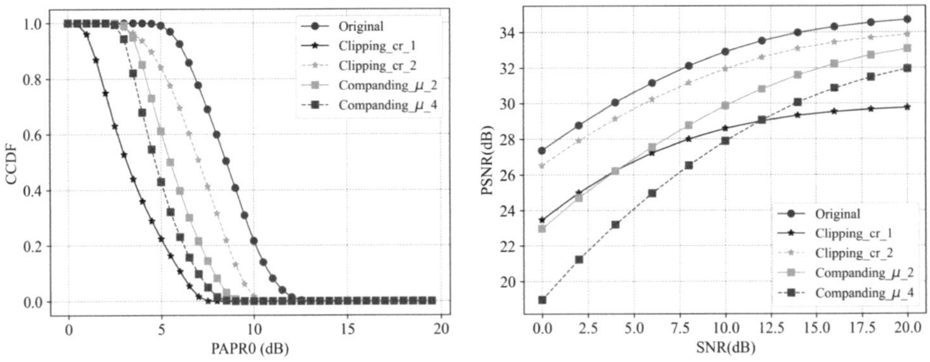


**Fig. 11.8** Effect of clipping and companding on the PAPR (left) and the PSNR (right) of a DJSCC-based OFDM system

- **SLM and PTS** Under the assumption that the phase factor sequences of SLM and PTS are transmitted without errors, no distortion is introduced to the signal. Therefore, only the PAPR reduction performance of these methods is analyzed, as shown in Fig. 11.9. Both methods effectively suppress PAPR in DJSCC, with the level of suppression increasing as $V$ increases.
- **PAPR Loss** Figure 11.10 illustrates the performance of DJSCC after incorporating PAPR into the loss function. A reduction in the PAPR of DJSCC is observed with this method; however, it also leads to a decline in PSNR performance. Furthermore, both the PAPR of DJSCC and the reconstruction quality of the images degrade as the penalty factor $\lambda$ increases.
- **Clipping with Retraining** Figure 11.10 demonstrates the PAPR and PSNR performance achieved by incorporating the clipping process into the DJSCC training. A reduction in the PAPR of DJSCC is observed with this method, although it results in a degradation of PSNR performance. The decline in PSNR is less pronounced compared to the conventional clipping method at the same CR, particularly when the CR is larger.

The PAPR and PSNR performance of all the methods discussed in the previous section are evaluated and presented in Fig. 11.11. Among the signal distortion PAPR reduction techniques, clipping with retraining achieves the best PSNR performance, particularly at low SNRs. In contrast, within conventional SSCC systems using digital modulation, the com-

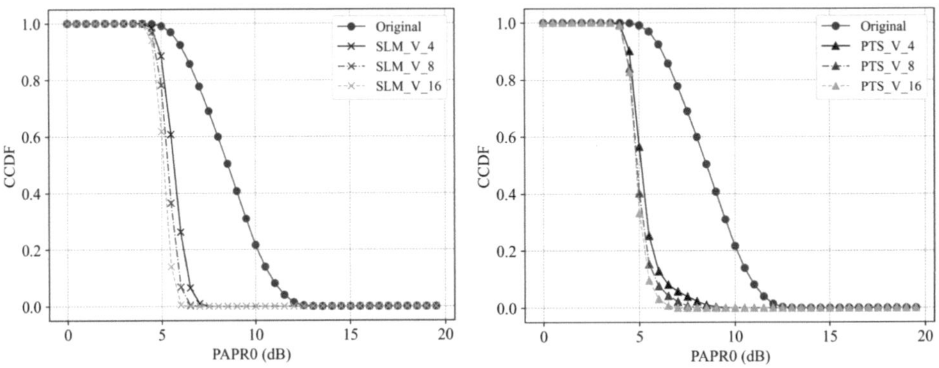


**Fig. 11.9** Effect of SLM (left) and PTS (right) on the PAPR of a DJSCC-based OFDM system

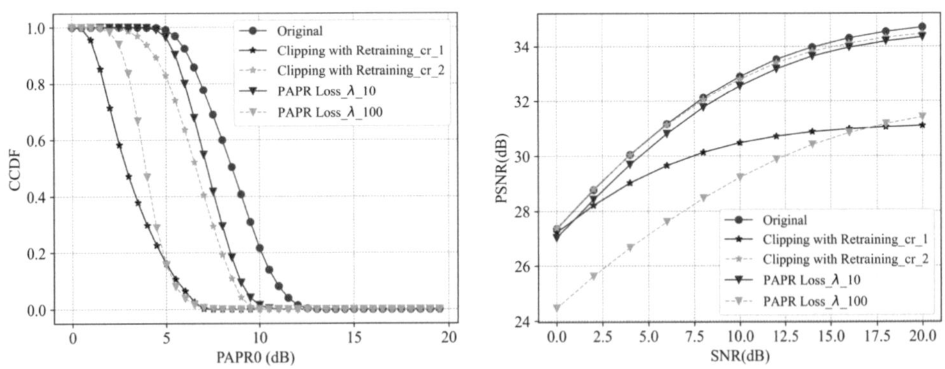


**Fig. 11.10** Effect of the PAPR loss and clipping with retraining on the PAPR (left) and the PSNR (right) of a DJSCC-based OFDM system

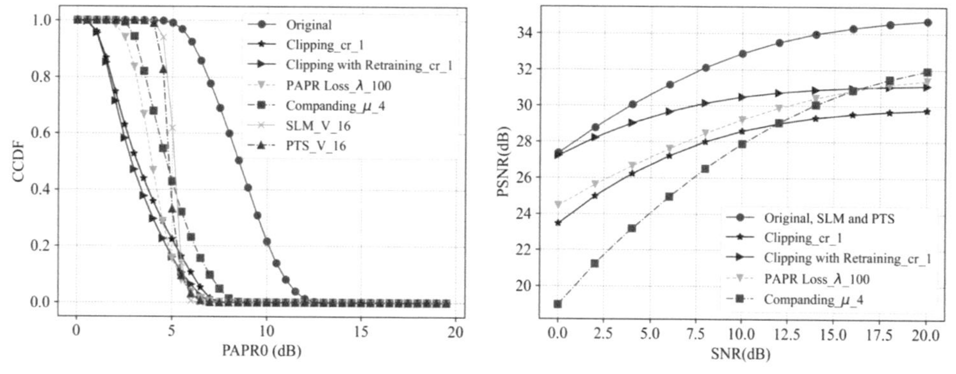


**Fig. 11.11** Performance comparison of all PAPR reduction techniques in the DJSCC-based OFDM system

panding method outperforms the clipping method, especially when higher-order modulations are employed [5].

An additional advantage of clipping with retraining is that it does not require modifications to the network model and imposes minimal computational complexity. Signal non-distortion PAPR reduction techniques, such as SLM and PTS, effectively reduce the PAPR in DJSCC without compromising signal reconstruction quality. This makes them attractive solutions for systems where maintaining image fidelity is critical.

In conclusion, the choice of PAPR reduction technique in DJSCC depends on the specific requirements of the application. Techniques like clipping with retraining offer a balance between computational efficiency and performance enhancement, making them suitable for scenarios with limited computational resources. On the other hand, non-distortion techniques like SLM and PTS are ideal for applications demanding high reconstruction quality.

## 11.4 Summary

This chapter has explored the integration of DJSCC with OFDM systems, highlighting the significant PAPR challenge that arises beyond the traditional discrete-time analog transmission setting. Also, this chapter gives an overview of both traditional and deep learning-based PAPR reduction methods, including clipping, companding, SLM, PTS, and deep learning-driven techniques such as PAPR loss and clipping with retraining. By applying these methods, the excellent image reconstruction performance of DJSCC can be preserved while keeping PAPR within an acceptable range in the OFDM system. This represents a significant step toward making DJSCC feasible for real-world semantic communication systems.

## Problems

11.1 What is the primary advantage of OFDM over conventional MCM?
A. Higher PAPR
B. Lower computational complexity in modulation/demodulation
C. Improved spectral efficiency through subcarrier orthogonality
D. Reduced susceptibility to narrowband interference

11.2 Which mathematical operation is used for OFDM modulation at the transmitter?
A. FFT
B. IFFT
C. DCT
D. Hilbert Transform

11.3 The CP in OFDM primarily mitigates:
A. Multipath interference
B. Phase noise
C. Carrier frequency offset
D. Quantization errors

11.4 Which PAPR reduction technique preserves signal quality but requires side information?
A. Clipping
B. $\mu$-law companding
C. SLM
D. Deep learning-based clipping

11.5 In DJSCC-OFDM systems, why is PAPR higher compared to traditional 16QAM?
A. Continuous amplitude signals from neural network encoders
B. Larger number of subcarriers

C. Higher bandwidth efficiency
D. Absence of cyclic prefix

11.6 Which parameter balances distortion and PAPR reduction in the combined loss function $L = L_1 + \lambda L_2$?
A. Learning rate
B. Weighting factor $\lambda$
C. Clipping ratio $\rho$
D. Companding constant $\mu$

11.7 The $\mu$-law companding function:
A. Compresses large amplitudes and expands small ones
B. Discards signal peaks above a threshold
C. Applies random phase rotations
D. Uses neural networks for PAPR prediction

11.8 What is the key difference between SLM and PTS techniques?
A. SLM uses clipping while PTS uses companding
B. SLM applies phase factors to entire signals while PTS partitions them
C. PTS requires retraining but SLM doesn't
D. SLM works only with discrete constellations

11.9 Explain why OFDM signals inherently have high PAPR, and how increasing subcarriers affects this.

11.10 Calculate the PAPR (in dB) for an OFDM signal with maximum instantaneous power of 25W and average power of 4W, using Eq. 11.10 from the text.

11.11 Compare the trade-offs between clipping and $\mu$-law companding in terms of PAPR reduction and signal distortion.

11.12 Describe how the clipping with retraining method improves upon conventional clipping in DJSCC-OFDM systems.

11.13 Analyze why SLM and PTS are classified as probabilistic PAPR reduction methods.

11.14 Given an OFDM system with $N = 64$ subcarriers and symbol period $T = 100\mu s$, calculate the subcarrier spacing $\Delta f$.

11.15 Propose a scenario where PAPR loss would be preferred over SLM/PTS despite its PSNR degradation, justifying your choice.

## References


1. M. Liu, W. Chen, J. Xu, B. Ai, A comprehensive study of papr reduction techniques for deep joint source channel coding in ofdm systems, in *IEEE International Conference on Communication Technology (ICCT)* (IEEE, 2023), pp. 1395–1399
2. Y. Shao, D. Gunduz, Semantic communications with discrete-time analog transmission: a PAPR perspective. IEEE Wirel. Commun. Lett. **12**(3), 510–514 (2022)
3. M. Kim, W. Lee, D.-H. Cho, A novel PAPR reduction scheme for OFDM system based on deep learning. IEEE Commun. Lett. **22**(3), 510–513 (2017)
4. M. Yang, C. Bian, H.-S. Kim, Deep joint source channel coding for wireless image transmission with OFDM, in *ICC 2021-IEEE International Conference on Communications* (IEEE, 2021), pp. 1–6
5. J. Kim, Y. Shin, An effective clipped companding scheme for PAPR reduction of OFDM signals, in *2008 IEEE International Conference on Communications* (IEEE, 2008), pp. 668–672

# Semantic Communication with Multi-antenna System

# 12

**Abstract**

This chapter provides an overview of recent advances in applying semantic communication techniques to MIMO systems, with a focus on both open-loop and closed-loop MIMO scenarios. In the open-loop scenario, existing studies have investigated the integration of semantic communication with linear detection techniques to maintain robust performance without CSI at the transmitter. In the closed-loop scenario, research efforts have explored the combination of semantic communication with precoding strategies, including both implicit approaches based on attention mechanisms and explicitly designed with importance-aware semantic communication systems, where the transmitted semantics are aligned with MIMO subchannel quality. In addition, adaptive CSI feedback mechanisms can help improve feedback efficiency, typically by adjusting the feedback overhead based on the prediction of reconstruction quality.

**Keywords**

Multi-antennas • Diversity • Multiplexing • Channel state information • Precoding

## 12.1 Introduction to Multi-antenna System

Multi-Input Multi-Output (MIMO) refers to a technology that uses multiple antennas at both the transmitter and receiver to improve communication performance, as shown in Fig. 12.1. In a MIMO system, the key idea is to transmit multiple signals simultaneously over the same frequency band, utilizing the spatial diversity provided by multiple antennas. This enables the system to exploit the spatial dimension of the communication channel to enhance data throughput, improve link reliability, and increase spectral efficiency. MIMO systems are capable of both spatial diversity and spatial multiplexing. Spatial diversity helps combat fading by transmitting the same signal over different paths, while spatial multiplexing increases the data rate by transmitting multiple independent data streams simultaneously [1].

The MIMO channel can be characterized by a complex matrix $\mathbf{H}$, where each element represents the channel gain between a specific transmitter antenna and a specific receiver antenna. Mathematically, the MIMO channel model can be expressed as:

$$\mathbf{y} = \mathbf{H}\mathbf{x} + \mathbf{n}, \tag{12.1}$$

where $\mathbf{y}$ is the received signal vector, $\mathbf{H}$ is the $M \times N$ channel matrix (with $M$ being the number of receiver antennas and $N$ being the number of transmitter antennas), $\mathbf{x}$ is the transmitted signal vector, and $\mathbf{n}$ is the noise vector. The matrix $\mathbf{H}$ models the wireless channel between the transmitter and receiver. Each element of $\mathbf{H}$ represents the path gain from a specific transmitter antenna to a receiver antenna. The channel matrix depends on the physical environment, the relative positions of the transmitter and receiver, and other factors such as multipath fading and interference.

In MIMO systems, the wireless communication channel is typically modeled as a random process due to factors such as multipath fading, interference, and environmental noise. The general assumptions for MIMO channels are as follows:

- **Slow Fading:** The channel is considered to change slowly over time, with the path gains remaining approximately constant over several symbol periods.
- **Flat Fading:** It is assumed that the channel exhibits flat fading, meaning the channel's frequency response is constant over the signal's bandwidth.
- **Channel State Information (CSI):** It represents the information about the wireless channel that carries signals between a transmitter (e.g., base station or access point) and a receiver (e.g., user equipment or device). CSI needs to be estimated at the receiver and usually quantized and feedback to the transmitter. It is often assumed that the receiver has perfect knowledge of the CSI in research,


W. Chen and Z. Qin, *Semantic Communications*, Textbooks in Telecommunication Engineering,
https://doi.org/10.1007/978-3-032-11005-3_12

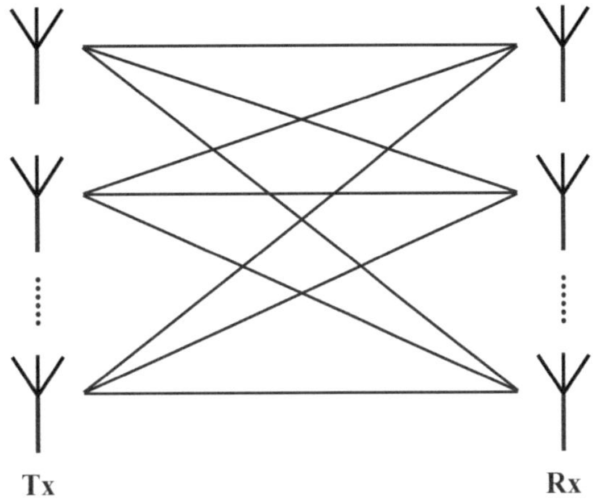


**Fig. 12.1** MIMO channel model

which is crucial for optimizing the communication strategy. In some cases, CSI at the transmitter is also considered, though acquiring it in practice is more challenging.
- **Statistical Model:**The channel coefficients are treated as random variables, often modeled as independent and identically distributed (i.i.d.) random variables with specific statistical properties, such as those following a Rayleigh fading model or a Rician fading model, depending on the channel conditions.

CSI plays a pivotal role in optimizing communication performance in wireless systems, particularly MIMO systems. The availability of CSI allows both the transmitter and receiver to adapt their transmission strategies to current channel conditions, thus maximizing the SNR and minimizing interference under certain transmit power limitations. The effect of CSI on communication performance can be described in two primary scenarios:

- **Open-loop MIMO scenarios**: When only the receiver has knowledge of the channel state, the transmitter must rely on statistical information or an outdated estimate of the channel, which limits its ability to optimize the transmission. However, the receiver can still perform adaptive techniques such as space-time coding to mitigate the impact of channel impairments.
- **Closed-Loop MIMO scenarios**: When both transmitters and receivers have perfect knowledge of the channel conditions, the system can fully exploit the available degrees of freedom in the channel. This includes precoding at the transmitter and combining at the receiver, which together optimize the power and spatial distribution of the transmitted signal, leading to the best possible channel capacity.

Understanding the difference between Open-Loop MIMO scenarios and Closed-Loop MIMO scenarios is crucial when exploring how to enhance communication performance using Deep JSCC in MIMO systems. In the next sections, we will discuss how semantic communication can be tailored to these two different CSI scenarios.

- Semantic communication over Open-Loop MIMO: This approach leverages the limited channel information available at the receiver to optimize communication, though it cannot achieve the same performance gains as when full CSI is available.
- Semantic communication over Closed-Loop MIMO: With complete channel knowledge, both the receiver and the transmitter can coordinate their efforts to optimally design the source and channel codes, significantly improving efficiency and robustness in communication.

## 12.2 Semantic Communication Over Open-Loop MIMO

In this section, we focus on open-loop MIMO systems, where only the receiver has access to CSI, and the transmitter lacks any knowledge of the channel state. Due to the absence of feedback, open-loop systems typically allocate power equally across each transmit antenna. This scenario is common in many practical systems where feedback is either costly, impractical, or unavailable, such as in large-scale wireless networks, satellite communication, or IoT systems.

As shown in Fig. 12.2, we consider a MIMO communication system with $N_t$ transmit antennas and $N_r$ receive antennas. We assume a block Rayleigh fading channel which remains constant during the transmission of a single-source message and changes for the next transmission.

The basic communication model for an open-loop MIMO system is as follows, which can be expressed as:

$$\hat{s} = f_\theta(s). \tag{12.2}$$

Let $s$ represent the source data at the transmitter, such as an image, video, or message that needs to be transmitted. The encoder at the transmitter maps the source data $s$ into a semantic feature $\hat{s}$, where $f_\theta(\cdot)$ is the semantic encoder function, parameterized by $\theta$. The encoder aims to generate a transmission signal $\hat{x}$ that optimally represents the semantic

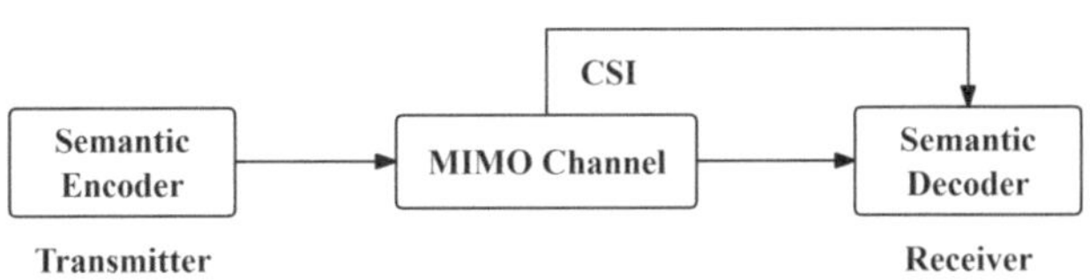


**Fig. 12.2** Semantic communication over open-loop MIMO

content of the source data. Parameters $\theta$ are learned during the system training phase, ensuring that the encoder is tailored to preserve the semantic integrity of the transmitted data.

Then, to adjust the semantic feature for the $N_t \times N_r$ MIMO system, the semantic feature is reshaped and power-normalized to the signal to be transmitted. Specifically, the output of the semantic encoder is transformed as follows:

$$x = \text{Reshape}(\sqrt{PN_tN_k}\frac{\hat{s}}{\|\hat{s}\|_2}), \tag{12.3}$$

where $P$ represents power constraint at each antenna; $N_k$ denotes the number of channel used by the transmitted signal; Reshape$(\cdot)$ is a function that changes the shape of the input. The reshaped output $\boldsymbol{x} \in \mathbb{C}^{N_t \times N_k}$ represents the signal to be transmitted. After passing through the MIMO channel, the process of the signal to be transmitted can be represented as:

$$\boldsymbol{y} = \boldsymbol{H}\boldsymbol{x} + \boldsymbol{n}, \tag{12.4}$$

where $\boldsymbol{y}$ is the received signal at the receiver, $\boldsymbol{H}$ is the channel matrix, and $\boldsymbol{n}$ represents additive noise. At the receiver, CSI is available, which enables the receiver to process the received signal and recover the original transmitted signal.

In MIMO systems, space diversity and spatial multiplexing are two fundamental techniques that impact the overall system performance [2].

- **Diversity**: This technique aims to improve the reliability of communication by exploiting multiple antennas to create independent copies of the signal across different paths. Using space-time coding, the signal is spread across multiple antennas, enhancing the robustness of the system to fading and interference. Space diversity typically improves the performance of system in terms of signal reliability but may limit the transmission rate.
- **Multiplexing**: In contrast, spatial multiplexing increases the communication capacity by transmitting multiple independent data streams simultaneously over the same frequency band, using multiple antennas. This technique allows for higher data rates but may be more susceptible to noise and interference, especially in highly dynamic or interference-prone environments.

In the following sections, we will discuss semantic communication schemes based on space diversity multiplexing techniques respectively.

### 12.2.1 Open-Loop MIMO with Diversity

Space-time coding (STC) is a powerful technique used to achieve diversity in MIMO systems by transmitting multiple copies of the signal across both spatial and temporal dimensions [3]. This allows the receiver to combine signals from different antennas and time slots, improving the chances of recovering the original signal even in the presence of fading or noise. In semantic communication, where the focus is on transmitting semantic information rather than raw data, space-time coding plays a critical role in ensuring that semantic content is reliably delivered to the receiver, even under challenging channel conditions. With STC, semantic communication systems can maintain high levels of transmission reliability, thereby improving the overall performance of the system.

Space-time block codes (STBCs) are one of the main types of STC methods, which operate on blocks of data (similar to block codes) and offer diversity gain, but do not provide coding gain.

This chapter will focus on STBCs as an example to explore semantic communication with diversity techniques, using the well-known Alamouti scheme to illustrate the concepts in detail [4].

From the history of coding development, the Alamouti scheme was proposed initially for two-antenna systems, providing full diversity in the transmission system [5]. Due to its simple matrix structure and potential for scalability to more antenna configurations, it is possible to realize full diversity by using the Alamouti scheme with straightforward encoding.

The principle of the Alamouti STBC encoder is shown in Fig. 12.3. In the Alamouti space-time coding, we map the symbols to be transmitted onto the transmit antennas according to the following coding matrix:

$$[x_1\, x_2] \rightarrow \begin{bmatrix} x_1 & -x_2^* \\ x_2 & x_1^* \end{bmatrix}, \tag{12.5}$$

where $x_1^*$ represents the complex conjugate of $x_1$.

The output of the encoder is two successive transmission periods, where the encoder sends the symbols through two transmit antennas. In the first transmission period, $x_1$ is transmitted from the first antenna and $x_2$ from the second antenna. In the second transmission period, $-x_2^*$ is transmitted from the first antenna and $x_1^*$ from the second antenna, which is a simple form of space-time coding. It is evident that this method performs encoding in both the spatial and temporal domains. The inner product of the two transmission sequences is zero, meaning they are orthogonal. The encoding vector has the following properties:

$$\boldsymbol{x} \cdot \boldsymbol{x}^H = \begin{bmatrix} |x_1|^2 + |x_2|^2 & 0 \\ 0 & |x_1|^2 + |x_2|^2 \end{bmatrix} = (|x_1|^2 + |x_2|^2)\boldsymbol{I}, \tag{12.6}$$

where $\boldsymbol{I}$ is a $2 \times 2$ identity matrix.

At the receiver, the received signals $y_1$ and $y_2$ over two consecutive symbol periods can be expressed as:

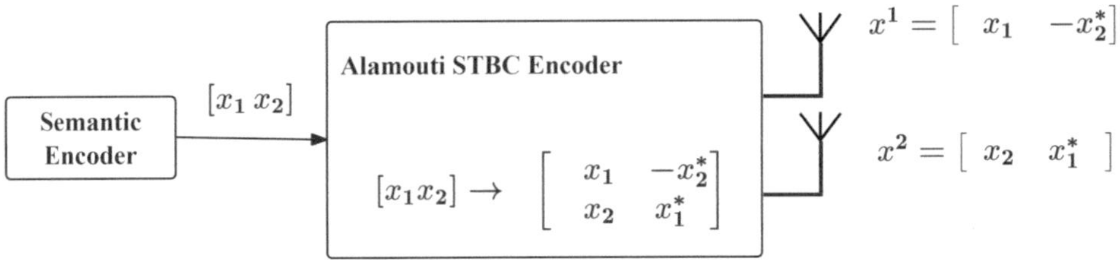


**Fig. 12.3** The principle of Alamouti STBC encoder

$$y_1 = h_1 x_1 + h_2 x_2 + n_1, \tag{12.7}$$
$$y_2 = -h_1 x_2^* + h_2 x_1^* + n_2, \tag{12.8}$$

where $\boldsymbol{h}_1$, $\boldsymbol{h}_2$ are the first and second columns of $\boldsymbol{H}$, $\boldsymbol{n}_1$ and $\boldsymbol{n}_2$ represent additive noise.

Then we introduce intermediate variables $m_1$ and $m_2$ as

$$m_1 = h_1^* y_1 + y_2^* h_2, \tag{12.9}$$
$$m_2 = h_2^* y_1 - y_2^* h_1. \tag{12.10}$$

By substituting the expressions of $\boldsymbol{y}_1$, $\boldsymbol{y}_2$ into (12.10), the simplified forms of $m_1$ and $m_2$ can be obtained as:

$$m_1 = (h_1^* h_1 + h_2^* h_2) x_1 + h_1^* n_1 + n_2^* h_2, \tag{12.11}$$
$$m_2 = (h_1^* h_1 + h_2^* h_2) x_2 + h_2^* n_1 - n_2^* h_1. \tag{12.12}$$

The estimated symbols $z_1$ and $z_2$ are obtained as:

$$z_1 = m_1/(||h_1||_2^2 + ||h_2||_2^2 + \theta^2), \tag{12.13}$$
$$z_2 = m_2/(||h_1||_2^2 + ||h_2||_2^2 + \theta^2). \tag{12.14}$$

Then the recovered semantic features with noise are fed into the semantic decoder to complete the communication task.

In this integrated framework, semantic encoding is first performed to extract meaningful features from the source data. These features are then mapped to the Alamouti coding structure, enabling robust transmission over unreliable wireless channels. At the receiver, semantic decoding is combined with the diversity gains provided by the Alamouti scheme, ensuring reliable recovery of semantic content even in challenging channel environments.

The Alamouti scheme introduces space-time diversity by adding redundancy across spatial and temporal dimensions, effectively mitigating channel fading and improving wireless transmission reliability. In the Alamouti scheme, each symbol output by the semantic encoder can be accurately received and decoded by the receiver, avoiding symbol interference and ensuring that the correct symbols are input to the semantic decoder. This process significantly enhances the reliability and accuracy of semantic communication, especially in complex MIMO channel conditions. By using Alamouti coding, the communication system ensures that each symbol is properly transmitted and decoded, enabling the effective training and deployment of the semantic communication model under multi-antenna MIMO channel conditions.

By incorporating this diversity technique into semantic communication systems, the resilience of transmitted semantic information under low SNR conditions can be significantly enhanced.

### 12.2.2 Open-Loop MIMO with Multiplexing

In modern wireless communication systems, multiplexing is a key technique that significantly enhances the data rate by transmitting multiple independent data streams simultaneously through multiple antennas. By exploiting the spatial dimension, multiplexing allows a system to achieve a linear increase in throughput relative to the number of antennas, without requiring additional bandwidth or power.

However, implementing spatial multiplexing poses challenges, particularly at the receiver end, where an accurate signal detection becomes critical. Due to the interference between spatially multiplexed streams, advanced signal processing techniques are required to separate the streams and recover the transmitted data effectively.

The process of spatial multiplexing involves encoding independent data streams and transmitting them through multiple antennas. At the receiver, the received signals, affected by the MIMO channel and noise, must be processed to recover the original streams. The signal detection process is a vital step in this workflow, aiming to mitigate interference between streams and reduce noise effects.

Two common detection algorithms often used in spatial multiplexing are the ZF detector and the MMSE detector, as shown in Fig. 12.4. These methods differ in their trade-offs between complexity, noise suppression, and interference management.

The ZF detector aims to completely eliminate inter-stream interference by inverting the MIMO channel matrix. The output of the ZF detector can be expressed as:

$$\hat{\boldsymbol{x}}_{\mathrm{ZF}} = (\boldsymbol{H}^H \boldsymbol{H})^{-1} \boldsymbol{H}^H \boldsymbol{y}, \tag{12.15}$$

where $\boldsymbol{H}$ is the channel matrix, $\boldsymbol{y}$ is the received signal vector, $(\boldsymbol{H}^H \boldsymbol{H})^{-1} \boldsymbol{H}^H$ is ZF detector matrix, and $\hat{\boldsymbol{x}}_{\mathrm{ZF}}$ is the estimated transmitted signal vector.

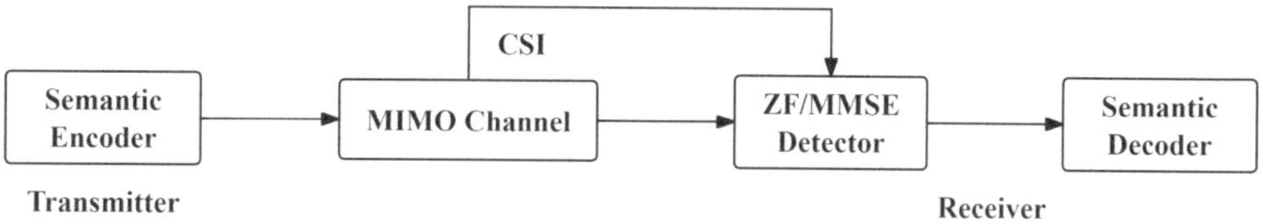


**Fig. 12.4** Semantic communication with detector at receiver

Although ZF effectively removes interference, it amplifies noise, especially when the channel matrix is ill-conditioned. The MMSE detector balances interference suppression and noise amplification by incorporating noise variance into the detection process. The MMSE detection formula is:

$$\hat{x}_{\text{MMSE}} = (\boldsymbol{H}^H \boldsymbol{H} + \sigma^2 \boldsymbol{I})^{-1} \boldsymbol{H}^H \boldsymbol{y}, \tag{12.16}$$

where $\sigma^2$ is the noise variance, $\boldsymbol{I}$ is the identity matrix, and $(\boldsymbol{H}^H \boldsymbol{H} + \sigma^2 \boldsymbol{I})^{-1} \boldsymbol{H}^H$ is MMSE detector matrix. Compared to ZF, MMSE achieves better performance in noisy environments by reducing the impact of noise, albeit at the cost of increased computational complexity.

Spatial multiplexing introduces higher data transmission rates by simultaneously transmitting multiple data streams over multiple antennas. However, this gain in throughput comes with the challenge of increased signal detection complexity at the receiver. Interference between spatially multiplexed streams requires advanced signal processing techniques to ensure reliable recovery of the transmitted data.

When integrated with semantic communication systems, spatial multiplexing enables enhanced performance in high SNR conditions. The semantic encoder extracts meaningful features from the source data, and these features are transmitted using spatially multiplexed streams. At the receiver, robust signal detection algorithms, such as ZF or MMSE detection, are applied, followed by semantic decoding. This combination ensures efficient utilization of spatial resources while achieving superior semantic transmission performance in high SNR environments.

In this section, we discuss the implementation of semantic communication in an open-loop mode, focusing on systems that leverage diversity and multiplexing techniques. In the diversity scenario, the introduction of spatial and temporal redundancy enhances transmission reliability, making it particularly suitable for semantic communication systems operating under low SNR conditions. On the other hand, spatial multiplexing achieves higher transmission rates and efficiency, offering superior performance at high SNR levels.

In the next section, we will explore the implementation of semantic communication systems in a closed-loop mode, introducing the crucial concept of precoding to further optimize communication performance.

## 12.3 Semantic Communication Over Closed-Loop MIMO

Closed-loop semantic communication refers to a communication framework where the transmitter has access to CSI, enabling it to adapt the transmission process based on the channel conditions, as shown in Fig. 12.5. In this setup, the receiver typically estimates the CSI and feeds it back to the transmitter. This feedback loop allows the transmitter to optimize both the encoding of semantic information and the wireless transmission strategy, making the communication system more efficient and resilient to varying channel conditions.

The key advantage of closed-loop systems over open-loop systems lies in the availability of CSI at the transmitter, which enables the use of precoding approaches. Precoding is an extension of beamforming designed to support multi-stream transmission in multi-antenna wireless communication systems. In traditional single-stream beamforming, the same signal is transmitted from all antennas, weighted appropriately in terms of phase and gain to maximize the received signal power at the output. In contrast, as shown in Fig. 12.6, precoding enables the transmission of multiple independent data streams from the antennas, each with its own set of optimized weightings. This approach aims to maximize the overall link throughput at the receiver, leveraging the spatial multiplexing capabilities of multi-antenna systems while effectively managing interference between the streams. The Singular Value Decomposition (SVD) precoding scheme is a widely used precoding technique that can minimize channel interference, and we can reach the channel capacity of the MIMO system.

***Proof*** The channel capacity $C_1$ of a MIMO system is given by the well-known formula:

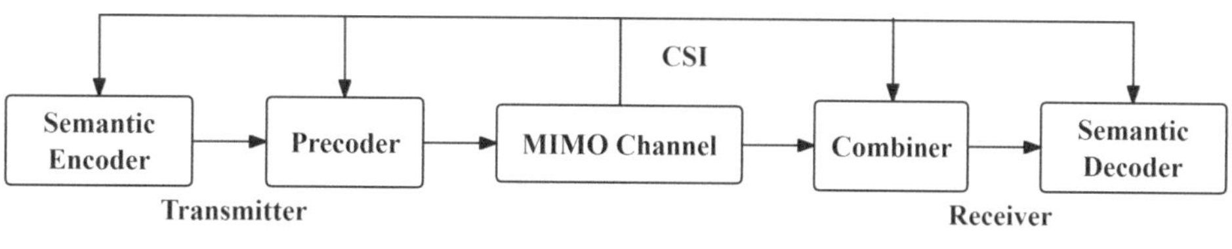


**Fig. 12.5** Semantic communication over closed-loop MIMO

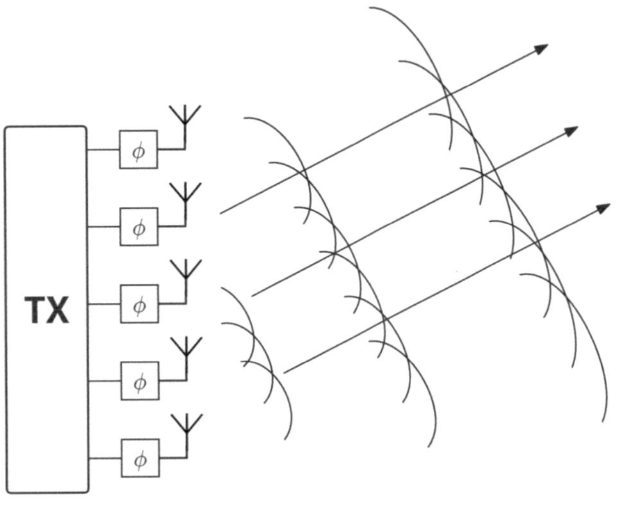


**Fig. 12.6** Precoding in multi-antenna system

$$C_1 = \log_2 \det(\mathbf{I}_{N_r} + \frac{\rho}{N_t}\mathbf{H}\mathbf{H}^H), \tag{12.17}$$

where $\mathbf{H}$ is the $N_r \times N_t$ channel matrix, $N_r$ and $N_t$ are the number of receive and transmit antennas, respectively, $\rho$ is the total transmit power, $\mathbf{I}_{N_r}$ is the identity matrix. Substituting $\mathbf{H}\mathbf{H}^H = \mathbf{U}\Sigma\Sigma^H\mathbf{U}^H$, and noting that the determinant is invariant under unitary transformations, we have:

$$C_1 = \log_2 \det\left(\mathbf{U}\left(\mathbf{I}_{N_r} + \frac{\rho}{N_t}\Sigma\Sigma^H\right)\mathbf{U}^H\right). \tag{12.18}$$

Simplifying further using the property of unitary matrices ($\det(\mathbf{U}) = 1$):

$$C_1 = \log_2 \det\left(\mathbf{I}_{N_r} + \frac{\rho}{N_t}\Sigma\Sigma^H\right). \tag{12.19}$$

Since $\Sigma\Sigma^H$ is a diagonal matrix with eigenvalues $\sigma_i^2$, we expand the determinant:

$$C_1 = \sum_{i=1}^{\min(N_r,N_t)} \log_2\left(1 + \frac{\rho\sigma_i^2}{N_t}\right). \tag{12.20}$$

When CSI is available at both the transmitter and receiver, SVD can be applied to the channel matrix $\mathbf{H}$:

$$\mathbf{H} = \mathbf{U}\Sigma\mathbf{V}^H, \tag{12.21}$$

where $\mathbf{U}$ and $\mathbf{V}$ are unitary matrices, $\Sigma$ is a diagonal matrix containing the singular values $\sigma_i (i = 1, \ldots, \min(N_r, N_t))$. By applying precoding at the transmitter using $\mathbf{V}$ and decoding at the receiver using $\mathbf{U}$, the MIMO system can be transformed into $\min(N_r, N_t)$ parallel and independent subchannels, each with a gain of $\sigma_i^2$. The capacity then becomes:

$$C_2 = \sum_{i=1}^{\min(N_r,N_t)} \log_2\left(1 + \frac{\rho\sigma_i^2}{N_t}\right) = C_1. \tag{12.22}$$

This matches the capacity after SVD precoding.

The integration of CSI into MIMO systems not only enhances channel capacity but also results in subchannels of varying quality. If this property can be fully exploited, better communication performance can be achieved by incorporating CSI into the encoder and decoder design of closed-loop semantic communication systems. By allocating transmission power according to the quality of each subchannel, more critical information can be transmitted over higher-quality subchannels, optimizing the system's overall performance. Intuitively, the communication system should try to transmit the most important information through the best subchannels.

However, the introduction of CSI into closed-loop semantic communication systems requires ingenious design to overcome the challenges pronounced in MIMO scenarios, where the complexity of CSI utilization increases. In the following, we outline two major challenges.

- **High Dimensionality of CSI**: In MIMO systems, CSI is represented as a high-dimensional channel matrix, with dimensions growing proportional to the number of transmit and receive antennas. As the system scales, such as in massive MIMO configurations with hundreds of antennas, the dimensionality of CSI increases dramatically. This poses substantial challenges in terms of computational complexity, memory requirements, and the design of neural networks capable of effectively utilizing high-dimensional CSI. Efficiently integrating this high-dimensional CSI into semantic encoding, decoding, and precoding modules requires novel approaches to reduce complexity while preserving the meaningful structure of the channel information.
- **Time-Varying of MIMO Channels**: MIMO channels are highly dynamic, with channel states constantly changing due to mobility, environmental variations, and other factors. This time-varying nature of CSI introduces challenges in maintaining accurate and up-to-date channel knowledge at the transmitter. Feedback delays and inaccuracies can lead to outdated CSI, which reduces the effectiveness of semantic-aware encoding and adaptive precoding strategies. To address this issue, closed-loop semantic communication systems must be designed to be resilient to CSI estimation errors and capable of adapting to rapid channel fluctuations in real-time, all while maintaining efficient operation.

These challenges highlight the complexities of integrating CSI into closed-loop semantic communication systems.

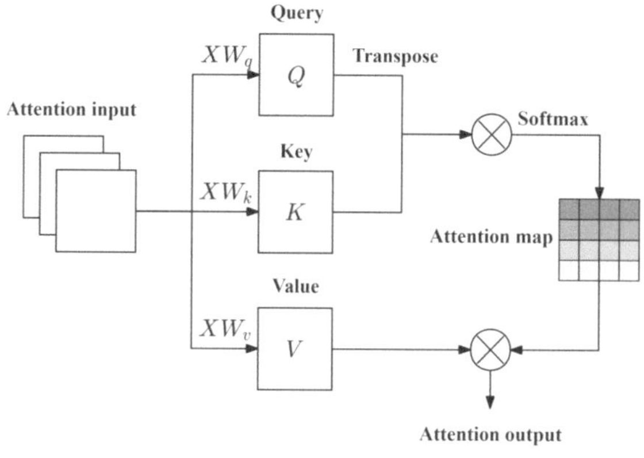


**Fig. 12.7** The illustration of attention mechanism

Overcoming these obstacles requires innovative designs that leverage the high-dimensional structure of CSI and develop robust mechanisms to adapt to time-varying channels. Successfully addressing these challenges will enable the effective use of CSI for optimized semantic communication, enhancing the system's adaptability and efficiency.

To tackle the aforementioned challenges, specific techniques can be adopted to efficiently integrate CSI into closed-loop semantic communication systems, enhancing their performance under complex MIMO scenarios.

The high dimensionality of CSI in MIMO systems can be effectively addressed using precoding methods such as SVD. By precoding, the original channel matrix $\mathbf{H}$ is decomposed into independent subchannels of varying qualities, as shown in Eq. 12.21, and the MIMO channel is effectively transformed into a set of parallel subchannels with different qualities. This decomposition allows the system to encode the CSI into a compact representation by focusing solely on the channel gains $\sigma_i^2$ of these subchannels. Consequently, these subchannel gains can be easily integrated into the design of closed-loop semantic communication systems to fully characterize the transmission process over the MIMO channel. By doing so, the complexity introduced by the high dimensionality of the CSI is significantly reduced, enabling the system to efficiently utilize channel information without the need to process the full CSI matrix [6].

To address the challenges posed by the varying channel conditions and take advantage of MIMO channel gain, researchers have proposed several methods to enhance the adaptability of semantic communication systems under MIMO scenarios. These methods aim to effectively utilize CSI and semantic features to optimize communication performance. Next, we explore two key approaches: adaptive transmission in semantic communication and adaptive CSI feedback leveraging semantic information.

### 12.3.1 Closed-Loop MIMO with Adaptive Transmission

The technical roadmap of adaptive transmission encompasses two main approaches: the implicit attention-mechanism-based semantic communication system and the explicit importance-awareness-based semantic communication system. In the following sections, we will introduce each of these approaches in detail.

**Closed-Loop Systems with Attention Mechanism:** Incorporating attention mechanisms into semantic communication systems allows dynamic adaptability to changing channel conditions by focusing on the most relevant features of input data. This enables efficient resource allocation based on the feature output and the channel conditions implicitly.

Attention is a machine learning method that determines the relative importance of each component in a sequence relative to the other components in that sequence, and this mechanism is widely used in deep learning to focus on specific parts of the input data [7]. An attention mechanism typically works as shown in Fig. 12.7, by assigning a weight to input feature, focusing on the most relevant parts of the input.

The attention input consists of feature representations. These are linearly transformed into three components using learned weight matrices $W_q$, $W_k$, and $W_v$. These transformations allow the network to generate representations in different subspaces for querying, referencing, and weighting. Then the dot product of the Query $(Q)$ and the Key $(K)$ is computed. This operation measures the similarity between the $Q$ and $K$ vectors. The result is a score matrix representing how much each query vector relates to each key vector. To improve numerical stability and interpretability, the scores are normalized using the Softmax function. This step converts the scores into a probability distribution, referred to as the Attention map. The Attention map is used as weights to scale the Value $(V)$ vectors. This scaling process emphasizes the most relevant parts of the $V$ based on the $Q-K$ similarity. The resulting weighted values are summed to produce the *Attention output* $\mathbf{A}_{\text{out}}$, which is a contextually enriched representation of the input.

The attention process can be mathematically expressed as follows:

$$\mathbf{A}_{\text{out}} = \text{Softmax}\left[(\mathbf{XW_q}) * (\mathbf{XW_k})^T\right] * \mathbf{XW_v}, \qquad (12.23)$$

where $x$ is the vector of the Attention input and $X$ is vectors stacked together as a matrix. The Softmax function normalizes the attention weights across all inputs.

In summary, the attention mechanism calculates a weighted representation of input features, focusing on the most relevant parts of the input based on learned $Q-K-V$ interactions. In semantic communication systems, this concept is extended to selectively focus on critical channel properties (e.g., CSI) and semantic features.

In the context of semantic communication, attention mechanisms allow the network to assign different levels of importance semantic features to various CSI components implicitly. In the work of Wu et al., the proposed DeepJSCC-MIMO model utilizes a vision Transformer-based architecture to exploit contextual semantic features and channel conditions through a self-attention mechanism [8].

In this work, the SVD method is used to decompose the MIMO system into several subchannels. Based on the channel gain and noise power of each subchannel, we derive the equivalent noise power for each subchannel. According to the predefined size of the features transmitted through the channel, we reshape the equivalent noise power of each subchannel into a channel heatmap. This heatmap is then concatenated with the input source, representing the equivalent noise term faced by each real encoder output element.

After concatenating the channel heatmap with the input source, the encoder network, built upon a vision Transformer-based architecture, processes the combined input through its self-attention layers. As shown in Fig. 12.7, these self-attention layers take the input features derived from the input source and the channel heatmap as attention inputs. This allows the model to adaptively focus on features of varying importance and account for the dynamic channel conditions, enabling a more effective alignment between the output features and the subchannels.

Finally, the processed features are transmitted over the MIMO channel, and the receiver reconstructs the transmitted information using a decoder also built upon a vision Transformer-based architecture. The entire system is trained end-to-end to ensure optimal performance across diverse channel conditions and user requirements.

Xie et al. consider the feedback CSI as a form of side information and generate an attention masking map by combining the computed semantic importance of features with the CSI feedback [9].

Since semantic elements with lower importance and those transmitted over poor subchannels inevitably impose a negative effect on the core semantic keypoints, transferring unequal protection into the semantic extraction process becomes essential. By emphasizing critical semantic elements while suppressing unimportant ones, this approach improves the robustness of semantic communication. To achieve this, they propose using protection mechanisms through random masking of semantic elements, which sharpens the semantic feature distribution. This sharpening effect ensures that the semantic distribution is less likely to be influenced by external channel noise or interference compared to the original distribution.

First, the semantic features pass through a gated residual network, which computes the semantic importance of each input feature patch. This step reduces the input dimension and provides guidance for the subsequent CSI fusion process. The refined semantic importance is then combined with the estimated CSI to generate the masking strategy $M_s$. According to a predefined ratio, elements in $M_s$ are set to either 0 or 1 based on their values, resulting in a mask matrix.

The masking strategy $M_s$, along with the *Attention Map* $\mathbf{A}_{\text{map}}$ in Fig. 12.7, is used to construct the *Masked Attention Map* $\mathbf{A}_{\text{masked}}$, which is expressed as:

$$\mathbf{A}_{\text{masked}} = M_s \times \mathbf{A}_{\text{map}}. \tag{12.24}$$

Here, $M_s$ ensures that less important semantic elements under poor channel conditions are effectively suppressed, while critical semantic features remain unaffected.

The CSI fusion masking model aims to produce an optimized attention weight distribution that is aware of varying channel states. By applying the masked attention map, intrinsic semantics are preserved in the transmitted signals, and unimportant semantic elements, particularly those subjected to strong channel interference, are masked. This approach avoids introducing negative effects on more important semantic features, ensuring the robustness of the transmitted information. This ultimately enhances the overall quality and efficiency of the communication system.

**Closed-Loop Systems with Importance Aware Mechanism:**
Another prominent method for introducing CSI in to semantic communication system design is using importance awareness to match important semantic information with better channel quality explicitly. In this approach, the focus is on leveraging the differences in subchannel quality that arise from precoding operations. After MIMO channels are precoded, subchannels of varying quality are created across the spatial and temporal domains. These differences in subchannel quality provide an opportunity for optimization by aligning the transmission of semantic features with the best available subchannels.

The importance awareness method involves quantifying the significance of the semantic features output by the encoder. Features are ranked based on their importance to the semantic task, and the most critical features are mapped to the highest quality subchannels as shown in Fig. 12.8. We take the feature allocation methods proposed by Zhou et al. as an example [10]:

Feature importance is computed by analyzing the contribution of each feature to the system's loss during the semantic communication process. They calculate feature importance by analyzing the gradients of the system loss $L(s, \hat{s})$ typically based on a task-specific metric such as MSE or a perceptual loss, to quantify the quality of reconstruction. The

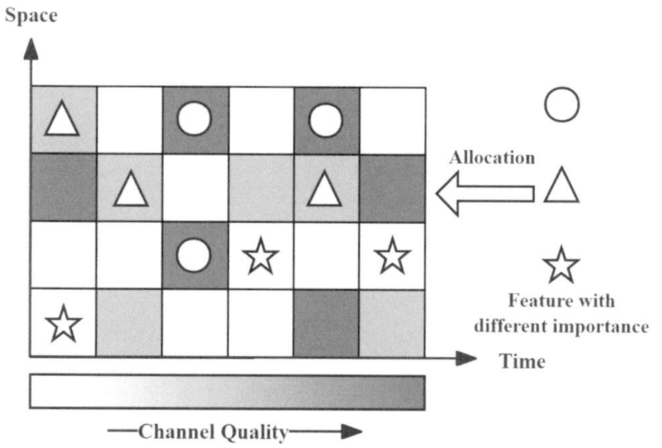


**Fig. 12.8** The illustration of feature allocation

loss measures the system's performance, while the gradients offer valuable insights into how the loss changes in response to variations in a specific feature. This relationship highlights the degree to which each feature impacts the overall system performance.

For each feature channel $k$ (from 1 to $c$, where $c$ is the total number of channels), we can calculate the gradient matrix $\Omega_k$, which reflects the sensitivity of the loss function $L$ to the feature $A_k$, as follows:

$$\Omega_k = \frac{\partial L(s, \hat{s})}{\partial A_k}, \tag{12.25}$$

where $A_k$ represents the $k$-th feature map. Then apply average pooling to $\Omega_k$ to obtain a scalar value $\omega_k$, which represents the importance of the $k$-th feature channel. Once the importance values $\omega_k$ for all channels are obtained, apply min-max normalization to map the values to a predefined range (e.g., $[0, 1]$).

In practice, the application of importance-aware methods faces challenges during the deployment phase. This is primarily because both $A$ and $L(s, \hat{s})$ are fundamental inputs to this approach. However, in communication systems, the system loss can usually only be observed after transmission. To address this issue, it is necessary to design a method that can calculate feature importance solely based on the features themselves. Furthermore, they employed an importance evaluator utilizing knowledge distillation techniques.

After computing the feature importance vector $\omega$ for multiple input samples, the system generates a new dataset that pairs the feature tensor $A$ with its corresponding importance vector $\omega$. This dataset is then used to train a neural network, referred to as the importance evaluator, through supervised learning. The importance evaluator is designed to predict the importance vector $\omega$ directly from the feature tensor $A$, effectively learning the mapping:

$$f_{\text{importance}} : A \to \omega. \tag{12.26}$$

The importance evaluator replaces the computationally expensive gradient-based approach during deployment, enabling real-time feature importance estimation.

During the deployment phase, the importance evaluator replaces computing importance based loss, providing a fast and efficient way to estimate feature importance. Based on $\omega$ calculated by the importance evaluator, the system allocates higher-quality subchannels (e.g., subchannels with higher signal-to-noise ratios) to the most critical features while mapping less important features to lower-quality subchannels. This process ensures that critical semantic information is transmitted with higher fidelity.

The importance-awareness-based semantic communication system bridges the gap between semantic feature extraction and channel-aware transmission. By quantifying feature importance and leveraging an importance evaluator for real-time predictions, the system achieves efficient and adaptive transmission. This approach is particularly advantageous in dynamic MIMO scenarios, where subchannel quality varies significantly.

### 12.3.2 Closed-Loop MIMO with Adaptive CSI Feedback

From the above, we can see that channel feedback for semantic communication with MIMO is crucial. In particular, the benefits of MIMO, such as high spectral efficiency, heavily rely on the base station (BS) acquiring accurate CSI. This requires the user equipment to send the CSI back to the BS through feedback links, as illustrated in Fig. 12.9. However, the large number of antennas at the BS in massive MIMO systems results in a high-dimensional CSI matrix, significantly increasing the feedback overhead.

Several methods have been proposed to reduce the overhead of channel feedback, such as vector quantization and codebook-based approaches. Additionally, many deep learning-based techniques leverage deep neural networks (DNNs) to compress the CSI matrix, and have achieved remarkable performance, such as CsiNet [11] and DJSCC methods [12–14]. However, despite their satisfactory performance, these methods have not accounted for the semantics of the sources in their designs, particularly when feedback CSI is applied in semantic communication.

However, it is capable for semantic communication framework to predict communication quality and adapting the length of the CSI feedback accordingly [15]. As shown in Fig. 12.10, in this framework, the receiver first transmits a very short codeword to the transmitter, which is then used to recover a coarse estimation of the CSI matrix, denoted as $\bar{H}$. This coarse estimation is utilized to predict the reconstruc-

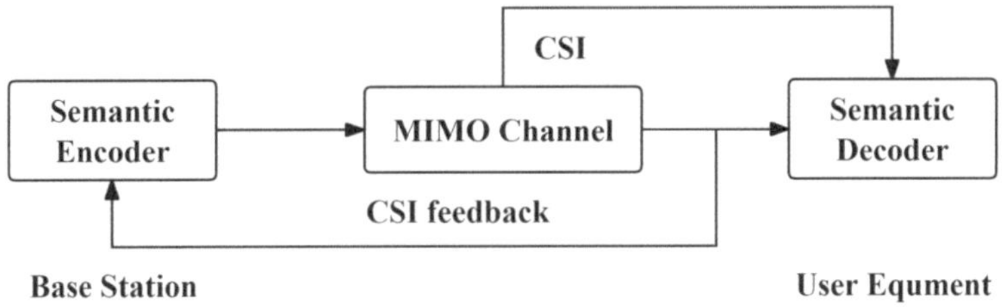


**Fig. 12.9** Semantic communication with feedback links

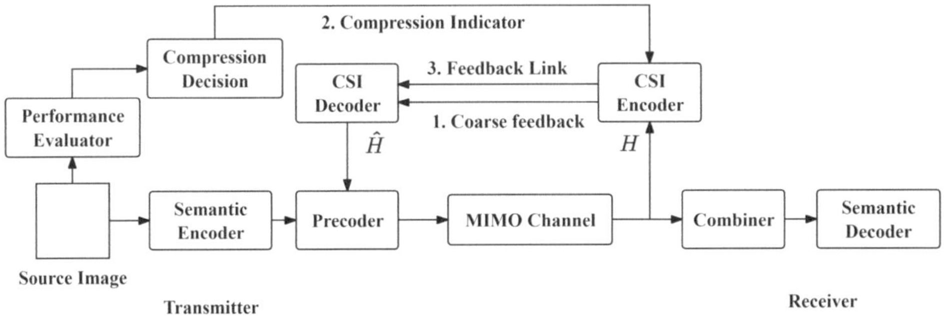


**Fig. 12.10** Semantic communication with adaptive CSI feedback

tion quality of the transmitted data. Despite the short codeword potentially introducing a slight increase in latency, its low feedback overhead ensures that the overall delay remains acceptable in practical applications.

Then, the transmitter evaluates the appropriate compression level for the CSI matrix based on the predicted reconstruction quality of the given image. It then sends a compression indicator to the receiver, specifying the required compression level. The receiver compresses the CSI matrix according to this indicator and feeds the compressed CSI codeword back to the transmitter via the feedback link for precoding.

To enable the transmitter to determine the compression level in advance, Zhang introduced a performance evaluator at the transmitter. This evaluator predicts the reconstruction quality, allowing the system to optimize the CSI compression process effectively.

The performance evaluator takes the transmitted images as input and outputs the predicted PSNR value. With the PSNR values achieved by semantic communication are set as the labels, the prediction task can be taken as a regression problem. Finally, the compression level of the CSI encoder at the receiver is determined by the compression indicator transmitted by the transmitter. After predicting the performance of the images, the transmitter determines the appropriate compression level and sends a compression indicator $r$ to the receiver. The receiver then compresses the estimated channel matrix $H \in \mathbb{C}^{N_t \times N_r}$ according to the indicated compression level and feeds the compressed CSI codeword back to the transmitter.

Using the received CSI codeword, the transmitter reconstructs the CSI matrix $\hat{H}$. It then performs precoding with $\hat{H}$ and transmits the precoded data symbols to the receiver. Finally, the receiver reconstructs the image based on the received symbol vector.

## 12.4 Summary

This chapter reviews recent advancements in semantic communication for MIMO systems, covering both open-loop and closed-loop scenarios. In open-loop MIMO, studies focus on integrating semantic communication with linear detection techniques to ensure robust performance without transmitter-side CSI. For closed-loop MIMO, research explores semantic-aware precoding, including implicit attention-based methods and explicit importance-aware designs that align transmission semantics with MIMO subchannel quality. Additionally, adaptive CSI feedback mechanisms are discussed, which optimize feedback efficiency by dynamically adjusting overhead based on predicted reconstruction quality. These approaches enhance spectral efficiency and robustness in semantic-aware MIMO communications.

## Problems

12.1 Which detector balances interference while suppressing noise?

A. Zero-forcing (ZF) detector.
B. Minimum mean square error (MMSE) detector.
C. Maximum likelihood (ML) detector.
D. Matched filter.

12.2 What is the main advantage of the Alamouti scheme?

A. Increase data transmission rate.
B. Provide full diversity gain.
C. Eliminate all noise interference.
D. Only suitable for single-antenna systems.

12.3 What is the main objective of spatial diversity in MIMO systems?

A. Increase data transmission rate.
B. Enhance signal reliability.
C. Reduce the number of antennas.
D. Simplify encoding complexity.

12.4 What is the main objective of spatial multiplexing in MIMO systems?

A. Increase data transmission rate.
B. Enhance signal reliability.
C. Reduce signal interference.
D. Minimize decoding delay.

12.5 Which characteristic is satisfied by the encoding matrix of the Alamouti scheme for two symbols?

A. Determinant of the matrix is zero.
B. Row vectors are orthogonal.
C. All matrix elements are real numbers.
D. Only suitable for single receive antenna.

12.6 In closed-loop MIMO systems, what is the primary role of SVD precoding?

A. Decompose the channel into independent subchannels.
B. Increase transmit power.
C. Eliminate all channel noise.
D. Reduce the number of antennas.

12.7 What is a drawback of spatial diversity in MIMO systems?

A. Increases data transmission rate.
B. Reduces reliability of signal transmission.
C. Sacrifices data rate to improve reliability.
D. Only works in high SNR scenarios.

12.8 What is a drawback of spatial multiplexing?

A. Reduces data rate.
B. Increases receiver detection complexity.
C. Reduces diversity gain.
D. Only suitable for low SNR.

12.9 Which method suppresses unimportant semantic features using a masking strategy?

A. SVD precoding.
B. Attention masking mechanism.
C. Alamouti coding.
D. ZF detection.

12.10 What is the typical power allocation strategy in open-loop MIMO systems?

A. Dynamic adjustment
B. Uniform allocation
C. Allocation only to the main antenna
D. Random allocation

12.11 Explain the core difference between open-loop and closed-loop MIMO.

12.12 What are the main differences in objectives between spatial diversity and spatial multiplexing?

12.13 How does SVD precoding improve MIMO capacity?

12.14 Why is an efficient CSI feedback mechanism necessary for closed-loop MIMO?

12.15 How does attention mechanism contribute to semantic communication?

## References


1. D. Tse, P. Viswanath, *Fundamentals of Wireless Communication*. (Cambridge University Press, 2005)
2. L. Zheng, D.N.C. Tse, Diversity and multiplexing: a fundamental tradeoff in multiple-antenna channels. IEEE Trans. Inf. Theory **49**(5), 1073–1096 (2003)
3. V. Tarokh, H. Jafarkhani, A.R. Calderbank, Space-time block codes from orthogonal designs. IEEE Trans. Inf. Theory **45**(5), 1456–1467 (1999)
4. C. Bian, Y. Shao, H. Wu, D. Gündüz, Space-time design for deep joint source channel coding of images over MIMO channels, in *2023 IEEE 24th International Workshop on Signal Processing Advances in Wireless Communications (SPAWC)* (IEEE, 2023), pp. 616–620
5. S.M. Alamouti, A simple transmit diversity technique for wireless communications. IEEE J. Sel. Areas Commun. **16**(8), 1451–1458 (1998)
6. W. Jiang, W. Chen, B. Ai, Deep joint source channel coding with attention modules over MIMO channels, in *2024 IEEE 99th Vehicular Technology Conference (VTC2024-Spring)* (IEEE, 2024), pp. 1–5
7. A. Vaswani, N. Shazeer, N. Parmar, J. Uszkoreit, L. Jones, A.N. Gomez, Ł. Kaiser, I. Polosukhin, Attention is all you need, *Advances in Neural Information Processing Systems*, vol. 30 (2017)
8. H. Wu, Y. Shao, C. Bian, K. Mikolajczyk, D. Gündüz, Deep joint source-channel coding for adaptive image transmission over MIMO channels. IEEE Trans. Wirel. Commun. (2024)
9. B. Xie, Y. Wu, Y. Shi, W. Zhang, S. Cui, M. Debbah, Robust image semantic coding with learnable CSI fusion masking over MIMO fading channels. IEEE Trans. Wirel. Commun. (2024)
10. K. Zhou, G. Zhang, Y. Cai, Q. Hu, G. Yu, A.L. Swindlehurst, Feature allocation for semantic communication with space-time importance awareness, arXiv preprint arXiv:2401.14614 (2024)
11. C.-K. Wen, W.-T. Shih, S. Jin, Deep learning for massive MIMO CSI feedback. IEEE Wirel. Commun. Lett. **7**(5), 748–751 (2018)
12. J. Xu, B. Ai, N. Wang, W. Chen, Deep joint source-channel coding for CSI feedback: an end-to-end approach. IEEE J. Sel. Areas Commun. **41**(1), 260–273 (2022)
13. Y. Guo, W. Chen, J. Xu, L. Li, B. Ai, Deep joint CSI feedback and multiuser precoding for MIMO OFDM systems. IEEE Trans. Veh. Technol. (2024)
14. Y. Guo, W. Chen, F. Sun, J. Cheng, M. Matthaiou, B. Ai, Deep learning for CSI feedback: one-sided model and joint multi-module learning perspectives. IEEE Commun. Mag. (2024)
15. G. Zhang, Q. Hu, Y. Cai, G. Yu, Scan: semantic communication with adaptive channel feedback. IEEE Trans. Cogn. Commun. Netw. (2024)

# Part IV

# Applications and Open Problems of Semantic Communication

# Real-World Applications 13

**Abstract**

This chapter presents the practical applications of semantic communication in diverse real-world scenarios, highlighting its transformative role in next-generation wireless systems. Four representative cases are discussed: immersive communications, satellite communications, autonomous driving, and smart grid systems. For each application, the fundamental framework and enabling technologies of semantic communication are analyzed, demonstrating how meaning-centered transmission can improve efficiency, reliability, and context-awareness. The chapter further explores how semantic communication reduces data redundancy, prioritizes task-relevant information, and supports intelligent decision-making in dynamic environments. Key challenges such as standardization, scalability, real-time processing, and integration with emerging technologies are also outlined, offering insights into future research directions.

**Keywords**

Semantic communication • Immersive communication • Satellite communication • Autonomous driving • Smart grid • Task-oriented transmission • Contextaware systems • Intelligent networks

## 13.1 Immersive Communications

(1) Introduction of Immersive Communication

Communication has seen enormous advances over the past 100 years including radio, television, mobile phones, video conferencing, and Internet-based voice and video calling. It strives to pull us together faster than transportation technology and globalization push us apart. This has permitted us to serve our customers far from where we work, work far from where we live, and live far from our friends and families [1].

Still, remote communication remains less natural and more fatiguing than face-to-face. We continue to meet in conference rooms; we continue to commute to work; we continue to fly to conferences, trade shows, and weddings. The reasons are clear. Colocation enables us to exhibit and interpret various nonverbal signals and cues, such as touch, proximity, position, direction of gesture, eye contact, and level of attention. It enables us to greet each other with a handshake or embrace, dine together, and appreciate objects in a shared environment [1]. Traditional communications do not allow us to perform many of these social interactions. Therefore, immersive communication is recognized as the most promising development.

Advancements in technology are progressing at a rate that outstrips those in the field of biology. It will soon become a viable proposition to deliver a greater number of bits per second to an observer than can be transmitted through the sensory cut-set separating the environment from the nervous system. The human visual system can absorb only about 10 Mb/s; the haptic system about 1 Mb/s; auditory and olfactory systems about 100 Kb/s each; and the gustatory system about 1 Kb/s [2]. In contrast, fiber-to-the-home is already delivering on the order of 10 Mb/s, while links in the Internet backbone are approaching 1 Tb/s. In light of the considerable advances that have already been made in the realm of information transmission, the central question is no longer how to convey information, but rather how to capture and render that information in a manner that aligns with the human system. Fundamentally, this concerns the desired nature of our remote communication and experiences.

Immersive communication can be defined as the exchange of natural social signals with remote individuals, as would be observed in face-to-face meetings, and/or the experience of remote locations (and having the individuals there experience you) in ways that suspend disbelief in the presence of the


W. Chen and Z. Qin, *Semantic Communications*, Textbooks in Telecommunication Engineering,
https://doi.org/10.1007/978-3-032-11005-3_13

participants. The concepts of transparency, fluidity of interaction and a sense of presence are all pertinent here. Visions of immersive communication are illustrated by such examples as Enders Game's warship system, Star Wars' Jedi council meetings, The Matrix's matrix, and Avatar's Pandora.

A fundamental aspect of immersive communication is the capacity for participants to engage with the remote environment, thereby detecting and evaluating that interaction through their senses. Visually, as we move, we perceive the world from diverse perspectives, the environment responds, and other individuals react, fostering a sense of integration into the scene. Similarly, we perceive the auditory aspects of the environment as we turn around, the environment adapts to us, and we perceive the results. This reciprocal interaction fosters a sense of immersion, whereby individuals perceive an illusion of being fully engaged with their surroundings. It is noteworthy that other senses, such as vestibular (balance and acceleration) and olfactory senses, are analogous to the aforementioned, though their technical conveyance is more challenging. The vestibular system, which governs balance and acceleration, is intricately linked to the visual system, and discrepancies between these two senses can be profoundly disconcerting. Touch and smell, while being the next most significant senses after vision and hearing, will not be the focal point of this discussion.

Immersive communication takes various forms, such as extended reality (XR), haptic communication, and holographic communication. Regardless of the immersive communication scenario, it involves the reconstruction and representations of 3D content. In terms of 3D data processing, a variety of 3D data representations and their characteristics are shown in Table 13.1.

Among the various 3D data representations, point cloud can provide a 6 degree-of-freedom (DoF) perspective and retain rich 3D details. Compared with mesh, light field and other technologies, the data volume is small, scalability is strong, algorithm pre-processing and post-processing is convenient, and acquisition, reconstruction and other devices are relatively mature. Therefore, point cloud is expected to become one of the mainstream three-dimensional data characterization of holographic business in the future.

A point cloud consists of an unordered set of points in 3D space, defined by three-dimensional coordinates and extensible attributes such as color, opacity, and reflectivity. Compared to modalities like text, speech, images, and videos, point cloud data presents significant challenges due to its large volume and the high difficulty of compression. Addressing the data compression of point clouds is a critical issue in the development of immersive communication systems.

(2) Semantic Communication Enabled Autonomous Driving

To tackle point cloud compression, MPEG has established two standards: Video-based point cloud compression (V-PCC) and geometry-based point cloud compression (G-PCC). V-PCC leverages mature video coding technologies and relies on the quality of conversions between 3D and 2D data. In contrast, G-PCC is designed to optimize compression based on the geometric characteristics of point clouds. Notably, neither of these standards incorporates deep learning techniques.

In recent years, numerous deep learning-based point cloud compression methods have emerged, primarily focusing on bit compression. These methods often employ entropy models for precise probability estimation, enabling advanced entropy coding for extreme bit reduction. However, these approaches are highly error-sensitive; even a single-bit error can render the decoder inoperative.

Developing a semantic codec tailored for point clouds requires in-depth exploration of neural network models specialized for point cloud feature extraction. The unprecedented demand for data transmission rate and the stringent requirements on latency and reliability create challenges for traditional codecs and networks to support immersive communications. Therefore, designing semantic communication codec frameworks, where semantically-unimportant information is highly-compressed and more tolerant to the channel noise, are one of the pivotal research directions supporting immersive communication.

The overall framework of the point cloud semantic communication (PTSC) is shown in Fig. 13.1. At the transmitter

**Table 13.1** 3D data representations and characteristics

| Representations | Degree of freedom | Bit-rate | Applications |
|---|---|---|---|
| RGBD video | 1 DoF | 1–10 Mbps | Simple visual experience, assisted computer vision tasks, etc. |
| Multi-view video | 2 more DoF | 10–100 Mbps | 3D movies, 3D games, etc. |
| 360° video | 3 DoF | 10–500 Mbps | Panoramic movies, XR, etc. |
| Voxel | 6 DoF | 0.1–10 Gbps | Autonomous driving, etc. |
| Point cloud | 6 DoF | 0.1–10 Gbps | Holography, 3D modeling, autonomous driving, XR, etc. |
| Mesh | 6 DoF | 0.1–10 Gbps | Holography, computer graphics, XR, etc. |
| Light field | 6 DoF | 0.1–10 Tbps | Holography |

**Fig. 13.1** PTSC framework

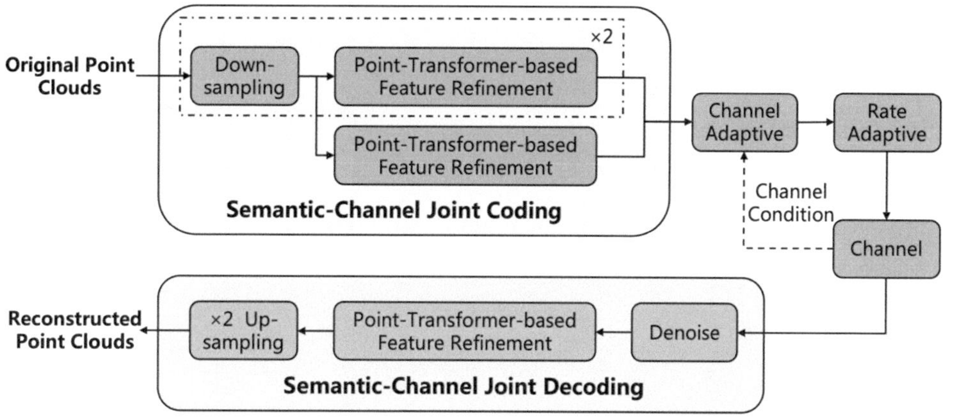


side, the original point cloud is first down-sampled, reducing the number of points but increasing the feature dimensionality to obtain semantic features of the point cloud. In the auxiliary branch, the down-sampled point cloud semantic features undergo feature refinement based on Point-Transformer. In the main branch, the refined features are further down-sampled and undergo another round of feature refinement based on Point-Transformer to obtain semantic features that preserve more geometric details. Afterward, the features output by both the auxiliary and main branches are combined and fed into the channel adaptation module and bit-rate adaptation module. The channel adaptation module adjusts the semantic features based on channel conditions to ensure that the output of the semantic-channel joint encoding can adapt to different channel conditions. The bit-rate adaptation module selects the semantic features generated by the auxiliary branch that need to be transmitted, ultimately producing the symbols for channel transmission.

At the receiver side, the received symbols are first denoised, then the semantic features are adjusted through Point-Transformer-based feature refinement, similar to the transmitter side, and finally, the reconstructed point clouds are obtained through offset-based up-sampling.

Semantic communication for point clouds mainly improves immersive communication performance through three aspects:

- A hierarchical semantic sampling method is designed to obtain point cloud semantic features at different granularities.
- A master-slave parallel joint semantic-channel encoder-decoder is proposed, which reduces the dimensionality of the semantic features while enhancing their robustness.
- A bitrate adaptation module and a channel adaptation module are introduced, enabling the transmitter to perform adaptive transmission based on the source semantics and channel conditions.

Based on the framework of PTSC in Fig. 13.1, numerical simulation experiments were conducted. The G-PCC of MEPG is selected as the encoder of the point cloud and LDPC is used as the channel coding scheme. The modulation and coding scheme (MCS) selects the parameters in Table 5.1.3.1-1 of 3GPPTS 38.214 version 16.2.0. for the physical downlink shared channel (PDSCH).

Under the D2 PSNR metric, the performance as shown in Fig. 13.2 (left) is obtained. Compared with the GPCC+5G LDPC scheme, PTSC has a performance advantage of more than 2 dB at all symbol numbers. The method can still accomplish effective coding and decoding under the condition of 0dB SNR, and the performance is better than the traditional method at 15 dB. In addition, the output dimension of the semantic-channel joint encoder, i.e., the number of channel symbols generated, is determined for each simulation point in the simulation experiments. By comparing the performance curves with and without RAM, it can be seen that the code-rate adaptive model effectively reduces the number of transmitted symbols under good channel conditions. For example, at a SNR of 15 dB, a model set to transmit 300 symbols actually transmits only about 250 symbols (top darker line with dots). Under poor channel conditions, the code-rate adaptive model does not choose to reduce the number of symbols transmitted in order to ensure reconstruction performance.

For different categories of point clouds, the results of the number of transmitted symbols under different channel conditions are shown in Fig. 13.2 (right). The method can select different numbers of transmitted symbols for different categories of point clouds and realize semantic adaption. Meanwhile, the proposed method is also able to adaptively select transmission symbols according to different channel conditions. As the channel conditions increase, the number of transmitted symbols produces a corresponding decrease.

(3) Challenges and Future Directions

To bring the semantic-aware wireless ultimate XR from theory to practice, the following challenges should be addressed [3]:

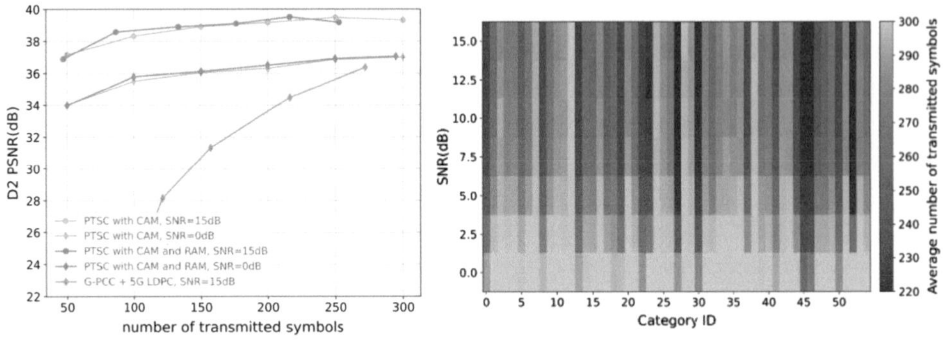


**Fig. 13.2** Performance curves of point cloud transmission system

A. Trade-Off Between Performance and Computing Costs

Existing semantic coding modules have shown great success in data size reduction. However, running semantic coding modules on head-mounted devices (HMDs) usually takes a long time due to the limited on-board computing resources. To reduce the running time, one potential way is to consider model distillation to find the best trade-off between the model performance and computing costs.

B. Task-Specific Network Generation

Most of the existing semantic sensing or communication modules are designed in a task-specific way. This will result in a great amount of semantic models to be stored at HMDs, which usually have limited storage space. One potential way to reduce the memory costs is to use meta-learning to train a network-generation network, which dynamically generates the optimal network architectures and parameters for each task in wireless ultimate XR.

C. Semantic-Aware Resource Allocation

In a practical communication system, users of a particular XR application need to share wireless resource with others. The data generated by each XR user also has various sizes and different semantic-importance levels. Therefore, it is challenging to find the optimal resource allocation policy in this complicated situation. To maximize the resource utilization efficiency, deep reinforcement learning could be considered.

## 13.2 Satellite Communications

(1) Introduction of Satellite Communication Networks

Wireless communications have undergone a series of transformative advancements, fundamentally changing the way individuals and devices connect, enabling seamless and efficient data exchange across various platforms. The introduction of 6G communication systems is poised to further revolutionize these interactions, offering a wealth of novel techniques and applications. Among the most promising innovations is the integration of satellite communications, which play a pivotal role in the evolution of global communication infrastructure. Satellite communication networks are particularly advantageous in scenarios where cellular infrastructure is limited or unavailable, such as in remote, rural, or disaster-stricken areas, where they can provide critical wireless connectivity with minimal reliance on ground-based facilities.

Satellite communication networks encompass a range of orbital categories, including Geostationary Earth Orbit (GEO), Medium Earth Orbit (MEO), and Low Earth Orbit (LEO). GEO satellites orbit at an altitude of 35,786 km, maintaining a fixed position relative to the Earth's surface. This allows for continuous coverage with just three satellites, providing a simple solution for global reach. However, the inherent propagation delay of approximately 120 ms to GEO satellites limits their applicability in latency-sensitive use cases. In contrast, MEO and LEO satellites, which are non-geostationary, operate at altitudes ranging from 7,000 to 25,000 km for MEO, and from 300 to 1,500 km for LEO. These satellites are fast-moving, with velocities that can exceed 10 km/s, necessitating the formation of constellations to maintain continuous coverage for ground users. LEO satellites are less susceptible to environmental factors or energy limitations, and can deliver persistent coverage across vast areas. Furthermore, being closer to the Earth's surface than both GEO and MEO, LEO satellites can offer lower-latency, higher-quality communication services. With advances in satellite manufacturing and launch capabilities, LEO satellites are poised to become a key driver in the evolution of satellite communication networks.

The growing prominence of satellite communication network (especially for LEO systems) is driven by their ability to complement terrestrial networks, particularly in regions where laying traditional infrastructure is cost-prohibitive or physically challenging. The integration of satellites into the communication ecosystem offers a promising solution for bridging the digital divide and enabling ubiquitous connectivity. Moreover, as the demand for mobile broadband and IoT applications continues to rise, satellite communication networks are expected to provide essential coverage in remote and underserved areas, supporting applications ranging from global navigation systems to real-time video communication and beyond.

However, several challenges remain for the successful deployment and operation of satellite communications. One of the primary obstacles is the limited spectrum available for satellite communications. Unlike terrestrial networks, where high-bandwidth fiber-optic cables and wireless technologies can be deployed in large quantities, satellite systems face inherent constraints due to the limited frequency spectrum allocated for satellite communications. The bandwidth available in satellite systems is critical for supporting high-throughput services, such as high-definition video streaming, large-scale data transfers, and the growing number of IoT devices.

The bandwidth limitation in satellite communication is largely due to the finite nature of the radio frequency spectrum, which is allocated by regulatory bodies like the International Telecommunication Union (ITU). Most satellite systems operate in frequency bands such as the L-band, C-band, Ku-band, and Ka-band. These frequency bands are shared among multiple users and services, which leads to congestion and interference, further limiting the available bandwidth for satellite communications. The most commonly used frequency bands for commercial satellite communication systems are often oversubscribed, resulting in bandwidth contention and reduced data rates, particularly during peak usage times. For instance, LEO satellites typically operate in the Ka-band, which offers higher bandwidth but is also more susceptible to weather conditions, such as rain fade, that can further reduce effective bandwidth. GEO satellites, while providing wider coverage areas, have limited bandwidth resources in their assigned frequency bands. For example, the Iridium satellite system, which operates in the L-band, has a relatively small available bandwidth of just 10.5 MHz, limiting its maximum data rate to around 2.4 kbps for both uplink and downlink transmissions. This narrow bandwidth severely constrains the satellite's ability to support bandwidth-intensive applications, such as real-time video conferencing, cloud computing, and large file transfers. The bandwidth limitation is also exacerbated by the fact that satellite communication systems often use shared resources among numerous users across vast geographical areas. This shared spectrum is typically allocated on a per-user basis, which means that the available bandwidth must be dynamically managed to prevent congestion and maintain service quality. As the number of users increases, especially in high-demand regions, the bandwidth available to each user decreases, leading to reduced speeds and potential service disruptions.

The second key challenge in satellite communication systems is the inherent delay introduced by the vast distances that signals must travel between Earth and space. Unlike terrestrial communication systems, which involve relatively short transmission paths, satellite links require signals to travel much greater distances, leading to significant latency. This delay is primarily due to the long propagation times associated with the signal's journey to and from satellites orbiting at altitudes ranging from LEO to GEO. For satellites in GEO, the round-trip signal propagation delay can be substantial, typically around 240 milliseconds. While this delay may seem minimal in everyday communication, it can significantly impact time-sensitive applications such as real-time video streaming, voice communication, and interactive services like online gaming. Furthermore, when considering the need for multiple signal hops (such as those required for satellite constellations or multi-link systems), this delay is compounded, resulting in higher latencies that can hinder the overall system performance. The round-trip signal delay for LEO satellites is typically between 10 and 20 ms, making them more suitable for applications that demand low latency, such as autonomous vehicle communications, remote surgery, and real-time financial trading.

The third critical challenge in satellite communication systems arises from the mobility of both the satellites and the ground terminals, which can lead to unstable channel conditions, frequent handovers, and severe Doppler effects. Unlike terrestrial networks, where communication links are relatively stable due to fixed locations of both transmitters and receivers, satellite systems operate in dynamic environments, where the relative motion between satellites and ground users can significantly affect the quality and reliability of the communication channel. The high speed of satellites, especially those in LEO, causes significant Doppler shifts. The Doppler effect leads to frequency changes in the received signal, which can degrade the signal quality, increase error rates, and lead to interruptions in the communication link. This effect is particularly problematic for high-speed applications, such as real-time video conferencing or live data streaming, where maintaining a stable link is crucial.

Moreover, the relative motion between satellites and mobile users on the ground creates a phenomenon known as "handover", where a user's communication needs to be transferred from one satellite to another. For LEO and MEO systems, handovers occur much more often due to the satellites' fast orbits. For instance, a LEO satellite typically provides coverage for only a few minutes before it moves out of range, requiring a switch to a neighboring satellite to maintain continuous communication. The handover process itself introduces delays and disruptions, as it involves re-establishing the communication link, re-aligning antennas, and possibly adjusting the signal parameters, all of which can lead to a temporary loss of connectivity. The combination of frequent handovers and unstable channel conditions due to mobility can significantly impact the quality of service in satellite communication systems.

(2) Semantic Communication-Enabled Satellite Communication Network

The increasing demand for efficient, reliable, and intelligent communication systems in satellite networks has highlighted the need for a shift from traditional communication paradigms to more advanced approaches. Semantic communication is regarded as a powerful way to solve the above challenges in satellite communication networks, which is able to offer numerous advantages that can significantly enhance the performance, efficiency, and scalability of satellite communication systems.

Traditional communication systems rely on the transmission of large amounts of raw data, which consumes substantial bandwidth, often leading to congestion and reduced data rates. Semantic communication, by focusing on the meaning of the transmitted information, allows for more efficient use of bandwidth. It enables the encoding and transmission of the most relevant or essential parts of the information, significantly reducing the amount of data that needs to be sent. This not only alleviates bandwidth congestion but also allows satellite networks to support more users and diverse applications, even in areas with limited spectrum availability.

Satellite communication networks are often deployed in challenging environments, such as remote regions, disaster-stricken areas, or in mobile contexts like aviation or maritime communications. In these settings, traditional communication methods can suffer from poor channel conditions, including high latency, signal attenuation, and interference. Semantic communication, which focuses on optimizing the transmission of meaningful content rather than raw bits, can adapt more effectively to these challenging environments. By prioritizing relevant information and utilizing AI-driven techniques to jointly optimize source and channel coding, semantic communication systems can ensure more robust and reliable transmission, even in the face of channel impairments or limited signal quality.

Latency is a significant challenge in satellite communications, especially for systems using higher orbits like geostationary satellites, where signal propagation times are relatively high. This delay can cause issues for time-sensitive applications, such as real-time video streaming, voice communication, and IoT devices. Semantic communication systems can help mitigate this issue by enabling more intelligent, context-aware processing of data. For instance, AI-driven techniques can prioritize and compress the most critical information, reducing the overall data load and, in turn, lowering the transmission time. This can result in a noticeable reduction in latency, making satellite networks more suitable for applications that require low-latency communication, such as autonomous vehicles, remote surgery, and interactive services.

As satellite communication networks evolve, especially with the deployment of large-scale LEO constellations, scalability and flexibility become increasingly important. Traditional communication methods may struggle to meet the demands of rapidly growing networks, diverse applications, and a large number of users. Semantic communication, by contrast, is inherently more scalable. It allows for the dynamic adaptation of communication protocols based on the context and needs of the user or application. This adaptability enables satellite systems to handle a diverse range of applications, from broadband internet services to mission-critical communications in remote areas, while optimizing resource allocation to ensure efficient use of network capacity.

The use of AI and machine learning is a cornerstone of semantic communication. These technologies enable the network to autonomously adjust transmission parameters, select optimal encoding schemes, and adapt to varying channel conditions. By integrating AI into satellite communication networks, semantic communication can provide more intelligent management of the communication process, allowing for real-time adjustments based on network performance, user demand, and environmental factors. This self-optimization capability is particularly valuable in satellite systems, where the operational environment is dynamic, and maintaining optimal performance under varying conditions can be difficult.

As satellite communication networks evolve with the advent of 6G and beyond, the need for innovative communication techniques becomes even more critical. Semantic communication is a key enabler for future-ready satellite communication systems. It not only supports the growing demand for high-efficiency communication in both terrestrial and space-based networks but also paves the way for novel applications such as deep space communication, seamless connectivity across satellite constellations, and the integration of satellite networks with other communication infrastructures like 5G, 6G, and beyond. By leveraging AI and semantic techniques, satellite systems can evolve to meet the needs of future communication challenges, ensuring continuous global connectivity for an expanding digital world.

(3) Challenges and Future Directions

Although semantic communication has the potential to achieve efficient and robust satellite communications, there are still some challenges that need to be addressed before implementation.

A. Limited Computational Resources

As AI becomes increasingly integrated into satellite communication systems within the framework of semantic communications, a new set of challenges has emerged. One of the most significant challenges is the computational resource limitations imposed by the complex AI models used for data encoding, decoding, and signal processing. In traditional satellite communication systems, the focus has been primarily on optimizing the transmission and reception of signals with minimal computational overhead. However, with the advent of semantic communication, which focuses on optimizing the transmission of meaning rather than raw data, the use of AI-driven algorithms such as deep learning models, reinforcement learning, and neural networks is becoming more prevalent. While these models can significantly enhance the efficiency and performance of satellite systems, they also place substantial demands on computational resources.

Semantic communication systems leverage AI to understand, predict, and optimize the content being transmitted, making them particularly useful for applications such as video conferencing, IoT, and real-time data analytics. AI models in these systems are tasked with interpreting the data's meaning and context, compressing it efficiently, and ensuring that the most relevant information is prioritized during transmission. However, these sophisticated AI models require significant computational power, which can strain the limited onboard resources available in satellite systems. Unlike terrestrial systems that can rely on centralized data centers and cloud computing, satellites operate in space with limited computational capacity, energy resources, and storage. This limitation poses a challenge for the deployment of resource-intensive AI models that are essential for the next-generation communication systems.

For instance, large-scale deep learning models, which are commonly used for semantic data processing, require significant processing power and memory, often exceeding the capabilities of satellite onboard computers. The need for real-time processing in dynamic environments further amplifies the challenge. In scenarios involving LEO satellites, where frequent handovers and high mobility require real-time adjustments to communication protocols, the computational burden becomes even more critical. The AI models must not only process and compress data efficiently but also respond to changing channel conditions, mobility, and user demands in a timely manner, all while operating within the constraints of limited computational resources.

B. Limited Energy Resources

Another issue is the trade-off between the complexity of AI models and the energy consumption of the satellite. High-performance AI models can consume considerable amounts of power, which is particularly problematic in satellite systems where energy resources are limited, often relying on solar panels and batteries for power. The challenge lies in balancing the computational power required for effective AI-driven semantic communication with the energy limitations of the satellite. If AI models are too computationally intensive, they can drain power reserves too quickly, jeopardizing the satellite's overall functionality and service reliability.

Furthermore, the data processing required by AI models for semantic communication can introduce significant latency, especially in satellite systems with limited computational resources. The need for rapid data processing and decision-making in satellite communication systems, particularly those using LEO constellations that require quick handovers, may be compromised by the time taken to run AI models on limited hardware. This latency could degrade the overall user experience, especially for real-time applications such as interactive communication or mission-critical services in remote areas.

To overcome these challenges, several approaches are being explored. Edge computing techniques, where computational tasks are offloaded from the satellite to nearby ground stations or other satellites in the constellation, can help mitigate the strain on onboard resources. Additionally, lightweight AI models, such as model pruning or quantization, are being developed to reduce the computational load while maintaining performance. Collaborative processing across satellite constellations, where multiple satellites share computational tasks, can also alleviate the resource constraints. Furthermore, innovations in satellite hardware, such as the development of more efficient AI accelerators or specialized chips designed for AI processing, hold promise for enhancing the computational capabilities of satellite systems without exceeding power and resource limitations.

C. Lack of Standards

A primary challenge in standardizing semantic communication for satellite systems is the lack of a unified framework for semantic encoding and decoding. Traditional communication systems focus on transmitting raw bits or symbols, but semantic communication emphasizes the transmission of meaningful information by encoding the context and intent of the data. This shift requires the development of new coding schemes, modulation techniques, and transmission protocols that are specifically designed to optimize semantic content. However, there is currently no widely accepted standard for the semantic encoding of information, making it difficult to ensure compatibility across different satellite communication platforms.

In response to these challenges, efforts are underway to develop standards for semantic communication in satellite systems. Initiatives such as the International Telecommunication Union's (ITU) work on next-generation satellite systems and the 3rd Generation Partnership Project (3GPP) efforts to define satellite-terrestrial integration protocols are vital to shaping the future of satellite communications. However, significant work remains in defining common frameworks, protocols, and performance benchmarks for AI-driven, semantic communication-based satellite systems. This work will require close collaboration between academia, industry, and standardization bodies to ensure that semantic communication technologies are fully integrated into the global satellite communication ecosystem.

In conclusion, while the integration of AI in semantic communication has the potential to revolutionize satellite communication systems by improving data efficiency and enabling more intelligent transmission of information, it also introduces significant challenges related to computational resource limitations. Addressing these challenges will require advancements in both AI model design and satellite system architecture, ensuring that AI-driven semantic communication can be effectively implemented in space-based environments with limited computational and energy resources.

## 13.3 Autonomous Driving

(1) Introduction to Autonomous Driving

Autonomous driving, often referred to as self-driving or driverless vehicles, represents a transformative development in the field of transportation. This technology allows vehicles to operate without human intervention, relying on sophisticated systems that process data from the environment and make real-time decisions to control the vehicle. Autonomous driving is at the intersection of various fields, including artificial intelligence, robotics, computer vision, and machine learning.

The basic process of autonomous driving begins with the vehicle perceiving its environment through a combination of advanced sensors, such as LiDAR, radar, cameras, and ultrasonic sensors. These sensors work in tandem to gather data on surrounding objects, road conditions, and traffic signals, providing the vehicle with a comprehensive understanding of its environment. The vehicle's onboard systems then process this sensory information to create a model of the surroundings, which is critical for safe navigation. Additionally, the vehicle uses high-definition maps and GPS data for localization, helping it determine its precise position within the environment. This is particularly important in complex scenarios where accurate spatial awareness is necessary to make decisions regarding route planning, obstacle avoidance, and other driving tasks.

Once the environment is understood, the vehicle's decision-making and path planning algorithms take over. These systems analyze the data and generate an optimal path, making real-time decisions based on factors such as traffic flow, road rules, and the behavior of other road users. The vehicle then executes this planned trajectory through its control systems, which manage actions like steering, braking, and acceleration to ensure that the vehicle follows the intended path while maintaining safety. These processes occur simultaneously and continuously, enabling the vehicle to respond dynamically to changes in its environment, such as pedestrians crossing the street, other vehicles adjusting speed, or unexpected road conditions. This seamless integration of perception, decision-making, and control is the foundation of autonomous driving.

Autonomous vehicles rely heavily on vast amounts of sensor data to navigate safely and efficiently. Traditional communication systems transfer raw sensor data, such as high-resolution LiDAR point clouds or camera images, which demands substantial bandwidth and computational resources. However, as autonomous driving systems progress, the need for more efficient communication has become apparent. Semantic communication represents a revolutionary shift in communication paradigms, where the emphasis is not on transmitting raw data, but on conveying the meaning behind the data. This focus on understanding and delivering task-relevant information is particularly beneficial in the context of autonomous driving, where real-time decision-making and data prioritization are crucial for safety and performance.

In a semantic communication system, only the most meaningful and context-specific information is transmitted, significantly reducing the amount of data exchanged. For example, instead of transmitting a full HD video stream from the vehicle's cameras, the system might only communicate essential information such as the location of detected objects (e.g., "pedestrian 30 m ahead" or "traffic light: red"), along with relevant classifications. This prioritization ensures that the data sent over the network is not only relevant but also time-sensitive, which is critical for real-time operation. Furthermore, this approach minimizes latency by avoiding the transmission of unnecessary data, such as background scenery or irrelevant visual details. For instance, in cases where an obstacle appears suddenly, semantic communication ensures that crucial data—like the detection of the obstacle's position, type, and trajectory—reaches the vehicle or control center with minimal delay, thereby enabling faster response times.

Beyond enhancing the efficiency of data transmission, semantic communication can also enable more intelligent network management. By allowing systems to prioritize certain types of data based on context (e.g., safety-critical information such as traffic conditions or sudden obstacles), the communication network can dynamically allocate bandwidth to meet the demands of real-time decision-making. Additionally, integrating semantic communication with other vehicle-to-everything (V2X) technologies could further enhance the coordination between vehicles, infrastructure, and pedestrians, allowing for more cohesive and synchronized traffic management. Overall, the adoption of semantic communication could significantly improve the reliability, scalability, and safety of autonomous driving systems, facilitating a more efficient exchange of information while ensuring that critical decisions are made with minimal latency and maximum accuracy.

(2) Semantic Communication Enabled Autonomous Driving

Autonomous driving relies on several critical processes, including perception, transmission, computation, and control. This section explores the current and potential applications of semantic communication across these processes, highlighting its role in optimizing data exchange, improving system efficiency, and enhancing safety.

A. Perception
Semantic communication plays a pivotal role in enhancing cooperative perception for autonomous vehicles by facilitating the fusion of data from multiple sources, such as vehicles and road infrastructure. This fusion improves environmental understanding and safety by reducing information overload and prioritizing the transmission of contextually relevant data. One such example is the VIPS system [4], which employs a graph-based approach to achieve real-time, decimeter-level perception fusion. By representing objects as graph structures that encode both their semantics and spatial relationships, VIPS ensures data consistency despite communication delays, enabling a more accurate and reliable understanding of the vehicle's surroundings. Through this method, VIPS extends the vehicle's perception range and improves fusion accuracy, all while reducing the volume of data transmitted. This showcases how semantic communication can optimize bandwidth and enhance the reliability of autonomous driving systems.

In addition to cooperative perception, semantic communication also improves tasks like semantic segmentation for Bird's Eye View (BEV) perception. Traditional single-agent systems, often reliant on camera-based sensors, may struggle with issues such as occlusions and the detection of distant objects. By integrating vehicle-to-vehicle (V2V) communication, semantic communication enables vehicles to share sensory data in real-time, significantly improving perception accuracy. The CoBEVT framework [5] exemplifies this by utilizing a transformer-based architecture with a fused axial attention module, which effectively integrates multi-agent and multi-view data for BEV semantic segmentation. This approach achieves state-of-the-art performance, demonstrating the potential of semantic communication to enhance the perception capabilities of autonomous vehicles.

B. Transmission
The transmission requirements of connected autonomous vehicles (CAVs) are substantial, involving both vehicle-to-vehicle (V2V) and vehicle-to-infrastructure (V2I) communication. These interactions form the backbone of Autonomous Driving Networks (ADNs). Semantic communication enhances ADNs by improving communication efficiency and addressing challenges such as high data volume and stringent latency constraints. One key advancement in this area is the use of federated learning [6] to optimize communication in resource-limited environments. By incorporating semantic communication frameworks within federated learning, CAVs can achieve more efficient data transmission while preserving privacy. This approach has shown promising results in optimizing both collaboration and model personalization within ADNs, thus enhancing overall network intelligence.

Moreover, research on federated learning in ADNs has addressed several key challenges, such as the need for privacy preservation and the efficient handling of multimodal data. These challenges are especially important in autonomous driving, where timely and secure transmission of critical information is vital for safe operation. By focusing on task-oriented data transmission, semantic communication systems enable CAVs to prioritize important information while minimizing unnecessary data exchange, ensuring efficient and reliable communication.

C. Computation
Task-oriented semantic communication is particularly valuable in autonomous driving systems, where it enables the transmission of only relevant information for specific tasks, thus enhancing computational efficiency. This approach reduces the volume of data to be processed, lowers latency, and ensures robust performance even in low signal-to-noise ratio (SNR) environments. For example, a convolutional autoencoder framework [7] has been developed to semantically encode road traffic signs for transmission between vehicles, allowing CAVs to share critical information even under challenging weather conditions. By using task-specific decoders for both image reconstruction and classification, this approach outperforms traditional communication methods, reducing bandwidth usage by up to 89% while improving classification accuracy.

Additionally, systems like VIS-SemCom [8] leverage semantic communication to reduce redundancy and prioritize critical data in image segmentation tasks for driving safety. By utilizing advanced architectures such as the Swin Transformer, VIS-SemCom enhances segmentation accuracy and employs multi-scale semantic extraction to emphasize key objects in the driving environment. Experimental results have shown significant reductions in data transmission (up to 70%) while achieving improvements in both coding efficiency and segmentation performance. These advancements highlight the potential of semantic communication to optimize data transmission, reduce resource consumption, and enhance the reliability of autonomous vehicle perception systems.

D. Control
In the realm of autonomous driving control, semantic communication can significantly enhance system accuracy and

responsiveness. By transmitting only task-relevant information, this approach reduces communication overhead and ensures that critical data is delivered in real-time, supporting timely decision-making. A source-channel joint training framework [9], for example, has been developed to optimize both source and channel coding in a task-oriented manner, demonstrating its effectiveness in maintaining robust driving performance under noisy communication conditions. By improving bandwidth utilization and reducing latency, this framework supports precise vehicle control, even in challenging environments.

The integration of semantic communication into autonomous driving control systems not only improves accuracy but also enhances the overall reliability of the system. By focusing on the transmission of relevant, high-priority data, semantic communication ensures that essential control information such as real-time adjustments to speed, steering, and braking reaches the vehicle with minimal delay. This capability is crucial for enabling safe and efficient operation in dynamic driving environments, positioning semantic communication as a key enabler for the future of connected and autonomous vehicles.

(3) Challenges and Future Directions

While semantic communication holds significant promise for autonomous driving, several challenges must be addressed to fully realize its potential.

One of the primary challenges is standardization, which involves developing common semantic languages and protocols for communication between vehicles and infrastructure. Without a universal framework, vehicles from different manufacturers or regions may struggle to exchange critical information effectively.

Another key challenge is error handling, as semantic communication systems must be robust enough to handle issues such as packet loss or transmission errors. In the highly dynamic environment of autonomous driving, even minor disruptions in communication can lead to safety risks. Therefore, ensuring reliability in the face of potential network failures is essential.

Scalability is also a significant concern. As the number of connected vehicles increases, the communication system must be able to accommodate high traffic densities without overwhelming the network. This requires advanced mechanisms for efficiently managing resources and prioritizing data transmission, especially in urban environments with dense traffic.

Finally, real-time processing is a critical issue. While semantic communication offers substantial transmission efficiency gains, it must be carefully balanced with the latency of semantic extraction. In autonomous driving, any delay in processing could compromise the vehicle's decision-making capabilities, potentially affecting safety and operational efficiency. Therefore, optimizing the latency of semantic extraction and ensuring timely transmission are vital for maintaining real-time performance.

Looking ahead, the future of semantic communication in autonomous driving will likely focus on integrating advanced AI models for dynamic and context-aware semantic extraction. These models would enable the system to adapt to varying driving conditions and evolving requirements, further enhancing communication efficiency. Additionally, as the technology matures, semantic communication will continue to evolve, playing an increasingly critical role in enabling fully autonomous and connected mobility. It is expected to optimize various aspects of autonomous driving, from perception and decision-making to control, making it a cornerstone of next-generation transportation systems.

## 13.4 Smart Grid

(1) Introduction to Smart Grid

Smart grid is a kind of intelligent electricity network that uses digital technology to facilitate electricity monitor, control, generation, distribution, and consumption. In order to ensure safety and economy in the process of new energy transformation, it is necessary to improve the flexibility of demand-side resources to complete energy supply guarantees. Virtual power plant is a clean, low-carbon, safe and efficient solution for aggregating demand-side flexible resources. It aggregates massive user-side resources such as flexible loads, electric vehicles and distributed energy storage through information network technology to help improve the power grid. It has balanced adjustment capabilities, promotes the consumption of new energy, and has significant low-carbon characteristics. International experience shows that virtual power plants are one of the key strategies to promote the safe consumption of new energy. In scenarios such as extreme cold wave climate, unstable power electronic equipment, and sudden unit failures, virtual power plants can provide solid support for power systems dominated by new energy sources. In May 2023, the Chinese National Energy Administration released the "Electricity Demand Side Management Method (Draft for Comments)". The document clarifies that by 2025, the power demand side response capability will reach 3–5% of the maximum electricity load. In 2030, it plans to form large-scale, real-time demand response capabilities through virtual power plants and other means.

The core of the virtual power plant lies in its communication technology, which is essentially a complex cyber-physical coupling system composed of numerous distributed resources and communication terminals. The communication data that virtual power plants need to transmit includes measurement and status collection data, resource control data, user transaction data, etc. With the access to massive distributed resources, communication requirements have also exploded, requiring tens of millions of connection capabilities, milliseconds and below delay requirement, extremely low jitter, extremely high timing accuracy and transmission reliability. Moreover, different virtual power plant services have greatly different requirements for data transmission, so advanced support by emerging communication technologies that are reliable, stable, and efficient is an urgent need for virtual power plants. The distributed resources of virtual power plants are usually distributed in the distribution network and the last mile of the city, and communication connections of large-scale resources are mainly achieved through wireless access networks (such as cellular communications). For example, in China, the demonstration project in Jiangsu Province initially successfully connected 1,726 industrial control users, 2,715 non-industrial flexible control users, and 208,000 active demand response users, most of whom accessed through wireless communication networks to participate in interactions. However, even a structurally robust transmission network will show a certain degree of vulnerability due to cyber-physical coupling. For example, the system may be paralyzed due to man-made cyber attacks. However, more common situations are accidents caused by the communication network itself. For example, State Grid Jiangsu Electric Power Company's optical cable faults accounted for 87% of all communication faults from 2016 to 2020. The use of wireless communication networks to access distributed resources puts forward higher requirements for the operational reliability of virtual power plants: After a photovoltaic power station in Shanxi Province adopted a wireless communication solution, the system reported hundreds of thousands of incidents within three months. Communication interruption alarm records, the power grid company also prompts caution in using wireless communications and may even disable wireless communications to collect and control distributed resources.

Therefore, there is an urgent need to re-examine the reliable and economical operation of virtual power plants from the perspective of cyber-physical systems and take the impact of the communication side into consideration. At present, the operation of power cyber-physical systems usually regards the communication network as a fixed and unfavorable external factor. As a result, the operation strategy of the virtual power plant can only passively adapt to the performance limitations of the communication network, while ignoring the active collaborative regulation on the communication side. Possibly, this limits the further improvement of virtual power plant performance. For virtual power plants, it is crucial to consider the impact of information systems on their operation and control. In addition, existing research on power cyber-physical coupling mainly focuses on the structural coupling of communication networks and power networks, analogizing information transmission with power flow, ignoring the differences between information, and also ignoring the importance, relevance, and semantics of information. Related content. Ignoring information semantics will make it difficult for virtual power plants to distinguish the value of heterogeneous information for various tasks, becoming a bottleneck that restricts virtual power plants from completing their various services reliably and efficiently.

In recent years, with the development of wireless communication technology, its capacity has gradually approached the Shannon limit. Semantic communication is an effective way to break through this bottleneck. Compared with traditional communication, semantic communication significantly reduces transmission bandwidth requirements and improves communication efficiency by understanding, extracting, and transmitting the semantic features of information rather than simply accurately transmitting bits or symbols. Similarly, virtual power plants not only need to accurately and quickly transmit information, but more importantly, accurately enable the information to play its role. Therefore, when transmitting heterogeneous data from massive devices, it should analyze the semantics and value of heterogeneous information to express the meaning. As the goal is to distinguish which information is more important to it, give this important information higher communication priority and communication resource allocation weight during communication access and transmission, so as to ensure the reliability of virtual power plant services to the greatest extent and quality. It has become a consensus in the industry to consider information semantics and optimize communication service quality according to the special needs of the power industry.

(2) Framework of Semantic-aware Smart Grid Systems

This section draws on the idea of semantic communication and proposes a semantic-aware cyber-physical collaborative control model for smart grid systems specifically for power systems and virtual power plants. First, an information semantic-aware cyber-physical system joint optimization model is established. Based on this model, an analytical representation method of information semantics is abstracted to ensure that the communication system collaboratively supports the operation of virtual power plants.

For a cyber-physical system, the ideal unified optimization problem is modeled as follows:

$$\begin{aligned} &\min_{\mathbf{u}^{\mathrm{p}},\mathbf{u}^{\mathrm{c}}} \mathcal{F}^{\mathrm{p}}(\mathbf{x}^{\mathrm{p}},\mathbf{u}^{\mathrm{p}})+\mathcal{F}^{\mathrm{c}}(\mathbf{x}^{\mathrm{c}},\mathbf{u}^{\mathrm{c}}) \\ &\text{s.t. } (\mathbf{x}^{\mathrm{p}},\mathbf{u}^{\mathrm{p}})\in\Theta^{\mathrm{p}},\ (\mathbf{x}^{\mathrm{c}},\mathbf{u}^{\mathrm{c}})\in\Theta^{\mathrm{c}} \\ &(\mathbf{x}^{\mathrm{c}},\mathbf{x}^{\mathrm{p}})\in\Phi^{\mathbf{x}},\ (\mathbf{x}^{\mathrm{c}},\mathbf{u}^{\mathrm{p}})\in\Phi^{\mathbf{u}}. \end{aligned} \tag{13.1}$$

Here, $\mathbf{x}$ represents the state variable, $\mathbf{u}$ represents the control variable. More specifically, $\mathbf{x}^{\mathrm{p}}$ and $\mathbf{x}^{\mathrm{c}}$ represent the state variable vectors of the physical system and the information system, $\mathbf{u}^{\mathrm{p}}$ and $\mathbf{u}^{\mathrm{c}}$ represent the control variable vectors of the physical system and the information system, respectively. The model distinguishes the differences in data transmitted by different devices, rather than treating the data as a homogeneous data flow. It can directly or indirectly express the different impacts of each device's transmission information side variables on the optimal objective function value on the physical side. It explains the differences in the importance and relevance of different equipment information and physical operation goals, that is, taking into account the semantics of the information and the role of the information. For different distributed resources, the information transmitted carries different status values and control amounts, which has significant differences in system decision-making and also has different impacts on the decision-making objectives of the virtual power plant. Information is not just a simple bit stream, it affects decision-making and regulatory effects. This is the semantics of information.

The objective function of the optimization model mainly includes the global operating cost of the entire cyber-physical system. $\mathcal{F}^{\mathrm{p}}(\mathbf{x}^{\mathrm{p}}, \mathbf{u}^{\mathrm{p}})$ represents the cost function of the physical system, including the control cost of the virtual power plant. $\mathcal{F}^{\mathrm{c}}(\mathbf{x}^{\mathrm{c}}, \mathbf{u}^{\mathrm{c}})$ represents the cost function of the information system, including but not limited to the cost of channel resource usage. In a centralized control virtual power plant, the virtual power plant can centrally collect the status information and some decision parameters of distributed resources, and then centrally solve the operation control problems such as standby optimization and control instruction allocation. For centralized coordination and decentralized joint virtual power plants, since it is difficult to directly obtain the user's utility information, the status and technical parameters of distributed resources. It is necessary to rely on distributed optimization methods to solve the collaborative control optimization problem through multiple rounds of iterations until a convergent solution is found.

Physical-side constraint $(\mathbf{x}^{\mathrm{p}}, \mathbf{u}^{\mathrm{p}}) \in \Theta^{\mathrm{p}}$ describes the functional relationship between the state variables of the physical system and the control variables, and their combination must comply with the constraints. The physical constraints mainly cover the resource limitations of various distributed resources, including upper and lower bounds of power, upper and lower bounds of energy, and limitations on internal states. Information-side constraint $(\mathbf{x}^{\mathrm{c}}, \mathbf{u}^{\mathrm{c}}) \in \Theta^{\mathrm{c}}$ mainly describes the changing rules of information side state variables such as changing resource wireless access, bandwidth allocation, and routing path. Coupling constraints $(\mathbf{x}^{\mathrm{c}}, \mathbf{x}^{\mathrm{p}}) \in \Phi^{\mathbf{x}}$, $(\mathbf{x}^{\mathrm{c}}, \mathbf{u}^{\mathrm{p}}) \in \Phi^{\mathbf{u}}$ cover the coupled operation restrictions between the physical system and the cyber system. Among them, $(\mathbf{x}^{\mathrm{c}}, \mathbf{x}^{\mathrm{p}}) \in \Phi^{\mathbf{x}}$ represents the coupling relationship between the cyber system state variables and the physical side state variables. $(\mathbf{x}^{\mathrm{c}}, \mathbf{u}^{\mathrm{p}}) \in \Phi^{\mathbf{u}}$ represents the coupling relationship between the cyber system state variables and the physical side control variables.

(3) Challenges and Future Directions

A. Analysis of the Multi-link Paths and Mechanisms of Cyber-Physical Coupling

At present, the research on the multi-link processing of information in physical systems is still shallow. For example, some state data is used as the input of the prediction model, and its output results are used as the basis for multi-stage decision-making, forming a complex multi-stage and multi-link information processing chain. In this case, the analysis of the impact of information systems on physical operation decisions is more complicated. In addition, the feedback effect between multiple information may also produce more coupling phenomena, including the mutual coupling of multi-stage data transmission, and the analysis method of its impact mechanism faces huge challenges.

B. Methods, Mechanisms, and Means of Communication Collaborative Regulation

The research on communication collaborative regulation needs to be further deepened. For example, the collaborative planning of information systems can be carried out more finely and the layout of information transmission points can be optimized. At the same time, the economic incentive mechanism of communication collaboration is also an important area of future research. Since most cellular communication control methods need to cooperate with communication operators to form a sustainable business model, it is essential to reasonably price and design incentive mechanisms for the performance improvement brought by communication collaboration.

C. Communication Control Requirements for Multiple Types of Services in Virtual Power Plants

The application scope of communication collaboration also needs to be further expanded. For example, in the application of technology-based virtual power plants, it is possible to consider incorporating the monitoring status of lines and transformers into the decision-making model, and realize differentiated sensor data collection strategies by coordinating the acquisition frequency of equipment parameters through communication. In addition, the more rational personal strategic behavior of users is also considered to explore the game and generalized equilibrium problems between multiple users. The realization of communication coordination in user games can effectively improve the scalability and efficiency of regional transactions.

D. Application Extension of Information-Physical Collaborative Control Theory

The semantic-based information-physical collaborative control theory proposed in this paper has broad application potential and can be extended to other research on power systems. For example, this method can be used to evaluate the importance of different information to the operation and dispatching goals of the power grid, and to focus on protecting this information to defend against network attacks. This method can also be used to evaluate the coupling covariance coefficient of different resource delays on the under-damped mode of wide-area control, and to ensure the delay performance of key controllers through the network slicing technology of wired networks, thereby effectively suppressing system oscillations.

## 13.5 Summary

In this chapter, we presented several practical applications to illustrate the potential and applicability of semantic communication in realistic scenarios. Specifically, we examined four representative cases: immersive communications, satellite communications, autonomous driving, and smart grid systems. For each scenario, the fundamental framework and principles underlying semantic communication were systematically analyzed. Additionally, key challenges currently faced in the implementation and integration of semantic communication technologies were discussed, along with insights into future research directions. Overall, this chapter underscores the significance of semantic communication in advancing intelligent and context-aware wireless communication systems, highlighting its role as a pivotal enabler of next-generation technological development.

## References


1. J.G. Apostolopoulos, P.A. Chou, B. Culbertson, T. Kalker, M.D. Trott, S. Wee, The road to immersive communication. Proc. IEEE **100**(4), 974–990 (2012)
2. M. Zimmermann, The nervous system in the context of information theory, in *Human Physiology* (Springer, 1989), pp. 166–173
3. B. Zhang, Z. Qin, Y. Guo, G. Y. Li, Semantic sensing and communications for ultimate extended reality (2022). arXiv:2212.08533
4. S. Shi, J. Cui, Z. Jiang, Z. Yan, G. Xing, J. Niu, Z. Ouyang, Vips: real-time perception fusion for infrastructure-assisted autonomous driving, in *Proceedings of the 28th Annual International Conference on Mobile Computing And Networking* (2022), pp. 133–146
5. R. Xu, Z. Tu, H. Xiang, W. Shao, B. Zhou, J. Ma, Cobevt: cooperative bird's eye view semantic segmentation with sparse transformers, in *Proceedings of the 6th Conference on Robot Learning* ed. by K. Liu, D. Kulic, J. Ichnowski. Proceedings of Machine Learning Research, vol. 205 (PMLR, 14–18 Dec 2023), pp. 989–1000
6. Z. Qin, G.Y. Li, H. Ye, Federated learning and wireless communications. IEEE Wirel. Commun. **28**(5), 134–140 (2021)
7. E. Eldeeb, M. Shehab, H. Alves, A multi-task oriented semantic communication framework for autonomous vehicles (2024)
8. J. Lv, H. Tong, Q. Pan, Z. Zhang, X. He, T. Luo, C. Yin, Importance-aware image segmentation-based semantic communication for autonomous driving (2024)
9. Y. Diao, Z. Meng, X. Xu, C. She, P.G. Zhao, Task-oriented source-channel coding enabled autonomous driving based on edge computing, in *IEEE INFOCOM 2024-IEEE Conference on Computer Communications Workshops (INFOCOM WKSHPS)* (2024), pp. 1–6

# Open Problems

# 14

**Abstract**

This chapter outlines several key open problems that remain critical for advancing semantic communication systems. A universal mathematical framework is still lacking, since the notion of semantics depends heavily on context and task. Current approaches also rely on deep learning, which improves efficiency but consumes substantial computing resources, raising the need for lighter models and better integration of cloud, edge, and device computing. Privacy and security pose new risks, as transmitting condensed, high-level features may expose sensitive information. Moreover, ensuring robust performance under dynamic and uncertain conditions remains a key difficulty. Generative models show promise but face challenges in safety, reliability, and evaluation. Finally, high-mobility scenarios demand mobility-aware semantic strategies. Together, these issues define the path toward practical and trustworthy semantic communication in future networks.

**Keywords**

Mathematical framework • Computing resource • Privacy and security • Adaptive communication • Generative model • High-mobility scenario

This chapter outlines several key open problems that remain critical for advancing semantic communication systems.

(1) Mathematical Framework for Semantic Communication

Shannon's foundational information theory offers a rigorous and general mathematical framework for defining information and determining theoretical limits, which has been highly successful in guiding the design of conventional communication systems. Inspired by Shannon's work, researchers have sought to extend these principles toward semantic information theory. For example, notions such as semantic entropy have been proposed by substituting statistical probability with logical probability, and semantic rate-distortion theory has been investigated by incorporating distortions in semantic features. Despite significant progress in this area, there is no universal, operational definition of "semantics" as well as semantic information theory that can be directly optimized in communication systems. Unlike bit-error rate or capacity, semantic distortion is inherently task and context-dependent. Therefore, there may not be a single unified framework capable of modeling all semantic communication scenarios, especially those involving multimodal data. It is likely that definitions of semantic entropy or channel capacity will need to be tailored to the particular transmission task and the underlying knowledge base, adapting to different tasks accordingly.

(2) Computing Resource Consumption

Semantic communication is largely achieved by embedding AI techniques within wireless systems. Most existing approaches rely on DNNs for semantic encoding and decoding, which can deliver high communication efficiency but at the cost of substantial computing resources. This creates a difficult trade-off between the gains in network performance and the practical constraints of device computing capacity and energy consumption. Cloud-edge-end computing network could be a potential framework to integrate and reorganize heterogeneous computing resources. Lightweight AI models are also open directions to reduce the size of parameters and complexity, which remain to be fully developed. Therefore, designing semantic communication systems that deliver robust, high-quality semantic processing while operating within realistic computing constraints remains an important open research problem for 6G and beyond.

(3) Private and Secure Semantic Communication


W. Chen and Z. Qin, *Semantic Communications*, Textbooks in Telecommunication Engineering,
https://doi.org/10.1007/978-3-032-11005-3_14

Information privacy and security are fundamental concerns in all wireless communication systems, especially in the context of semantic communication systems. Unlike traditional systems that transmit raw bitstreams, semantic communication focuses on sending only task-relevant semantic features. By transmitting only the most important, high-level information, these systems may become vulnerable if that condensed information is intercepted or eavesdropped. A successful attacker could gain direct insight into the underlying meaning of the message with far less data. Moreover, there is a fundamental trade-off between data security and transmission efficiency. Stronger encryption and security measures can protect privacy but may increase latency or lead to bandwidth limitations. Addressing this trade-off in a principled way, especially under varying transmission tasks and QoS requirements, remains an open research question. Therefore, developing semantic communication systems that can robustly safeguard privacy and security without undermining their efficiency advantages is a key open problem for next-generation wireless networks.

(4) Adaptive and Robust Semantic Communication Under Dynamic Conditions

Despite substantial advances in bandwidth-adaptive and task-oriented semantic communication, a significant open problem persists: the design of universal frameworks capable of dynamically adapting to unpredictable variations in both source characteristics and channel conditions. In real-world scenarios, such as mobile wireless links and real-time multimedia streaming, the conventional assumption of fixed and known distributions is seldom valid. Future research must therefore focus on developing robust semantic coding and resource allocation strategies that autonomously adjust to these fluctuations, ideally without relying heavily on retraining or complete model redesign. Additionally, there remains an urgent need to establish rigorous theoretical performance bounds under statistical uncertainty and to systematically quantify the impact of model mismatch on end-to-end reliability.

(5) Challenges Faced by Generative Models in Semantic Communication

While generative models exhibit impressive capabilities for producing contextually coherent and semantically rich content, their application within semantic communication frameworks introduces two critical challenges. First, the field currently lacks standardized semantic evaluation metrics, hindering systematic and reproducible assessment of the communicative relevance and fidelity of generated outputs. Second, existing generative models often fall short of meeting the stringent safety, reliability, and latency requirements essential for mission, critical applications, such as autonomous driving, emergency response, and industrial automation, where even minor errors may have severe consequences. Bridging these gaps is crucial to ensure the dependable deployment of generative semantic communication systems in safety-critical and real-time contexts.

(6) Semantic Communication in High-Mobility OFDM

Scenarios OFDM remains central to contemporary wireless communication systems; however, its performance can be severely compromised in high-mobility environments—including vehicular networks, UAV-based communication, and high-speed railway systems—due to Doppler spread and pronounced ICI. A key open problem is to develop semantic communication frameworks that remain robust and efficient under such rapidly varying channel conditions. Promising research directions include dynamically integrating semantic importance into channel estimation and resource allocation, thereby prioritizing the transmission of semantically critical information even in the presence of severe channel impairments. Designing and validating mobility-aware semantic transmission strategies thus represents an essential yet underexplored frontier for advancing next-generation wireless communication systems.

# Index





W. Chen and Z. Qin, *Semantic Communications*, Textbooks in Telecommunication Engineering,
https://doi.org/10.1007/978-3-032-11005-3_

## S



## T



## V



## W

MIX
Papier aus verantwortungsvollen Quellen
Paper from responsible sources
FSC® C105338

If you have any concerns about our products,
you can contact us on
**ProductSafety@springernature.com**

In case Publisher is established outside the EU,
the EU authorized representative is:
**Springer Nature Customer Service Center GmbH**
**Europaplatz 3, 69115 Heidelberg, Germany**

Printed by Libri Plureos GmbH
in Hamburg, Germany